P9-DNC-259

Recent Advances in Carbohydrate Bioengineering

Recent Advances in Carbohydrate Bioengineering

Edited by

H.J. Gilbert

University of Newcastle upon Tyne, Newcastle upon Tyne, UK

G.J. Davies

University of York, York, UK

B. Henrissat

Centre Nationale de la Recherche Scientifique, Marseille, France

B. Svensson

Carlsberg Laboratory, Copenhagen, Denmark

The proceedings of the 3rd Carbohydrate Bioengineering Meeting held at the University of Newcastle upon Tyne on 11–14 April 1999.

Special Publication No. 246

ISBN 0-85404-774-3

A catalogue record for this book is available from the British Library

Published by The Royal Society of Chemistry,
Thomas Graham House, Science Park, Milton Road,
Cambridge CB4 0WF, UK

For further information see our web site at www.rsc.org

Printed and bound in Great Britain by MPG Books Ltd, Bodmin, Cornwall

Preface

In common with all areas of the biological sciences, glycobiology has benefited greatly from the developments, over the last 20 years, in molecular biology, structural biology, bioinformatics and analytical biochemistry. These technologies have transformed our understanding of the catalytic, structural and biological role of carbohydrate modifying enzymes. This information is now being exploited in the generation of novel carbohydrates designed for use in the pharmaceutical, food and agriculture industries. In recognition of the rapid developments in glycobiology over the last few years a conference on carbohydrate bioengineering was held in Elsinore, Denmark in 1995. The 2nd conference on this subject was in La Rochelle, France in 1997 and the 3rd Carbohydrate Bioengineering meeting took place in Newcastle upon Tyne, England, from 11th–14th April 1999

The Newcastle meeting attracted 250 delegates from 31 countries from Europe, Asia and America with interests ranging from carbohydrate chemistry, carbohydrate modifying enzymes and the biotechnological exploitation of carbohydrates. In addition to 36 lectures, both invited and selected, there were over 100 poster presentations. This book contains papers of the oral presentations. The first paper is the keynote address in which the classifications of both catalytic and non-catalytic enzyme modules, based on structural similarities, were presented. The resulting databases are pivotal to much of the research directed at understanding the structural and biochemical properties of carbohydrate modifying enzymes.

In Section 2 the papers discuss how a glycosidase-catalysed approach to oligosaccharide synthesis can be utilised to synthesise a variety of oligosaccharides of considerable biological interest. In common with the other Carbohydrate Bioengineering meetings, carbohydrate modifying enzymes was one of most widely covered topics. In Section 3 the biochemistry of these enzymes was discussed. The first paper in this section provides an excellent review of the use of physical-organic probes in understanding glycosidase catalysis. Here, the pioneering work on the engineering of glycosyl hydrolases to catalyse glycosidic bond formation, rather than cleavage, is discussed, and the use of the oligosaccharides generated, both as therapeutic agents and as mechanistic enzymes inhibitors, is reviewed. The section also contains papers on a wide range of enzymes including hemicellulases, pectinases and amylases.

Section 4 focuses on how the solving of the three dimensional structures of glycoside hydrolases and transferases has made a major contribution to our understanding of the mechanism of action of these enzymes. The following section deals with the rapidly expanding field of non-catalytic polysaccharide binding modules. Topics covered include a phylogenetic analysis of cellulosome structure from different micro-organisms, a review of the diversity in structure and function of non-catalytic carbohydrate binding modules, and an insight into the structural basis of their ligand specificity.

Section 7 contains papers that review the use of enzymes in the generation of industrially important polysaccharides and monosaccharides. In the final section the papers review the use of protein engineering to improve the industrial utility of carbohydrate modifying enzymes, and to increase our understanding of the mechanism of action of these biocatalysts. The section discusses the use of this technology to improve the biophysical and catalytic activities of amylases, glycosyltransferases and xylose isomerase. In the final papers the use of site-directed mutagenesis to probe the mechanism of action of both catalytic and non-catalytic modules of carbohydrases is discussed.

The meeting organisers would also like to take this opportunity to express their thanks to various individuals and institutions, who contributed to the success of the meeting. We greatly appreciate the financial support provided by, *Bio-Rad Laboratories Ltd, Dupont Ltd, Megazyme International Ltd, Newcastle Breweries, NovoNordisk A/S, Qiagen Ltd* and *The Wellcome Trust*. We are greatly indebted to **Heather Martin** for organising the meeting, dealing with all enquiries, and preparing the abstract booklet and this book, while maintaining her sanity and good humour. We also wish to thank all the authors for the preparation of their manuscripts, and to the delegates who attended the meeting.

Newcastle, July 5th, 1999

Gideon J. Davies, Harry J. Gilbert, Bernard Henrissat, Birte Svensson; The Editors

Contents

1 Keynote Address

2 Structure and Synthesis of Carbohydrates and Carbohydrate Analogues

3 Post-translational Glycosylation of Proteins

4 Biochemistry of Carbohydrate Modifying Enzymes

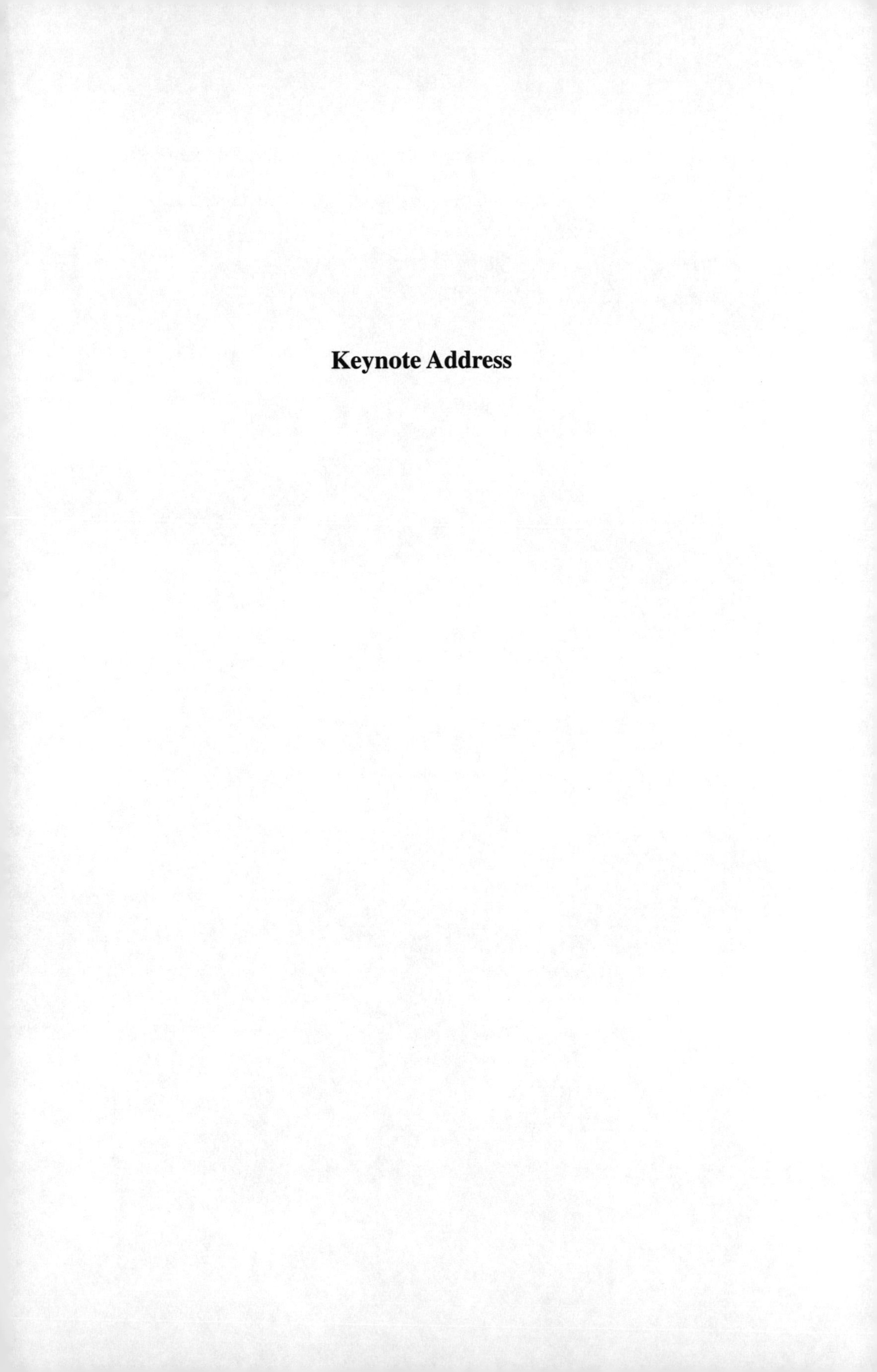

Keynote Address

CARBOHYDRATE-ACTIVE ENZYMES: AN INTEGRATED DATABASE APPROACH

P. M. Coutinho and B. Henrissat

Architecture et Fonction des Macromolécules Biologiques, AFMB-CNRS IFR1
31 Chemin Joseph Aiguier
F-13402 Marseille cedex 20
France

1 SUMMARY

Carbohydrate-active enzymes often display a modular structure featuring a catalytic module attached to one or several ancillary non-catalytic modules (NCMs). In glycoside hydrolases the function of NCMs is often but not always substrate-binding. The various modules, catalytic or not, of carbohydrate-active enzymes can be identified and classified in various families by careful sequence comparisons. CAZyModO (Carbohydrate-Active enZYmes Modular Organisation, http://afmb.cnrs-mrs.fr/~pedro/DB/db.html), a WWW server describing the modular structure of carbohydrate-active enzymes has been created to provide access to updated classifications of glycosidases and transglycosidases, glycosyltransferases, polysaccharide lyases, and carbohydrate esterases and their associated modules in families. A complete analysis of over 1,750 glycoside hydrolase sequences has been accomplished. This effort will be pursued until all modular descriptions of these enzymes can be obtained. The classification of these modules is a necessary step for the effective annotation of protein modules of carbohydrate-active enzymes and related proteins in sequence databases.

2 INTRODUCTION

Carbohydrates are an essential component of life as structural and energy storage components, as stabilisation, recognition, signalling, and communication agents. The multiple roles of carbohydrates arise from the stability and the stereochemical diversity glycosidic bonds. Half-lifes of up to 5 to 8 million years can be found for the non-catalytic hydrolysis at 25°C and neutral pH of polysaccharides like cellulose and glycogen, glycosidases being able to accelerate the reaction by a factor of 10^{17}.[1] The unique branching structure of carbohydrates has also an inherent information potential in biological recognition processes which enzymes, and proteins in general, can recognise, adapt to, and act upon.[2] They constitute therefore three-quarters of the biomass on earth and carbohydrate-active enzymes are the key for their synthesis, degradation, and modification.

The classification of carbohydrate-active enzymes into families started with a classification of cellulases into six distinct families based on amino acid sequence similarities.[3] This led to the classification of glycosides and transglycosidases in general, also referred as glycoside hydrolases (or GHs),[4-6] as no direct relation between substrate specificity and sequence similarities existed. Contrarily to the traditional EC classification system based primarily on substrate specificity, this new classification reflected structural features of the enzymes as sequence similarities are correlated by structural similarities.[7] The growing number of glycosidase structures resolved thereafter validated the approach.[8-9] The exponential growth in sequences in public databases lead to a steady growth in the number of GH families with 300 sequences and 35 families in 1991 and over 2,800 sequences in 74 families as of April 99. This classification system had received immediate acceptance by the scientific community as: (i) molecular biologists could rapidly assign their new sequences to families; (ii) microbiologists could screen micro-organisms for new gene sequences with primers corresponding to conserved regions within a family;[10] (iii) structural biologists could decide to determine the structure of a family with no known representative at the 3-D level; (iv) and enzymologists could investigate and rationalise their mechanistic data in structural terms.[11] More recently, similar classifications have been proposed for glycosyltransferases (GTs),[12] polysaccharide lyases (PLs), and carbohydrate esterases (CEs). Given the increasing importance of glycobiology and the elevated number of GTs found in eucaryotes,[13] these families are likely to provide the framework for structural and functional research. At the example of GHs, the fruitfulness of the classifications of carbohydrate-active enzymes will become even more apparent as soon as more structural and genomic information becomes available.

Among the best characterised carbohydrate-active enzymes, many GHs and particularly those involved in the hydrolysis of polysaccharides, are known to display a modular structure featuring a catalytic module (or unit) attached to: (i) one or several ancillary non-catalytic modules (NCMs) whose precise functions are often not known, and/or (ii) catalytic modules corresponding to GHs or other enzymes exhibiting complementary properties. The non-catalytic modules of cellulases (and other GHs) can be identified in many cases by sequence comparisons,[14] an analysis that can now be extended not only to GHs but to the carbohydrate-active enzymes in general. For GHs acting on insoluble polysaccharides such as cellulose, chitin, starch and β–1,3-glucan, several NCMs promote the attachment of the enzyme onto the polysaccharide matrix and thereby facilitate the degradation of crystalline polysaccharides. Among the NCMs with known functions, the cellulose-binding domains are those probably best characterised in terms of structure and function. These have already been classified in several families,[14-16] but other NCMs, like dockerins and cohesins involved in cellulosome assembly,[17-18] and the growing number of modules of unknown function lacked an integrated classification until now. Just as the catalytic modules which can be classified into several families based on amino acid sequence similarities, the NCMs also form a number of different families. A classification of such modules, independent and complementary to that of catalytic modules, is needed for a more meaningful description of carbohydrate-active enzymes, particularly of GHs whose sequence information has reached a critical mass.

The enzyme classifications, even when based on sequence similarities, cannot be dissociated from mechanistic considerations. In the GH classification the factors to take in account are: (i) the axial or equatorial orientation of the bond to be hydrolysed as a consequence of the anomeric configuration; (ii) the relative orientation between the hydrolysed bond in the substrate and in the resulting product, as the reaction is inverting or retaining,[19] which is accompanied by a different spatial distance between the two catalytic residues.[20] The combination of these two factors subdivides GHs into four groups, but the more significant factors are from structural nature. In fact, with the increasing sequence and structural information GH families have been grouped into clans or "superfamilies".[6,21-22] Within a clan, the fold, the catalytic residues, and mechanism are conserved, a fact which argues in favour of a divergence of the families during evolution. In the absence of mechanistic data, enzyme families within a given clan can be inferred as being inverting or retaining when the information exists for other clan members. Equally important is the inference of catalytic residues within a clan. In the GT classification, there is only one significant mechanistic factor in family division: the relation between orientation of the axial-linked nucleotide phospho-sugar and the newly synthesised glycosidic bond of the product; GTs are then either inverting or retaining. In PLs families, the same orientation of the cleaved glycosidic bond is presently found within a family. The axial or equatorial orientation of the glycosidic bond in the acetylated or methylated sugar seems also conserved within the CEs families, some CEs families interestingly extending into other non-carbohydrate acting esterases. However, these last two classes of carbohydrate-active enzymes have still a too reduced number of sequence and structural entries to allow generalisations.

As mentioned earlier, the traditional EC classification based primarily on substrate specificity is not able to clearly discriminate many carbohydrate-active enzymes. Also there are many cases of overlapping specificities and of activities presently not described. Furthermore, there are both inverting and retaining GHs in some EC enzyme classes as the anomeric relation between the substrate and the product is not systematically used in the description. For example, β–xylosidases are found with the EC 3.2.1.37 in retaining families GH3, GH39 and in the inverting family GH43 (besides family GH52 where the type of mechanism is still unknown), leading to different products. EC classification will be only used as a complement in enzyme descriptions.

The main issues with maintaining carbohydrate-active family classifications are: (i) how to make them available to the scientific community; (ii) how to disclose the new families and new family members; and (iii) how to keep up with the ever increasing number of sequences. The World Wide Web (WWW) is obviously the best medium for the distribution of family classifications.[6] For carbohydrate-active enzymes, the first milestone was achieved in 1996 with the availability of the GH classification on ExPASy (http://www.expasy.ch/cgi-bin/lists?glycosid.txt).[6] This useful document suffered however from containing only annotated SwissProt entries,[23] missing therefore the large number of entries already in GenBank,[24] and the information relative to resolved three-dimensional structures in the PDB.[25] Other drawbacks were the irregular updates, the impossibility to do family-by-family browsing and unavailability of family classification of other carbohydrate-active enzymes such as GTs. To overcome some of these problems, and to prepare the way for a server also providing descriptions of the modular structure of carbohydrate-active enzymes, a new WWW server named CAZy (Carbohydrate Active enZYmes, http://afmb.cnrs-mrs.fr/~pedro/CAZY/db.html) was created,[26] giving access to the classification of GHs, GTs and PLs in families based on sequence similarities.

CAZy displays online information relative to the family classification using hypertext to give access to sequence, structure, and associated information. Families can be directly accessed, each enzyme entry having direct links to the NCBI taxonomy browser,[24] ENZYME,[27] GenBank,[24] SwissProt/TrEMBL,[23] the PDB,[25] with a few other complementary links to other public databases (PIR, PRF, Genpept, patents, etc) for specific entries. Each family is annotated with information concerning the known activities of the respective enzymes, known PROSITE motifs,[28] catalytic mechanism and known catalytic residues (in general conserved throughout a family and so inferred from only a few studied cases), known taxonomic range, etc., allowing a rapid perception of all its common characteristics. Presently only available for GHs, the more remote structural and mechanistic relationships corresponding to clans are given. Links to HOMSTRAD[29] structure-based alignments are also provided for families having structures resolved for different members, and links to other relevant online references are likely to be added in the future. Tabular information on clan membership and on the EC classification of the various families are also available in CAZy.

The automatic annotation of proteins used in present day public databases are known to lead to: (i) the introduction and propagation of misannotations; (ii) the underprediction of functional properties where multifonctionality or multimodularity are rarely considered, and; (iii) the overprediction of properties, as the best protein hits do not necessarily share the same or similar functions.[30] As examples of typical problems found with carbohydrate active enzymes: (i) GH family 60 has been recently eliminated as its inclusion in the classification was due to a probable misannotation of a metalloproteinase as a cellulase; (ii) genome annotations like "putative glycosyl transferase" corresponds to all 31 GT families and therefore of limited utility for researchers; (iii) it is not rare to find proteins named after sequence similarities to regions not related to the functionality in question, a typical case being proteins named after homologies limited to NCM regions.

The creation of CAZy constitutes in fact only the -essential- first step in the implementation of a server dedicated to both family description and discovery of new family members. For carbohydrate-active enzymes, it is desirable to have expert functional and modular annotation, that can be used as a basis for the annotation of new proteins where a "module" is an independently folding structural and functional unit of catalytic or other nature. With the increasing number of sequences and resolved structures available, the delineation of catalytic modules, NCMs and other modules is significantly simplified. Singularly individualised modules that constitute carbohydrate-active enzymes can be used to semi-automatically and conservatively annotate new proteins as members of enzyme (and NCM) families. With sets of annotated modules, the sequences and structures of carbohydrate-active enzymes from the GenBank and SwissProt/TrEMBL databases or from the PDB from the updated classifications of GH, GT, PL, and CE families can be now characterised in all their extension. The new server created to complement CAZy is designated CAZyModO (Carbohydrate-Active enZYmes MODular Organisation). It includes the functionalities of CAZy together with a graphical description of the modular structure of the enzymes and related proteins.

3 METHODOLOGY

In the analysis of multimodular protein sequences, gapped BLAST and PSI-BLAST searches[31] were performed against both a local extensive library of modules and the non-redundant protein database at NCBI in search of complementary sequences, typically those presenting a $E < 0.001$. In the absence of three-dimensional data and when the similarity to known modules was relatively low, hydrophobic cluster analysis[32,33] (HCA) was used as a complement to refine the limits of the modules and of the linker regions. The remaining segments of the protein sequences were again used as templates for further BLAST analysis. Upon identification and annotation of the modules, the remaining segment(s) of the protein were further analysed, in an iterative process. Several types of new modules have been be found: (i) NCMs with already assigned functions; (ii) catalytic modules of carbohydrate-active or other enzymes; (iii) common regions found in other carbohydrate-active enzymes away from the catalytic unit. The later regions were designated unknown 'X' modules.

This iterative approach already allowed a completion of carbohydrate-active enzymes families, but also the creation and/or completion of known carbohydrate-binding module families. These include previously identified families of cellulose binding domains (CBDs),[15] chitin binding domains (ChBDs), starch binding domains (SBDs), lectin-like domains, etc. Several CBDs and ChBDs are known to bind to polysaccharides other than cellulose or chitin, respectively, a dramatic example being the overlap between families CBD12 and ChBD3. This is a problem similar to that found initially in the classification of GHs. Again, as no clear border between substrate specificities exist, a more generic carbohydrate-binding module (CBM) classification will be used. To facilitate the transition to a new nomenclature, the same family number will be maintained for families CBD1 to CBD13,[15] that will now be designated as CBM1 to CBM13.

For each enzyme analysed in CAZyModO a linear modular description is given. An example for a typical extracellular glycosidase is [(1-30)SIGN] [(31-40)PROP] [(41-240)GH*x*] [(241-260)LNK] [(261-350)CBM*y*] [(351-400)X*z*] that corresponds to primary structure subdivided into segments corresponding to a signal peptide (SIGN), a propeptide segment (PROP) that will be lost upon maturation, a glycosidase family *x* catalytic module (GH*x*), a linker region (LNK), and carbohydrate-binding module of family *y* (CBM*y*), and a module of unknown function belonging to family *z* (X*z*).

4 RESULTS AND DISCUSSION

The full extention of the enzymes covered by CAZy(ModO) is given in Table 1. CAZy comprises 4385 proteins as of April 1999, CAZyModO providing modular descriptions for only 41% of these. Modular descriptions are provided for full-length proteins. For protein fragments, they are provided only when significant features are described. Given that the descriptions are only made manually, the different families have received an unequal treatment, with some having descriptions for almost every single member while no descriptions exist for others. Efforts are being made to have expert descriptions for at least some representatives in each family. Due to an unequal description for different types of enzymes, 62% of GHs have a modular description, but only 1% of GTs have one. The large number of resolved GH three-dimensional structures corresponds in fact to only

Table 1 *Number of CAZy entries and modular organisation (ModO) descriptions in April 1999. Total number of entries are given for the public databases as well as the non-redundant (nr) number of PDB entries.*

	GH	*GT*	*PL*	*CE*	*Total Enzymes*	*CBM*	*Known NCMs*	*X Modules*	*Total NCMs*
Families	74	31	9	11	125	24	7	45	76
Families with 3-D representative	28	1	1	1	31	10	3	6	18
Proteins	2805	1391	109	138	4433	412	110	299	665
GenBank	3735	1857	144	161	5207	463	137	349	819
SP-TrEMBL	2593	1120	114	118	3853	363	100	279	624
PDB (total)	751	4	6	2	763	65	8	29	74
PDB (nr)	99	1	4	1	105	24	5	9	30
ModO Descriptions	1756	17	30	34	1837	412	110	299	665
ModO Coverage (%)	62	1	28	25	41	100	100	100	100

99 different proteins due to significant redundancy in some crystallographic studies as, for example, variations on bacteriophage T4 lysozyme account for a third of the total number of PDB entries in CAZy !

It is significant that CBMs account for almost two thirds of the annotated modules, and this proportion is likely to increase when some of the other known modules or some of the unknown X modules become functionally characterised as carbohydrate-binding.

The known binding properties and other designations of the different CBM families are described in Table 2. It is interesting to note that the same CBMs can be found adjacent to different enzyme families, exhibiting very different reactivities. The new nomenclature proposed here for these modules is not substrate-dependent, at the example of the enzyme classifications found in CAZy. This is particularly important as many of these binding modules have had their binding properties determined with a limited number of substrates. A particularly notorious example is the fact that CBM6 (formerly named CBD6) is only found in xylanases but has been only reported to bind to cellulose, a substrate rather similar to xylan but not tested to our knowledge. Given the large spectrum of possible carbohydrate targets, and the possible multiple specificities for each module, a less constraining nomenclature was needed, where structure rather than specificity prevailed. As mentioned earlier and to maintain a continuity in the classification of these binding modules, former CBD1 to CBD13 families[15] are now named CBM1 to CBM13. Family numbers have not yet been assigned for the remaining eleven CBM families, as efforts are presently being made to create a common nomenclature within the CBM community.

Other well described NCMs have been found in carbohydrate-active enzymes and related proteins and are described in Table 3. These include modules associated with cell adhesion, cellulosome assembly, or protein anchoring, and others of known structure with still unknown function in these enzymes. A typical example includes fibronectin III-like modules (FN3) which were probably acquired by bacteria from animals by horizontal transfer[34] and which most often interconnect catalytic domains and CBMs, suggesting a role as a linker, but other functions like cell-anchoring and/or carbohydrate-binding are not excluded.

Following the iterative analysis of carbohydrate-active enzyme sequences we find regions bearing only homology to other uncharacterised regions in other carbohydrate-active enzymes. These regions form presently 45 families, but this number is likely to increase as modular descriptions are made for all GH families and the remaining enzyme families. The description of these new 'X' module families will be made elsewhere.

A CAZyModO example for the well established GH family 6 is shown in Figure 1. Besides the "regular" family description and listing of known sequence and structure database entries already found in CAZy (Figure 1a), a graphical representation for the modular structure of the enzymes and proteins of this family is accessible (Figure 1b). This representation contains links to the modular representations of every single module family associated with GH6. Tools for the analysis and alignment of GH6 sequences are also provided.

The creation of module families is extremely useful when searching for functional module extension, fold estimation, potential global class or classes of enzymes, of new proteins. Its however critical to keep in mind that neither substrate specificity nor even activity, can be ascertained solely on amino acid sequence similarities.

Table 2 - *Carbohydrate-binding modules (CBMs) found in carbohydrate-active enzymes and their known binding substrates. Legend: GH - Glycosidases and Transglycosidases (or Glycoside Hydrolases); GT- Glycosyltransferases; CE - Carbohydrate Esterases;* ***bold*** *- available 3-D structures; * - family numbering to be assigned at a later date.*

Module Family	*Other Name(s)*	*Average Size*	*Associated Catalytic Modules*	*Substrates*	*Notes*
CBM1	CBD1	40	GH3, GH5, GH6, GH7, GH10, GH45, CE5	cellulose, chitin	fungal (with exceptions)
CBM2	CBD2	100	GH5, GH6, GH9, GH10, GH11, GH12, GH18, GH48, GH62, CE4	cellulose, chitin	bacterial
CBM3	CBD3	150	GH5, GH9, GH10, GH43, GH44, GH48	cellulose, chitin	bacterial
CBM4	CBD4	150	GH6, GH9, GH10, GH16	xylan, cellulose (amorphous but not crystalline)	bacterial
CBM5	CBD5	60	GH5	cellulose, chitin	structurally similar to CBM12
CBM6	CBD6	120	GH3, GH5, GH10, GH11, GH16, GH43, CE1, CE4	xylan, cellulose (amorphous), β–1,3-glucan	
CBM7	CBD7				deleted in CDB classification
CBM8	CBD8	150	GH9	cellulose	
CBM9	CBD9	170	GH10, CE4	cellulose	only present in xylanases
CBM10	CBD10	50	GH5, GH9, GH10, GH11, GH45, CE4	cellulose	
CBM11	CBD11	180-200	GH5, GH26, GH51	cellulose	
CBM12	CBD12 ChBD3	40-50	GH5, GH18, GH19	cellulose, chitin	structurally similar to CBM5
CBM13	CBD13 Lectin	150	GH10, GH16, GH27, GH62, GH64, GT27	xylan, mannose, galactose	3-fold repeat as in ricin and agglutinin
CBM*	-	200	GH5	cellulose	
CBM*	-	100	GH9	cellulose	
CBM*	-	150	GH10	cellulose	
CBM*	-	130-140	GH5, GH18	cellulose	
CBM*	Lectin	190	GH33	mono- or oligosaccharides	
CBM*		140	GH33	galactose	
CBM*	ChBD1	70	GH18	chitin	
CBM*	ChBD2	40	GH18, GH19	chitin	
CBM*	ChBD4	60-70	GH18	chitin	
CBM*	SBD	100	GH13, GH14, GH15	starch, cyclodextrins	
CBM*	SBD	100	GH13, GH15	starch	

Table 3 *Non-catalytic modules of known function or structure adjacent to carbohydrate-active enzymes. Legend as in Table 2.*

Module Family	*Average Size*	*Associated Catalytic Modules*	*Module Description*
COH	120	-	Cohesin. Receptors of DOC1 modules. Essential for cellulosome formation. Only indirect interaction with the catalytic module of enzymes.
DOC1	60	GH5, GH8, GH9, GH10, GH11, GH16, GH18, GH26, GH44, CE1, CE2, CE3, CE4	Dockerin 1. Promotes Ca^{2+}-mediated attachment to cohesins in the cell wall or in cellulosomes. Bacterial, found especially in Clostridiae
DOC2	40	GH5, GH6, GH11, GH26, CE3, CE6	Dockerin 2 . Putative fungal counterpart of DOC1, usually found in tandem repeats.
FN3	80-90	GH5, GH6, GH9, GH13, GH18, GH33, GH48	Fibronectin type 3-like module. Unknown function. Similar to eukaryotic fibronectin type 3 modules.
PTH	30-50	GH31	P-type trefoil-homology module. Potential interaction with mucins.
SLH	50-60	GH5, GH10, GH13, GH16, GH28	S-layer homology module. Interaction of enzymes with the cell surface in bacteria.
TSP3	30	GH5	Thrombospondin 3. Ca^{2+}-binding repeat putatively involved in cell-matrix interactions.

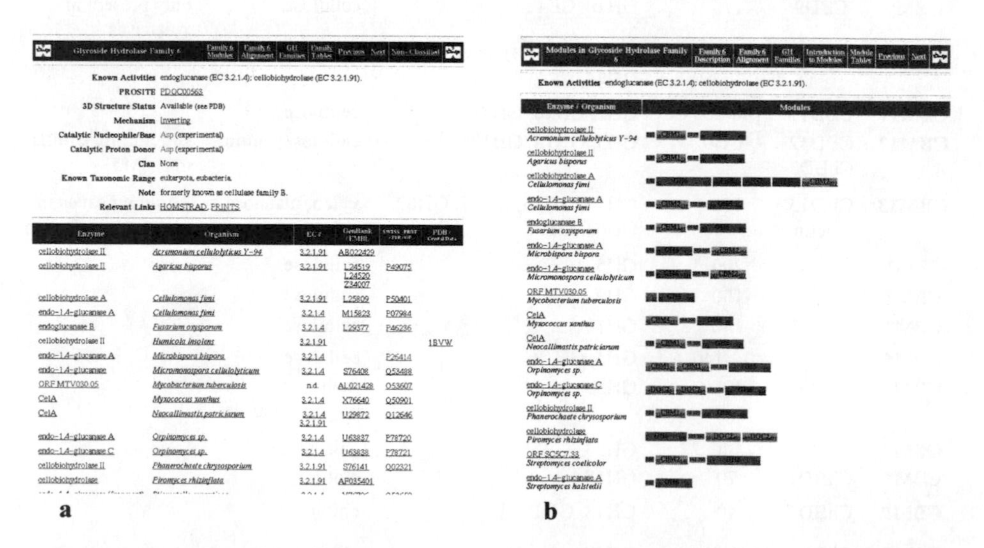

Figure 1. *Example of hypertext pages in CAZyModO for family GH6: (a) the updated classification for any given family of catalytic modules has links to sequence and structure information databases; (b) the modules associated with this family can be shown graphically with links to the families of adjacent NCMs and/or other catalytic modules.*

5 CONCLUSIONS

CAZyModO, a server dedicated to the classification of both catalytic and non-catalytic modules in carbohydrate-active enzymes and to their modular description has been created. It presently describes the classification of GHs, GTs, PLs, and CEs and associated modules and should be gradually extended to accommodate the wide spectrum of carbohydrate-active enzymes. Only then could an overall view of the complex interactions between different types of enzymes in each organism be made.

The sequence-based classification principles used earlier for defining families of carbohydrate-active enzymes are now used for the classification of CBMs. Again structure and not function is the basis of the classification. A thorough analysis of CBMs has been performed, together with that of known modules characterised elsewhere. This comprehensive analysis facilitates the identification of many previously undetected modules families of unknown function, paving the way for their functional analysis, a prerequisite to an improved knowledge of carbohydrate-active enzymes.

Acknowledgements

We wish to thank the Fundação Ciência e Tecnologia, Lisbon, Portugal and the TMR Program from the European Commission for financial support to P.M.C. and BIOTECH Program from the European Commision and INTAS for general support. This work involved discussions and help from many people from around the world in the last 10 years that we wish to thank: Kim V. Andersen, Amos Bairoch, Tristan Barbeyron, Edward Bayer, Pierre Béguin, Eric Blanc, Bernard Boutherin, R. Malcolm Brown, Jr., Vincent Bulone, Isabelle Callebaut, James A. Campbell, Marc Claeyssens, Gideon J. Davies, Michel Fèvre, Roberto Geremia, Harry J. Gilbert, Neil R. Gilkes, Micheline Guinant, Stefan Janecek, Frances Jurnak, Noel Keen Gérard Kleywegt, Bernard Kloareg, Jonathan Knowles, Jean-Paul Latgé, Stéphane Lavaitte, Laurence Lemesle-Varloot, Jean-Paul Mornon, Merja Penttilä, Anu Saloheimo, Inder Saxena, Martin Schülein, Serina Stretton, Henrik Stålbrand, Birte Svensson, Tuula Teeri, John Thompson, Peter Tomme, Anneli Törrönen, R. Anthony J. Warren, Takeshi Watanabe, Stephen Woodcock, J. Gregory Zeikus

References

1. R. Wolfenden, X. Lu and G. Young, *J. Am. Chem. Soc.* 1998, **120**, 6814
2. R. A. Laine, *Glycobiology*, 1994, **4**, 759.
3. B. Henrissat, M. Claeyssens, P. Tomme, L. Lemesle and J. P. Mornon, *Gene*, 1989, **81**, 83.
4. B. Henrissat, *Biochem. J.*, 1991, **280**, 309.
5. B. Henrissat and A. Bairoch, *Biochem. J.*, 1993, **293**, 781.
6. B. Henrissat and A. Bairoch, *Biochem. J.*, 1996, **316**, 695.
7. C. Chothia and A. M. Lesk, *EMBO J.*,1996, **5**, 823.
8. G. Davies and B. Henrissat, *Structure*, 1995, **3**, 853.
9. B. Henrissat and G. Davies, *Curr. Opin. Struct. Biol.*, 1997, **7**, 637.
10. H. Dalbøge and L. Lange, *Trends Biotechnol.*, 1998, **16**, 265.
11. J. Gebler, N. R. Gilkes, M. Claeyssens, D. B. Wilson, P. Béguin, W. W. Wakarchuk, D. G. Kilburn, R. C. Miller Jr, R. A. Warren and S. G. Withers, *J. Biol. Chem.*, 1992 **267**, 12559.

12. J. A. Campbell, G. J. Davies, V. Bulone and B. Henrissat, *Biochem J.*, 1997, **326,** 929.
13. R. Kleene and E. G. Berger, *Biochim. Biophys. Acta*, 1993, **1154**, 283
14. N. R. Gilkes, B. Henrissat, D. G. Kilburn, R. C. Miller Jr and R. A. J. Warren, *Microbiol. Rev.*,1991. **55**, 303.
15. P. Tomme, R. A. J. Warren and N .R. Gilkes, *Adv. Microb. Physiol.*, 1995, **37**, 1.
16. R. A. J. Warren, *Ann. Rev. Microbiol.*, 1996, **50**, 183.
17. E. A. Bayer, E. Morag, R. Lamed, S. Yaron and Y. Shoham, 'Carbohydrases from Trichoderma reesei and other microorganisms', M. Claeyssens, W. Nerinckx and K. Piens (Eds), The Royal Society of Chemistry, Cambridge, 1998, p. 39.
18. E. A. Bayer, H. Chanzy, R. Lamed and Y. Shoham, *Curr. Op. Struct. Biol.*, 1998, **8**, 548.
19. M. L. Sinnott, *Chem. Rev.*, 1990, **90**, 1171.
20. J. D. McCarter and S. G. Withers, *Curr. Op. Struct. Biol.*, 1994, **4**, 885.
21. B. Henrissat, I. Callebaut, S. Fabrega, P. Lehn, J. P. Mornon and G. Davies, *Proc. Natl. Acad. Sci. USA*, 1995, **92**, 7090.
22. J. Jenkins, L. L Leggio, G. Harris and R. Pickersgill, *FEBS Lett.*, 1995, **362**, 281.
23. A. Bairoch and R. Apweiler, *Nucleic Acids Res.*, 1999, **27**, 49.
24. D. A. Benson, M. S. Boguski, D. J. Lipman, J. Ostell, B. F. F. Ouellette, B. A. Rapp and D. L. Wheeler, *Nucleic Acids Res.*, 1999, **27**, .12.
25. J. L. Sussman, D. Lin, J. Jiang, N. O. Manning, J. Prilusky, O. Ritter and E. E. Abola, *Acta Cryst.*, 1998, **D54**, 1078.
26. P. M. Coutinho and B. Henrissat, 'Genetics, Biochemistry and Ecology of Lignocellulose Degradation', K. Ohmiya, K. Hayashi, K. Sakka, Y. Kobayashi and T. Kimura (Eds), Uni Publishers Co., Tokyo, 1999, p. 15.
27. A. Bairoch, *Nucleic Acids Res.*, 1999, **27**, 310.
28. K. Hofmann, P. Bucher, L. Falquet and A. Bairoch, *Nucleic Acids Res.*, 1999, **27**, 215.
29. K. Mizuguchi, C. M. Deane, T. L. Blundell and J. P. Overington, *Protein Sci.*, 1998, **7**, 2469.
30. T. Doerks, A. Bairoch and P. Bork, *Trends Genet.*, 1998, **14**, 248.
31. S. F. Altschul, T. L. Madden, A. A. Schaffer, J. Zhang, Z. Zhang, W. Miller and D. J. Lipman, *Nucleic Acids Res.*, 1997, **25**, 3389.
32. C. Gaboriaud, V. Bissery, T. Benchetrit and J. P. Mornon, *FEBS Lett.*, 1987, **224**, 149.
33. L. Lemesle-Varloot, B. Henrissat, C. Gaboriaud, V. Bissery, A. Morgat and J. P. Mornon,. *Biochimie* 1990, **72**, 555.
34. P. Bork and R. F. Doolittle, *Proc. Natl. Acad. Sci. USA*, 1992, **89**, 8990.

Structure and Synthesis of Carbohydrates and Carbohydrate Analogues

APPLICATION OF GLYCOSIDASES IN THE SYNTHESIS OF COMPLEX CARBOHYDRATES

D.H.G. Crout, P. Critchley, D. Müller, M. Scigelova, S. Singh and G. Vic

Department of Chemistry
University of Warwick
Coventry CV4 7AL, UK

The astonishing growth of interest in carbohydrates has been stimulated by the many biological studies that have revealed hitherto unrecognised roles in nature for these substances. If anyone is in any doubt about the number and extent of the systems in which carbohydrates have been implicated, these will be dispelled by reference to the frequently quoted review by Varki.[1]

Prominent among biological studies have been those dealing with adhesion, and in particular, adhesion of micro-organisms to mammalian tissues as the first step in the process of infection. To quote Jann and Jann "The great majority of bacterial infections are initiated by the adhesion of pathogenic bacteria to cells and mucosal surfaces of the host. .....With very few exceptions, adhesion is carbohydrate specific."[2]

The carbohydrate receptors on mammalian tissue to which proteins of the micro-organisms bind are not there to serve this rôle. They have quite a different primary purpose which has been subverted by micro-organisms to their own use. The need of pathogenic micro-organisms to attach themselves to specific carbohydrates offers the possibility of intervention by chemical agents that will interfere with adhesion and thereby prevent the development of full-scale infection. The possibilities for evasion of such a therapeutic intervention are limited because micro-organisms can work only on the carbohydrate receptors that are present on the surface of mammalian tissues. For this reason, there is a continuing and active interest in anti-adhesion measures as an alternative to antibiotic therapy for preventing or mitigating the effects of microbial infection. The obvious starting point in a search for inhibitors of carbohydrate-protein binding is the carbohydrate itself, which, in soluble form, can compete with the natural ligand for the adhesion protein. This has stimulated much activity directed towards the synthesis of the oligosaccharide components of the natural glycoconjugate receptors.

Examples of adhesion carbohydrates are shown in Fig. 1. Globotriose **1** (α-D-Gal*p*-(1$\rightarrow$4)-β-D-Gal*p*-(1$\rightarrow$4)-D-Glc*p*) is the receptor (as the ceramide Gb3) for Shiga toxin and Shiga-like toxins (SLTs or verotoxins). Shiga toxin is produced by *Shigella dysenteriae*, the organism responsible for many millions of cases of dysentery *per annum*, a significant proportion of which are fatal. It differs by only one amino acid residue from verotoxin 1, the toxin of the pathogenic *Escherichia coli* 0157. Gb3 is found on human tissue and is particularly abundant on kidney cells. The SLTs act to bind the toxin to the tissue. The toxin subsequently enters the cell and causes cell death by inhibiting protein synthesis (*vide infra*).[3] The bacterium *Helicobacter pylori* is the causative agent of chronic gastritis leading to peptic ulcers and gastric carcinoma. A carbohydrate receptor for the adhesion protein is the sialylgalactose **2**. This carbohydrate, formulated as NE-0080 (Neose Technologies Inc.) when administered to piglets, protected them from infection by *H. pylori*.[4]

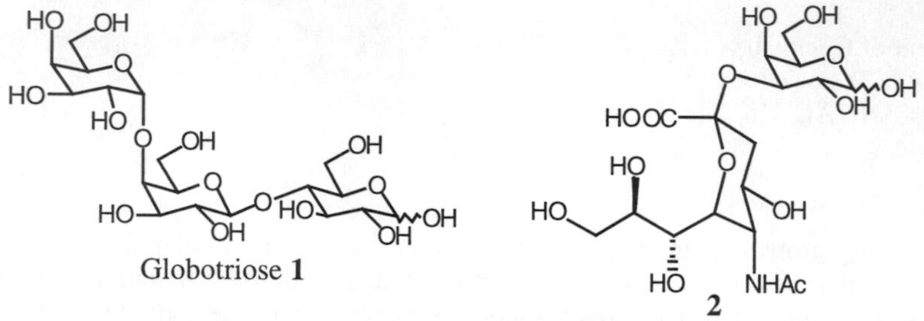

Globotriose **1**

2

Figure 1 *Carbohydrate receptors for adhesion proteins*

Shiga toxin and SLTs are AB_5 hexamers. The B chains are light (7.7 kDa). The A chain is larger (32 kDa). X-ray crystal structures of the B-pentamer[5], holotoxin[6] and holotoxin with bound globotriose[7] have been determined and an NMR structure of the B monomer has been obtained.[8] Following binding of the pentamer to Gb3, the A chain is translocated into the cell where, after proteolytic cleavage[9] it becomes an *N*-glycosidase. In the rat, the A chain catalyses hydrolysis of the adenosine residue at position 4324 of 28S ribosomal RNA, cleaving off the adenine residue.[10] This completely inactivates the ribosome, protein synthesis stops and cell death results.

The moderate to severe kidney damage observed in a proportion of cases is attributed to the high concentration of Gb3 on kidney cells. The lungs of cystic fibrosis patients are prone to infection with *Burkholderia cepacia*. The receptor for one adhesin from *B. cepacia* has been identified as Gb3. Patients with *B. cepacia* (20-30% in North America) have a poorer prognosis than patients free from infection.[11]

The remarkable selectivity of carbohydrate-protein interactions is illustrated in Fig. 2 which shows the gel electrophoresis (SDS-PAGE) of the total protein mixture from an organism producing Shiga toxin. The mixture was passed through a column of immobilised globotriose. Lanes A-C reveal proteins not retained by the column. After thorough washing, the column was eluted with guanidinium chloride to give Shiga toxin A and B chains as the only visible proteins (lanes D and E).

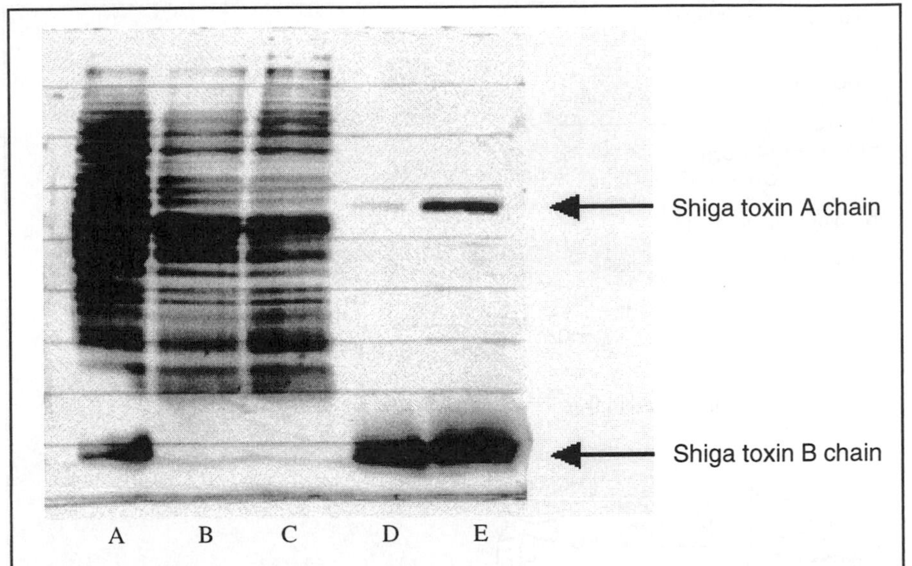

Figure 2 *SDS-PAGE of proteins eluted from a column of immobilised globotriose. A, B, C: Crude protein mixture D, E: 6 M Guanidinium chloride eluate*

The problem of working in the field of glycobiology are highlighted by reviewing the chemical synthesis of globotriose (Scheme 1),[12] which drew on previous work[13] for the synthesis of the glycosyl donor (Scheme 1a). Synthesis of the acceptor from lactose was straightforward, but required seven steps (Scheme 1b). It required sixteen steps to arrive at the trisaccharide with attached linker, in protected form (Scheme 1c). This is a major synthetic effort but conforms to the rule of thumb that on average, seven steps are required to attach a monosaccharide to an oligosaccharide under construction.

The length of the synthesis shown in Scheme 1 is largely determined by the multiplicity of protection-deprotection steps necessary to permit selective coupling of the saccharide units. Often, the effort required to produce oligosaccharides is the rate limiting step in glycobiology research. Against this background it was natural to turn to enzymatic methods of oligosaccharide synthesis as being potentially far more efficient. Two main approaches present themselves: the natural biosynthetic "Leloir" pathway and methods based on so-called "reverse hydrolysis" catalysed by glycosidases. The Leloir method requires a complex glycosyl donor (a sugar nucleotide derivative), an acceptor and the appropriate glycosyl transferase.[14]

The reactions are usually very efficient. Transfer is highly selective with respect to the linkage generated and control of configuration at the newly formed anomeric centre is absolute. Disadvantages are the complex nature of the donor and the inaccessibility of the enzymes, most of which are membrane-bound and few of which are readily available.

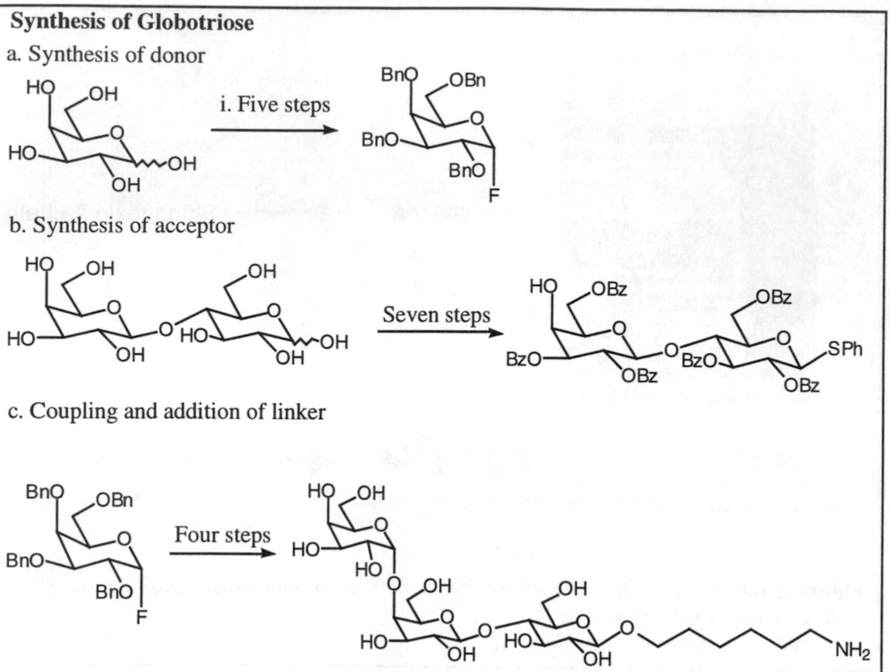

Synthesis of Globotriose

a. Synthesis of donor

i. Five steps

b. Synthesis of acceptor

Seven steps

c. Coupling and addition of linker

Four steps

Scheme 1

Nevertheless, the method has been widely applied, usually on a small scale, but occasionally on a multi-kilogram scale. The availability of the glycosyltransferases will undoubtedly improve, together with a corresponding increase in their use.[15]

The way in which glycosidases can be used to synthesise glycosides and oligosaccharides is illustrated in Scheme 2. Two acidic groups are implicated in catalysis. One protonates the leaving group, the other, in ionised form, attacks the anomeric centre to give a glycosyl-enzyme intermediate. Normally this intermediate is attacked by water (Scheme 2, ROH = H_2O) to give the hydrolysis product. However, if an alternative acceptor is present (ROH) this can intercept the intermediate to give a new glycoside (ROH = alcohol or oligosaccharide (ROH = saccharide). It is remarkable, but true, that although the acceptor, typically at 100 mM, is competing with water at 55 M, significant glycosyl transfer takes place to the acceptor ROH. This, of course, is simply a function of the relative values of k_{cat}/K_m for the acceptor ROH and water.

Preparative reactions can be carried out under thermodynamic or kinetic control. Under thermodynamic control, the reaction is allowed to evolve towards equilibrium, when the product mixture consists of substrate, hydrolysis product and glycosyl transfer product. Although this approach suffers from the disadvantage that a separation is always called for, its simplicity commends it many cases (*vide infra*). When XOH is a very activated leaving group, the composition of the product mixture is determined by the relative rates of the reaction of ROH and water with the glycosyl-enzyme intermediate.

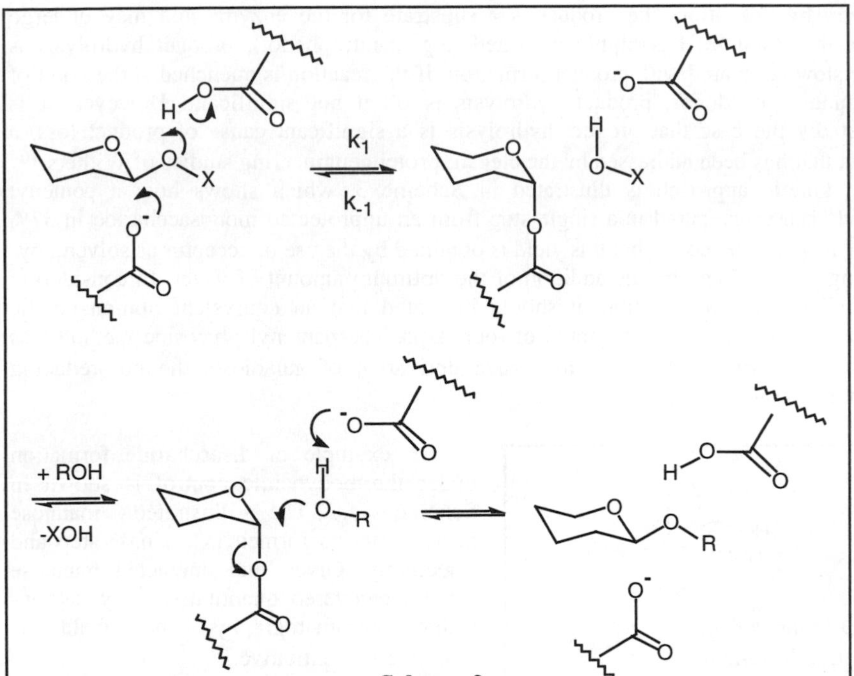

+ ROH

-XOH

k_1

k_{-1}

Scheme 2

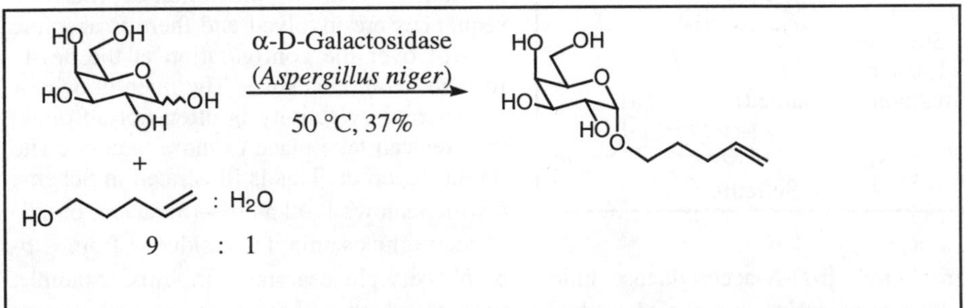

α-D-Galactosidase

(*Aspergillus niger*)

50 °C, 37%

: H_2O

9 : 1

$PhCH_2Br$/NaH/DMF

76%

Et_3SiOTf

NIS

40%

Scheme 3

However, by definition, the product is a substrate for the enzyme and may undergo hydrolysis. When XOH is highly activated (e.g. *p*-nitrophenol), product hydrolysis is usually slow compared with product formation. If the reaction is quenched at the point of disappearance of donor, product hydrolysis is often not significant. However, it is undoubtedly the case that product hydrolysis is a significant cause of product loss, a problem that has been addressed by the elegant protein engineering studies of Withers.[16]

The kinetic approach is illustrated in Scheme 3, which shows how a pentenyl glycoside can be prepared in a single step from an unprotected monosaccharide in 37% yield.[17] It should be noted that this yield is obtained by the use of acceptor as solvent, by operating at 50 °C and by the addition of the optimum amount of water. In considering the efficiency of this reaction, it should be noted that an equivalent non-enzymatic procedure would require a minimum of four steps. The pentenyl glycoside method was used in Fraser-Reid coupling[18] to give a derivative of galabiose, the non-reducing disaccharide component of globotriose.

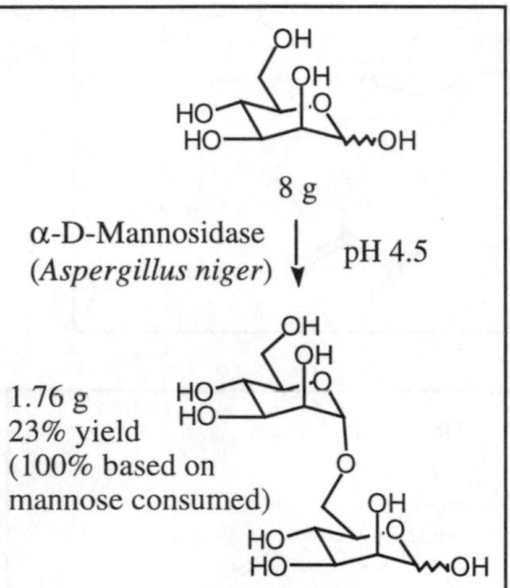

8 g

α-D-Mannosidase (*Aspergillus niger*) pH 4.5

1.76 g
23% yield
(100% based on mannose consumed)

Scheme 4

An example of disaccharide formation under thermodynamic control is shown in Scheme 4. The illustrated mannose disaccharide is formed in a single step and selectively. Given that unreacted mannose can be recovered quantitatively by carbon-Celite chromatography, the yield is effectively quantitative.[19]

The examples of Schemes 3 and 4 illustrate the advantages of the glycosidase method. No protection-deprotection sequences are involved and there is absolute control over the configuration at the newly formed anomeric centre. The main drawback is that regioselectivity is often not absolute; transfer can take place to more than one site on the acceptor. This is illustrated in Scheme 5 which shows 1→4 and 1→6 transfer of a β-*N*-acetylglucosaminyl residue from *p*-nitrophenyl β-D-*N*-acetylglucosaminide to *N*-acetylglucosamine. In this example, product purification is made easier by a stratagem which, although not general, can be applied in some cases. The product mixture was treated with a different *N*-acetylhexosaminidase (from *Canavalia ensiformis* [Jack bean]) that selectively hydrolysed the 1→6 product. By using carbon-Celite chromatography, whereby saccharides are separated by size, the 1→4 product, which was the only disaccharide in the final product mixture, was easily separated from the monosaccharide and was obtained in 55% yield.[20]

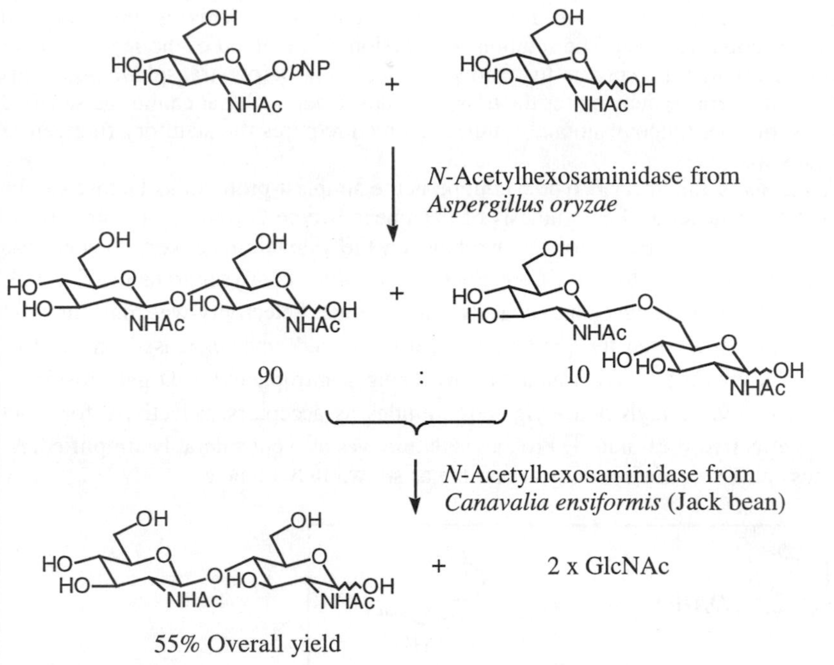

N-Acetylhexosaminidase from *Aspergillus oryzae*

90 : 10

N-Acetylhexosaminidase from *Canavalia ensiformis* (Jack bean)

+ 2 x GlcNAc

55% Overall yield

pNP = *p*-nitrophenyl

Scheme 5

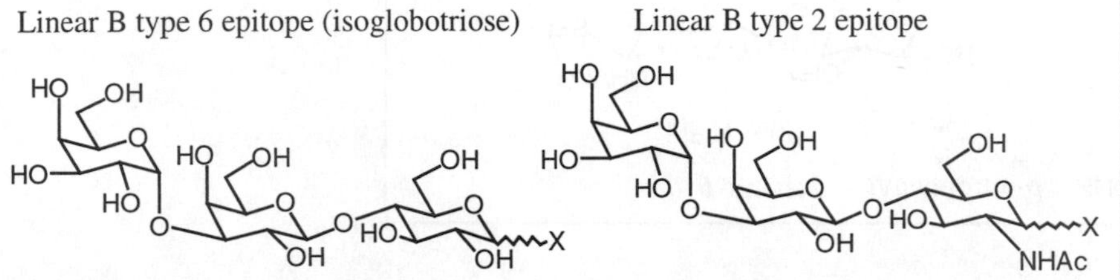

Linear B type 6 epitope (isoglobotriose)

Linear B type 2 epitope

Figure 3 *The trisaccharide epitopes on pig tissue responsible for the hyperacute rejection response in xenotransplantation.*

As a concluding example, we may consider the synthesis of isoglobotriose and its *N*-acetylglucosamine analogue (Fig. 3). These trisaccharides are of considerable biological interest. Isoglobotriose is the receptor for an adhesin from *Clostridium difficile*, an organism that occurs in the intestinal flora of about 3% of the elderly population. It is resistant to antibiotics and may flare up in such individuals following antibiotic therapy, producing severe colitis.[21] The two trisaccharides of Fig. 3 are found as epitopes on pig tissue. This gives rise to problems when xenotransplantation is attempted of pig organs into man or into old world monkeys. In man, approximately 1% of circulating immunoglobulins (antibodies) are specific for these carbohydrates. When they come into contact with the carbohydrate epitopes on introduced pig tissue, a dramatic immune response is initiated, resulting in rapid death of the tissue. This is the so-called hyperacute rejection response. The solution is perfusion of the blood of the recipient over immobilised carbohydrate epitope to remove the antibody before transplantation. This has led to a considerable demand for the trisaccharides, a demand that cannot be satisfied in a practical manner by conventional synthesis, which requires the statutory fourteen or so chemical steps.

Synthesis of the linear B type 6 epitope is the simplest problem as lactose can be used as a starting material. For synthesis of the linear B type 2 epitope, a source of *N*-acetyllactosamine was needed. For this we built on the valuable discovery by Japanese workers that the β-D-galactosidase from *Bacillus circulans* catalyses transfer of a β-D-galactosyl residue from lactose on to *N*-acetylglucosamine. Selectivity was high for 1→4 transfer but the 1→6 transfer product was also formed and increased in relative concentration with time.[22] We found that by using *p*-nitrophenyl β-D-galactoside as donor and thioalkyl or aryl- *N*-acetylglucosaminides as acceptors, selectivity for 1→4 transfer was effectively absolute.[23] Product isolation was also considerably simplified. *N*-Acetyllactosamine was made readily accessible as shown in Scheme 6.

β-Galactosidase (*Bacillus circulans*)

50% Yield

*p*NP = *p*-nitrophenyl **Scheme 6**

With a simple route to *N*-acetyllactosamine in hand, a further survey revealed that a strain of *Penicillium multicolor* produced an α-galactosidase that was highly selective for α1→3 transfer to both lactose and *N*-acetyllactosamine thioglycosides. Thus the linear B type 6 and type 2 epitopes were prepared in one and two enzymatic steps respectively from lactose or from *N*-acetylglucosamine and *p*-nitrophenyl α-D-galactoside (Schemes 6, 7).[24] Twelve to fourteen step chemical procedures are thereby reduced to one or two steps, powerfully illustrating the efficiency of the glycosidase-catalysed approach.

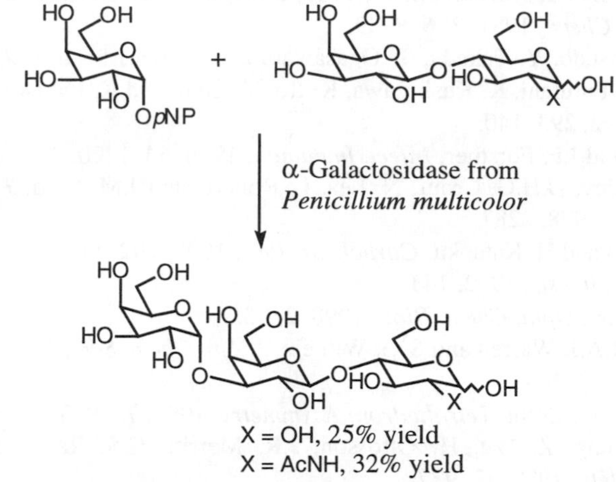

α-Galactosidase from *Penicillium multicolor*

X = OH, 25% yield
X = AcNH, 32% yield

*p*NP = *p*-nitrophenyl

Scheme 7

The syntheses described above, taken together with numerous other examples in the literature,[15] clearly show that the glycosidase-catalysed approach to oligosaccharide synthesis can now be considered as well-established and capable of providing a wide variety of oligosaccharides of biological interest.

This research was supported by the EPSRC, the BBSRC and by the Ciba Geigy Jubiläumsstiftung (to DM).

References

1. A. Varki, *Glycobiology*, 1993, **3**, 97.
2. K. Jann and B. Jann, 'Bacterial Adhesins' (Current Topics in Microbiology and Immunology 151) K. Jann and B. Jann (eds), Springer-Verlag, Berlin and Heidelberg 1990.
3. J.C. Paton and A.W. Paton, *Clin. Microbiol. Rev.*, 1998, **11**, 450.
4. D. Zopf, P. Simon, R. Barthelson, D. Cundell, I. Idanpaan-Heikkila and E. Tuomanen in 'Towards Anti-Adhesion Therapy for Microbial Diseases', I. Kahane and I. Ofek (eds), Plenum Press, New York and London, 1996, p.35.
5. P.E. Stein, A. Boodhoo, G.J. Tyrell, J.L. Brunton and J.R. Read, *Nature*, 1992, **355**, 748.
6. M.E. Fraser, M.M. Chernaia, Y.V. Kozlov and M.N.G. James, *Nature Struct. Biol.*, 1994, **1**, 59.
7. H. Ling, A. Boodhoo, B. Hazes, M.D. Cummings, G.D. Armstrong, J.L. Brunton and R.J. Read, *Biochemistry*, 1998, **37**, 1777.
8. J.M. Richardson, P.D.Evans, S.W. Homans and A .Donohue-Rolfe, structure 4ull.pdb in the Protein Databank.

9. S. Olnes, R. Riesbig and K. Eiklid, *J. Biol. Chem.*, 1981, **2546**, 8732; R. Reisbig, S. Olnes and K. Eiklid, *J. Biol. Chem.*, 1981, **256**, 8739.
10. Y. Endo, K. Tsurugi, T. Yutsudo, Y. Takeda, T. Ogasawara and K. Igarishi, *Eur. J. Biochem.*, 1988, **171**, 45; M. Furutani, K. Kashigawa, K. Ito, Y. Endo and K. Igarishi, *Arch. Biochem. Biophys.*, 1992, **293**, 140.
11. F.A. Sylvester, U.S. Sojjan and J.F. Forstner, *Infect. Immunol.*, 1996, **64**, 1420.
12. D. Müller, G. Vic, P. Critchley, D.H.G. Crout, N. Lea, L. Roberts and J.M. Lord, *J. Chem. Soc., Perkin Trans. 1,* 1998, 2287
13. K.C. Nicolaou, T.J. Caulfield and H. Kataoka, *Carbohydr. Res.,* 1990, **202**, 177.
14. J. Thiem and T. Wiemann, *Synthesis*, 1992, 141.
15. D.H.G. Crout and G. Vic, *Curr. Opin. Chem. Biol.*, 1998, **2**, 98.
16. L.F. Mackenzie, Q. Wang, R.A.J. Warren and S.G. Withers, *J. Am. Chem. Soc.*, 1998, **120**, 5583.
17. G. Vic, J.J. Hastings and D.H.G. Crout, *Tetrahedron: Asymmetry*, 1996, **7**, 1973.
18. B. Fraser-Reid, U.E. Udodung, Z. Wu, H. Ottosson, J.R. Merritt, C.S. Rao, C. Roberts and R. Madsen, *Synlett.*, 1992, **12**, 927.
19. M. Scigelova, S. Singh and D.H.G. Crout, unpublished.
20. S. Singh, J. Packwood and D.H.G. Crout, *J. Chem. Soc., Chem. Commun.*, 1994, 2227.
21. G.F. Clark, H.C. Krivan, T.D. Wilkins and D.F. Smith, *Arch. Biochem. Biophys.*, 1987, **257**, 217.
22. K. Sakai, R. Katsumi, H. Ohi, T. Usui and Y. Ishido, *J. Carbohydr. Chem.*, 1992, **11**, 553.
23. G. Vic, C.H. Tran, M. Scigelova and D.H.G. Crout, *Chem. Commun.*, 1997, 169.
24. M. Scigelova, S. Singh and D.H.G. Crout, unpublished.

PROTEIN ENGINEERING OF THE β-GLYCOSIDASE FROM THE HYPERTHERMOPHILIC ARCHAEON *SULFOLOBUS SOLFATARICUS*: MECHANISMS OF GLYCOSIDE HYDROLYSIS.

M. Moracci[a] , G. Perugino[a], A. Trincone[b], M. Ciaramella[a] and M. Rossi[a,c]

[a]Institute of Protein Biochemistry and Enzymology-CNR, Via Marconi 10, 80125, Naples, Italy. [b]Istituto per la Chimica di Molecole di Interesse Biologico-CNR, Via Toiano 2, Arco Felice 80072, Naples, Italy. [c]Dipartimento di Chimica Organica e Biologica, Universita' di Napoli 'Federico II', Via Mezzocannone 16, 80134 Naples, Italy.

1 INTRODUCTION

Glycosyl hydrolases follow two reaction mechanisms, one producing the overall retention and the other the inversion of the anomeric configuration of the substrate. Both mechanisms involve two carboxylic groups in the active site highly conserved in each family. In *inverting* enzymes (Scheme 1A) these residues function as an acid and a basic catalyst, respectively, and operate with a single displacement of the leaving group (1). Instead, *retaining* enzymes (Scheme 1B) follow a two-step mechanism with formation of a covalent glycosyl-enzyme intermediate. The carboxyl group in the active center functions as a general acid/base catalyst and the carboxylate as the nucleophile of the reaction (2, 3).

Family 1 of glycosyl hydrolases compiles enzymes which follow the *retaining* mechanism but show different substrate specificity: β-glycosidases with broad specificity, 6-phospho- β-gluco and galactosidases, myrosinases and lactases. One of the most hyperthermophilic members of Family 1 is the β-glycosidase from the Archaeon *Sulfolobus solfataricus* (Ssβ-gly; E.C. 3.2.1.x) which show optimal activity at temperatures over 85°C and remarkable thermostability: half life at 75°C > 24 hours (4). Ssβ-Gly shows broad substrate specificity for the hydrolysis reaction of several glycosides and β1-3, β1-4 and β1-6 glucose dimers. A remarkable exo-glucosidase activity was found against oligosaccharides with up to 5 glucose residues, which are hydrolyzed from the non-reducing end (5). This enzyme is also capable to efficiently synthesize different β-D-glycosides by transglycosylation (6).

Here, we report the identification of the residues involved in catalysis and the study of the modification of the reaction mechanism of Ssβ-gly by coupling site-directed mutagenesis, the action of external nucleophiles and the use ^{13}C-labelled substrates.

2 IDENTIFICATION OF THE ACTIVE SITE CARBOXYLIC GROUPS

The two carboxylic residues essential for catalysis in Ssβ-gly were searched among glutamic or aspartic acids conserved among mesophilic and thermophilic Family 1 glycosyl hydrolases. The glutamic acid 206 and 387 residues of Ssβ-gly have been found in two totally conserved motives called NEP and ENG, corresponding to the general acid/base catalyst and the nucleophile, respectively, and previously identified in the

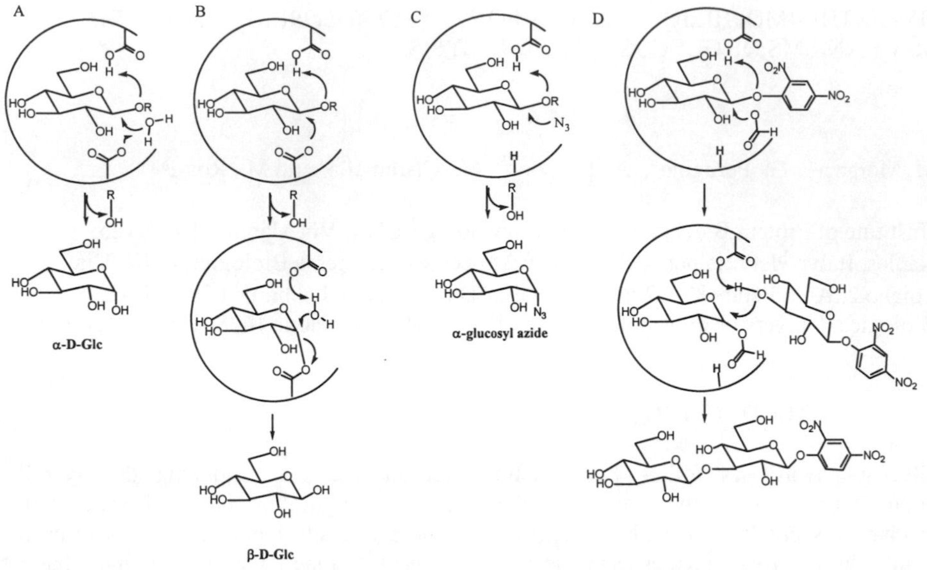

Scheme 1

active site of *Agrobacterium* β-glucosidase (7, 8). These residues were changed to isosteric glutamine by site-directed mutagenesis. In order to facilitate the purification of proteins expected to have impaired enzyme activity, wild type and mutant enzymes were expressed by fusing their amino-terminus to the *Schistosoma japonicum* glutathione S-transferase (9). Chimeric proteins expressed from this vector were efficiently cleaved, producing recombinant Ssβ-gly with a seven-residues extension at its amino-terminus. The recombinant enzyme does not differ significantly from the native one regarding pH optimum, thermostability, thermophilicity and kinetic parameters (data not shown); hence, this form of Ssβ-gly was used in the following characterisation and it will be defined 'wild type' throughout the paper.

2.1 Nucleophile

The E387Q mutant was totally inactive against all the substrates tested and the absence of activity could not allow reliable estimates of kinetic parameters.

The E387 is totally conserved in Family 1 and its crucial function has been confirmed in several enzymes from this family. Our data are consistent with the function of E387 as the attacking nucleophile (see also below).

2.2 Acid/base catalyst

The E206Q mutation strongly affects, but does not completely abolish, glycosidase activity: although the mutant is completely inactive on disaccharides (10), and lacks any transglycosylating activity (data not shown), it shows a residual activity on

2- or 4-nitrophenyl-β-D-glycosides. Kinetic constants were determined for E206Q mutant at 65°C and compared to the wild type (Table 1). A 10-30-fold decrease of the K_M values and a 60-fold decrease of the k_{cat} occurred upon mutation on 2- or 4-nitrophenyl-β-D-glucosides (2/4NpGlc), suggesting that the glucosyl-enzyme intermediate is accumulated during the reaction. k_{cat} values on 2- or 4-nitrophenyl-β-D-galacto and fucoside (2/4NpGal and Fuc) were less reduced. These results suggest that 2- or 4-Np-gluco and galactoside substrates make different interactions with the E206 residue in the Ssβ-gly active site.

Another indication that the general acid/base catalyst has been removed is the change in the activity versus pH upon mutation. Wild type Ssβ-gly shows a typical bell-shaped curve, with maximum activity at pH 6.5.

Table 1 *Kinetic constants of hydrolysis at 65°C by wild type and mutant E206Q*

Substrate	K_M (mM)		k_{cat} (s^{-1})	
	wild type	E206Q	wild type	E206Q
2-NpGal	0.95±0.08	1.53±0.1	295±6	34±0.8
4-NpGal	1.17±0.06	4.79±0.17	275±4	31±0.4
2-NpGlc	1.01±0.24	0.03±0.006	252±12	4.46±0.13
4-NpGlc	0.30±0.04	0.03±0.003	240±7	4.40±0.06
4-NpFuc	0.45±0.11	0.09±0.005	263±12	35±0.4

The E206Q mutation strongly affects the activity dependence on pH, confirming that the residual activity measured is due to the mutant itself and not to wild type contamination.

The low activity of E206Q on 4-NpGal shows the same thermal activation of the wild type, and no anomalous Arrhenius behavior over the whole temperature range was detected for both enzymes.

The results presented strongly suggest that glutamic acid 206 in Ssβ-gly is the general acid/base catalyst for the hydrolysis reaction. The studies reported so far on glycohydrolases mutated in the general acid/base catalyst demonstrate that the kinetic parameters can be affected at different extents (8, 11, 12). The reasons for these differences are not clear; it is possible that enzymes with different substrate specificities require the assistance of the general acid/base catalyst at different extent.

3 REACTIVATION OF COMPLETELY INACTIVE MUTANTS

It has been demonstrated that the reaction mechanism of glycosyl hydrolases can be converted by mutagenesis replacing residues involved (13, 14) or not involved (15) in catalysis. In the former case, the enzyme activity was restored by providing high concentrations of external nucleophiles and only by using 2,4-dinitrophenyl-based substrates which possess good leaving groups.

To evaluate whether the reaction mechanism of Ssβ-gly could be affected by mutation and subsequent reactivation with external nucleophiles, two new mutants of the active site nucleophile (E387A and E387G) were prepared. Extremely low activity on all substrates was recorded for the mutants E387A/G/Q; when sodium azide or formate were included in the reaction mixture as external nucleophiles a considerable rescue of the activity was obtained with E387A and G mutants on 2,4-dinitrophenyl-β-D-glucoside (2,4-DNp-β-Glc). Remarkably, the E387G mutant showed increased activity, with the same nucleophiles, also on 2-Np-β-Glc. No activity rescue could be obtained with the E387Q mutant on all the substrates tested under the same conditions. It was inferred that this result might be due to the different nature of the substituting amino acid: the larger side chain of glutamine, as compared with those of alanine and glycine, could not allow the external nucleophile to access the substrate in the active site. The residual activity observed in the three mutants without nucleophiles was unexpected since the glutamic acid 387 is absolutely required for the reaction. Experiments with a mechanism based inhibitor indicated that this small residual activity was due to small amounts of wild type enzyme in the mutant preparations (16). The origin of the wild type contamination is still obscure and probably occurred during the mutant enzyme preparation, however, it was demonstrated negligible for the purpose of this work (16).

3.1 Kinetic characterization of E387A and G mutants

The kinetic constants for hydrolysis at 65°C shown by wild type and mutants Ssβ-gly for 2,4-DNp-β-Glc and 2-Np-β-Glc substrates are reported in Table 2 and 3, respectively.

The maximal reactivation was observed with the E387G mutant on 2,4-DNp-β-Glc whose specific activity was 40% and 15% with azide and formate, respectively, if compared to the wild type enzyme without nucleophiles.

The affinity to the substrate appeared unchanged upon mutation, whereas the specificity constants with azide and formate were 38% and 20% respectively compared to the wild type value. In the case of the mutant into alanine, the calculated k_{cat} and k_{cat}/K_M values for the same substrate, with 2 M sodium azide, were 3.5% and 0.5% those of the wild type, respectively. Both the catalytic efficiency and the reaction rate enhancements on 2,4-DNp-β-Glc observed with the two mutants, and particularly with E387G, are remarkably high if compared with mesophilic glycohydrolases (13, 14).

Peculiarly, significant reactivation of the E387G mutant on 2-Np-β-Glc which has a relatively weak leaving group was observed (Table 3). Also in this case the affinity to the substrate did not change upon mutation, but the catalytic efficiency was 3.5% and 8.4% with azide and formate, respectively, if compared to the wild type without nucleophiles.

3.2 Identification of the products: reaction with sodium azide

In order to evaluate whether the reaction mechanism was affected by mutation, the analysis of the products of the reaction using anomerically ^{13}C labelled substrate catalyzed by wild type and mutant Ssβ-gly was performed using ^{13}C NMR spectroscopy. To ascertain the feasibility of the technique, the analysis was at first performed by using wild type Ssβ-gly for hydrolysis of 2,4-DNp-β-Glc in the absence of external anions at room temperature.

Table 2 *Kinetic constant for 2,4-DNp-β-Glc hydrolysis at 65°C of wild type and E387A/G mutants Ssβ-gly*[a]

Enzyme	K_M (mM)	k_{cat} (s^{-1})	k_{cat} / K_M (s^{-1} mM^{-1})
wild type	0.17±0.04	275±16	1617
wild type + 2M azide	1.64±0.20	428±20	261
wild type + 2M formate	0.61±0.06	367±12	602
E387A	NA[b]	NA	-
E387A + 2M azide	1.21±0.12	9.6±0.3	8
E387G	NA	NA	-
E387G + 2M azide	0.18±0.03	110±4.7	611
E387G + 2M formate	0.13±0.02	42±1.2	323

[a]At least part of these activities appears to be due to contaminating wild type enzyme (see text for details). Values are corrected for the spontaneous rate observed in the absence of enzyme. [b]NA (no measurable activity) means that, using concentrations of enzyme of 10 μg/mL in the assay, the rates of change in absorbance did not vary in the experimental conditions and were approximately the same as in the control without enzyme.

Table 3 *Kinetic constant for 2-Np-β-Glc hydrolysis at 65°C of wild type and E387G mutant Ssβ-gly*[a]

Enzyme	K_M (mM)	k_{cat} (s^{-1})	k_{cat} / K_M (s^{-1} mM^{-1})
wild type	1.01±0.24	538±11	533
wild type + 2M azide	0.98±0.10	480±13	490
wild type + 2M formate	0.50±0.07	425±11	850
E387G	NA[b]	NA	-
E387G + 2M azide	0.80±0.11	15±0.5	19
E387G + 2M formate	1.17±0.12	53±1.2	45

[a]At least part of these activities appears to be due to contaminating wild type enzyme (see text for details). Values are corrected for the spontaneous rate observed in the absence of enzyme. [b]NA (no measurable activity) see Table 2

Almost complete enzymatic hydrolysis is clearly observed after 20 h. The fact that β-glucose is the first compound formed during the hydrolysis of 2,4-DNp-β-Glc, was further direct indication of the *retaining* mechanism followed by this thermophilic enzyme in the hydrolysis reaction.

A similar procedure was followed for the reaction catalyzed by E387G mutant enzyme in the presence of sodium azide, taking care, by suitable blank experiment, to rule out the possibility of a chemical formation of azides. After 10 min at 75 °C the spectrum indicated the presence of a signal due to α-glucosyl azide. This spectrum also demonstrated that the reaction mixture did not contain any anomeric form of glucose and this led to the conclusion that E387G enzyme performs the hydrolysis of the substrate at high temperatures in the presence of sodium azide as an *inverting* glycosidase.

These findings indicate that the reaction catalyzed by the E387G mutant must proceed via direct attack of the azide to the α-side of the C1 of the substrate producing α-glucosyl azide (Scheme 1C). The attack is extremely selective since no β-glucosyl azide product could be detected by ^{13}C NMR spectroscopy experiments with the mutants. By contrast, the wild type enzyme, when assayed in the presence of azide, followed the mechanism showed in Scheme 1B: in this case the small anion competes with water in the second step of the reaction and forms β-glucosyl azide by glycosyl transfer reaction.

3.3 Identification of the products: reaction with sodium formate

In the reaction of E387G with sodium formate two interesting signals at 104.69 and 101.91 ppm originated from the anomeric carbon of the ^{13}C-labelled substrate after a 15 min reaction time (Figure 1a).

The two signals did not appear in two blank experiments where sodium formate or the biocatalyst were respectively absent. An experiment on analytical scale was then performed for the identification of the reaction products. Chemical identification of the reaction product indicate that the compound is a disaccharide derivative of 2,4-dinitrophenol in which the additional glucose molecule was β-linked to the 3-O-position of glucose. The disaccharide formed by the mutant survives in the reaction mixture and was consumed on addition of external wild type enzyme and incubation at room temperature. Spectra were taken at 5 min intervals showing a broadening of the signal at 104.69 and 101.91 (Figure 1b).

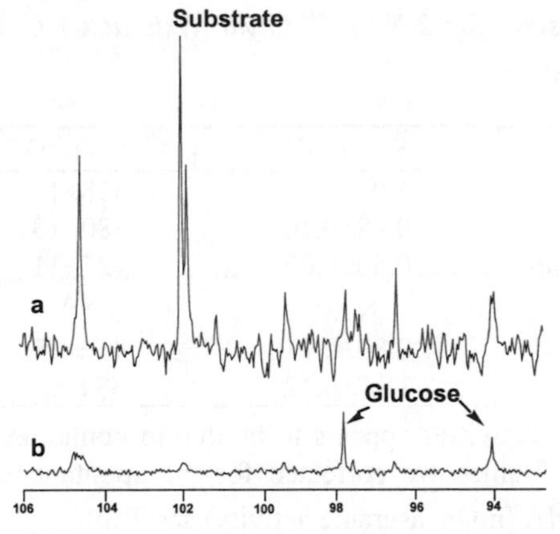

Figure 1 *Enzymatic reaction performed by E387G mutant in the presence of sodium formate.*

Preliminary results indicate that also in the presence of sodium formate and 2-Np-β-Glc as substrate, the mutant catalyzed the formation of a 3-O β-linked disaccharide derivative of the substrate. The role of sodium formate in the reaction catalyzed by Ssβ-gly mutant appears different from the role of azide described above. In the presence of formate, the anomeric configuration of the product indicates that the E387G behaves as a *retaining* enzyme. This is the first evidence of this kind since no product characterization in the reaction with formate as external agent was reported for mesophilic enzymes (13, 14). The restoration of the *retaining* reaction implies a two step mechanism. The formate would participate in the first step assisting the leaving of the phenolate group and then stabilizing the glucosyl unit as assistant nucleophile until it was transferred to a new substrate molecule (Scheme 1D). These results indicate that sodium formate has the optimal nucleophilicity for acting as a biomimicking agent since it is capable of restoring the function of the carboxylate group 387 removed by mutation, and therefore it makes the synthesis of β-products possible after enzymatic processing of α-glucosyl formate. This intermediate product has been in fact observed by a ^{1}H NMR monitoring of the reaction in a mesophilic β-glucanase (17). Hence, this mutant can be exploited as a thermophilic *glycosynthase* for the synthesis of disaccharides through the transglycosylation pathway described in Scheme 1D. A similar approach has been recently reported by using *Agrobacterium* β-glucosidase and *Bacillus* β-glucanase mutants and α-glycosyl fluoride as substrate (18,19).

4 CONCLUSIONS

The modification of the kinetic properties of the glycosyl hydrolases by protein engineering is an interesting challenge for both basic and applied research. With this approach, the *retaining* mechanism of Ssβ-gly has been characterized in detail through the identification of E206 and E387 as the the general acid/base and the nucleophile of the reaction, respectively. Moreover, we have demonstrated that completely inactive mutants in E387 could be reactivated by the addition of external nucleophiles. The mechanism of the catalyzed reaction resulted to be dependent on the nucleophilicity of external anions. In the presence of azide the mutant enzymes gave rise to product with *inverted* configuration (α-glycosyl-azide). Instead, surprisingly enough, in the presence of formate a retaining mechanism was followed and the observed product was a β-disaccharide which could not be hydrolyzed by the mutant.

5 ACKNOWLEDGMENTS

We thank Dr. S.G. Withers for kindly providing 2,4-DNp-β-Glc substrates used for the preliminary experiments described in this work; the staff of ICMIB-CNR NMR service (S. Zambardino and V. Mirra) for NMR spectra, Mr. Giovanni Imperato and Mr. Ottavio Piedimonte for the technical assistance. This work was supported by EC project: "Biotechnology of extremophiles" contract n. BIO2-CT93-0274 and by the EC FAIR project: "Enzymatic Lactose Valorization" contract n. 1048/96.

References

1. J. D. McCarter and S. G. Withers *Curr. Op. Struct. Biol.,* 1994,**4**, 885.
2. D. E. Koshland, *Biol. Rev.,* 1953, **28**, 416.
3. A. White and D. R. Rose*Curr. Op. Struct. Biol.,* 1997, **7**, 645.
4. M. Moracci, R. Nucci, F. Febbraio, C. Vaccaro, N. Vespa, F. La Cara and M. Rossi *Enzyme Microb. Technol.,* 1995, **17**, 992.
5. R. Nucci, M. Moracci, C. Vaccaro, N. Vespa and M. Rossi *Biotechnol. Appl. Biochem.,* 1993, **17**, 239.
6. A. Trincone, R. Improta, R. Nucci, M. Rossi and A. Gambacorta *Biocatalysis*, 1994, **10**, 195.

7. S. G. Withers, R.A.J. Warren, I. P.Street, K. Rupitz, J. B. Kempton and R. Aebersold *J. Am. Chem. Soc.,* 1990, **112**, 5887.
8. Q. Wang, D. Trimbur, R. Graham, R.A.J. Warren and S. G. Withers *Biochemistry*, 1995, **34**, 14554.
9. M. Moracci, L. Capalbo, M. De Rosa, R. La Montagna, A. Morana, R. Nucci, M. Ciaramella and M. Rossi, in *Carbohydrate Bioengineering*, S. B. Petersen, B. Svensson and S. Pedersen (eds.) Elsevier Science B.V., Amsterdam, 1995, 77.
10. M. Moracci, L. Capalbo, M. Ciaramella and M. Rossi, *Prot. Engineer.*, 1996, **9**, 1191.
11. E. Witt, R. Frank and W. Hengstenberg, *Protein Eng.,* 1993, **6**, 913.
12. A. M. MacLeod, T. Lindhorst, S. G. Withers and R. A. J. Warren *Biochemistry,* 1994, **33**, 6371.
13. Q. Wang, R. W. Graham, D. Trimbur, R. A. J. Warren, and S. G. Withers, *J. Am. Chem. Soc.*, 1994, **116**, 11594.
14. A. Mac Leod, D. Tull, K. Rupitz, R. A. J. Warren and S. G. Withers, *Biochemistry*, 1996, **35**, 13165.
15. R. Kuroki, L. H. Weaver and B. W. Matthews, *Nature Struct. Biol.,* 1995, **2**, 1007.
16. M. Moracci, A. Trincone, G. Perugino, M. Ciaramella and M. Rossi, *Biochemistry*, 1998, **37**, 17262.
17. J.L. Viladot, E. de Ramon, O. Durany and A. Planas *Biochemistry*, 1998, **37**, 11332
18. L. F. Mackenzie, Q. Wang, R. A. J. Warren and S. G. Withers, *J. Am. Chem. Soc.*, 1998, **120**, 5883.
19. C. Malet and A. Planas, *FEBS Lett.* 1998, **440**, 208.

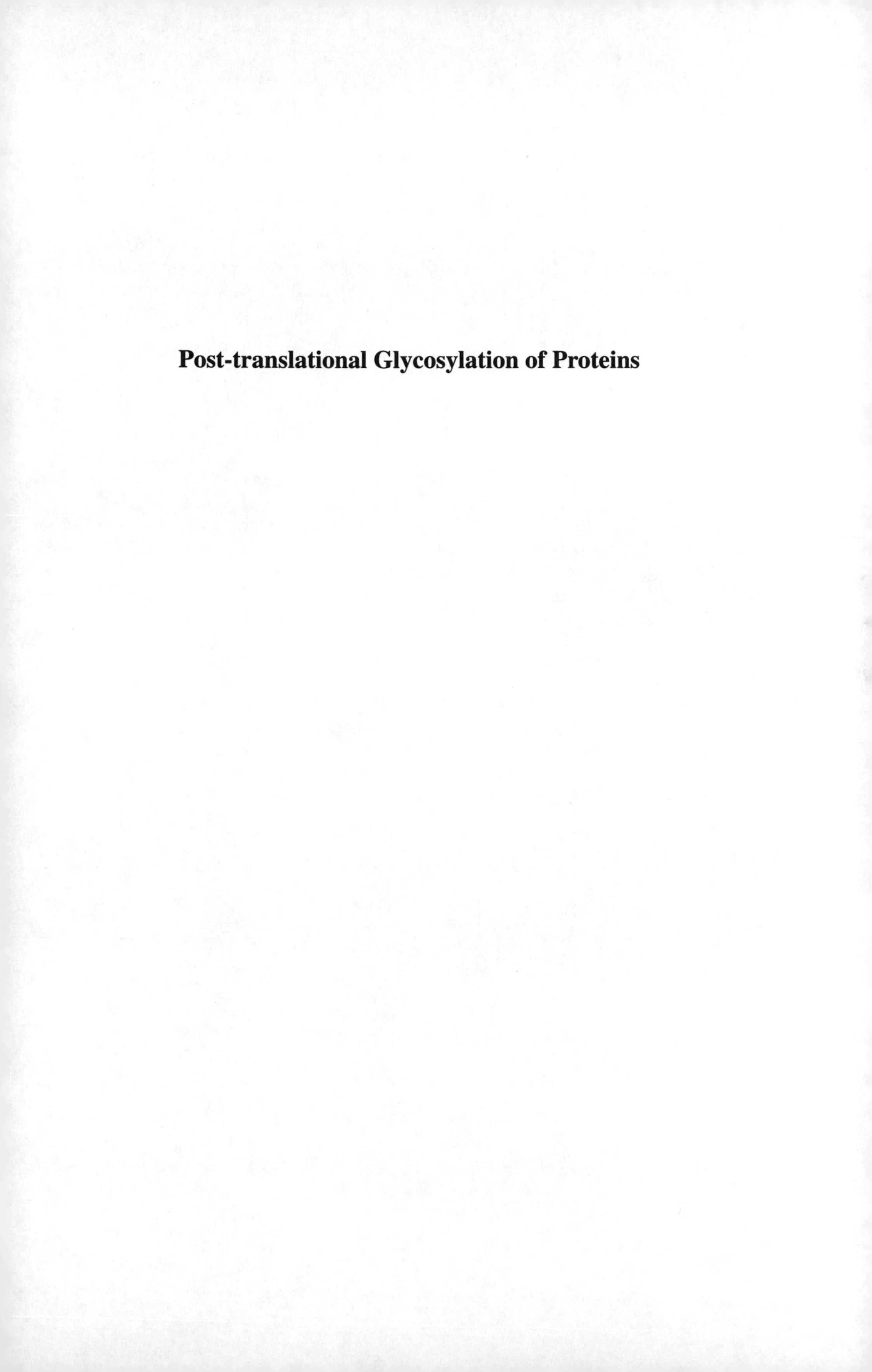

Post-translational Glycosylation of Proteins

OVEREXPRESSION OF α-D-MANNOSE-1-PHOSPHATE GUANYL-TRANSFERASE ENCODING GENE RESTORES THE VIABILITY OF THE *SACCHAROMYCES CEREVISIAE* MUTANTS AFFECTED IN EARLY STEPS OF GLYCOCONJUGATE FORMATION

Anna Janik, Joanna Kruszewska, Urszula Lenart, Monika Sosnowska and Grażyna Palamarczyk

Institute of Biochemistry and Biophysics, Polish Academy of Sciences
Pawińskiego 5a, 02-106 Warszawa, Poland

1 INTRODUCTION

GDP-mannose acts as a mannose donor in lipid and protein glycosylation. In the synthesis of N-linked glycans the oligosaccharide chain is linked to dolichyldiphosphate and DolPP-$GlcNac_2Man_9Glc_3$ (LLO) is formed. All the mannose residues of the glycan originate from GDPMan, either directly or via dolichylphosphate, the lipid intermediate. It is believed (for reviews see 1,2,3), that the first five mannose residues are added directly from GDPMan at the cytosolic side of the endoplasmic reticulum (ER) and the subsequent four from dolichylphosphate mannose (MPD) at the lumenal side of ER. MPD is synthesised from GDPMan and dolichylphosphate by MPD synthase. In yeast and filamentous fungi MPD is also a substrate for protein O-mannosylation, where it serves as a donor of the first mannose to be attached to hydroxyl groups of serine and threonine in the protein. The second and subsequent mannose residues are transferred directly from GDPMan (1, 3). MPD is also involved in the synthesis of the sugar part of glycosylphosphatidyl inositol anchor in yeast and other eukaryotes (3). MPD synthase, encoded in *Saccharomyces cerevisiae* by the *DPM1* gene, plays therefore a central role in three different pathways in the ER (4).

Studies *in vitro* on the biosynthesis of N-linked glycans in retina (5) lead to the conclusion that GDPMan, in addition to acting as a substrate for mannosyltransferase, acts in an indirect manner to stimulate the biosynthesis of non-mannose-containing GlcNac-lipids, a process which may participate in the regulation of glycoprotein biosynthesis. A similar observation was also made for thyroid tissue (6). We have reported earlier (7) that in filamentous fungus *Trichoderma reesei,* GDPMan stimulates mannosyl transfer from MPD to the hydroxyl groups of serine and threonine and thus in turn affects the rate of protein mannosylation.

Several lines of evidence has been presented, that the level of GDPMan affects the process occuring in the Golgi. compartment i.e. the outer mannose chain elongation of *Saccharomyces cerevisiae* glycoproteins (8, 9).

The results presented here indicate, that the avaiability of GDPMan affects also reactions ascribed to the endoplasmic reticulum, the early steps of glycoprotein formation i.e. assembly of the dolichol-linked oligosaccharide as well as mannosyl-phosphodolichol (MPD) formation.

1.1 Potential sources of GDPMan and its turnover in the cell

Biosynthesis of GDPMan like the other sugar nucleotides is the process occurring in the cytoplasm. The final reaction of sugar nucleotide assembly (Fig1) is preceded by the conversion of Man-6-phosphate to Man-1-P catalysed by phosphomannomutase (PMM). The evidence has been presented that in yeast the enzyme is coded by the *SEC53* gene, which is defective in the class B secretion mutant *sec53*, which seems to be blocked at an early stage in protein translocation into endoplasmic reticulum (10, 11). Alleles of *sec53* were also isolated as *alg4* in a [^{3}H] mannose suicide screen for glycosylation defective mutants (12). Mutants in PMM are blocked in N-glycosylation, O-mannosylation, GPI-anchoring and in the formation of mannosylphosphoceramides (4, 11-14), which seems to be a pleiotropic effect due to a block in the synthesis of GDPMan. The bulk of cellular Man-6-P might derive also by conversion of glucose [Fig 1, according to (15)]. The reaction is catalysed subsequently by phosphogluco- and phosphomannose isomerases. The final reaction in GDPMan synthesis (compare Fig.1) is transfer of mannose-1-phosphate into GTP which is catalysed by MPG transferase (EC 2.7.7.13). This enzyme is also known as GDP-α-D-mannose pyrophosphorylase. In the yeast *S.cerevisiae* MPG -transferase is encoded by the *MPG1* gene , the complete sequence of which has been deposited in the yeast genome data base by Shultz and Sprague (acc nb P41940). Alleles of *MPG1* were isolated also as the *PSA1* (17) a plasmid supressor of the *alg1* mutation in the yeast asparagine N-glycosylation pathway (3) and as the *VIG9* gene (18) by the screen of the collection of vanadate-resistant mutants (19). The gene is essential, encodes a 361 amino-acid protein with a high homology to NDP-hexose pyrophosphorylases, the enzyme that catalyse the formation of activated sugar nucleotides (17). We have recently cloned a cDNA fragment coding for MPG1 transferase in *Trichoderma reesei,* a filamentous fungus, by suppression of the yeast *Saccharomyces cerevisiae* mutation in the *DPM1* gene encoding MPD synthase. The nucleotide sequence of the 1.6 kb long cDNA revealed an ORF which encodes a protein of 364 amino acids. Sequence comparisons demonstrate 70% identity with the *S. cerevisiae* guanyl transferase gene (*MPG1)* and 75 % identity with the *Schizosaccharomyces pombe* homologue. (20). Overexpression of the *S. cerevisiae MPG1* gene in the wild type yeast resulted in the 3-fold increase of GDPMan concentration (9). Similarly a significant increase of MPG-transferase was observed upon expression of the *Trichoderma mpg1* gene in the yeast *dpm1*-mutant.(20) and the *VIG9* gene in the *vig9* mutant (18), substantiating further the assumption that the *MPG1/ VIG9* are the structural genes for MPG-transferase.

The cellular level of GDPMan might be also affected by the mutations in the genes encoding enzymes involved in the synthesis of GTP. One of them involves impairment in the transport of GDPMan from the site of its synthesis i.e., cytoplasm into the Golgi compartment in order to serve as a mannose donor for the extension of the outer mannan chain. Import of GDPMan into Golgi apparatus of *Saccharomyces cerevisiae*, analogously as in the animal cells, involves an antiport mechanism. GDPMan enters the lumen via a specific membrane-bound carrier protein and donates its mannose residue to the endogenous acceptor, releasing GDP, which is in turn is hydrolysed to GMP, whose exit from the Golgi compartment is coupled to the further import of GDPMan. This model was supported by the further results from Hirshberg Laboratory when the essential component of this antiport mechanism i.e. the Golgi GDP-ase, encoded by the *GDA* gene, was identified (8, 21)

More recentlly a defect in GTP synthesis, mutation in GMP-kinase encoded by the *GUK1* gene, has been described (9). This is the cytoplasmic step responsible for the conversion of GMP to GTP which in turn affects the concentration of GDPMan (Fig 2).

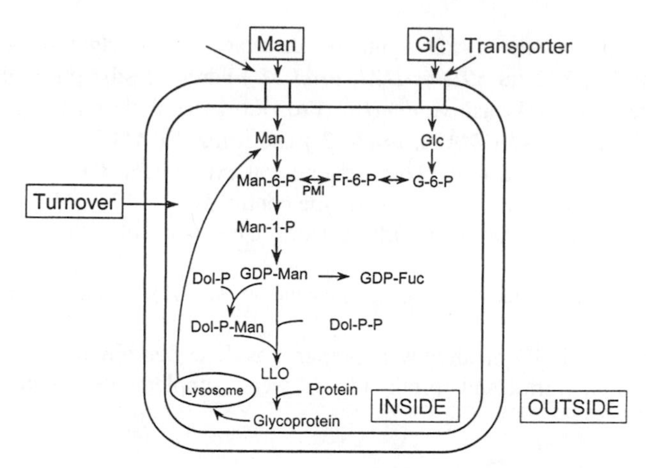

Figure 1 *Cellular mechanism of GDPMan synthesis and turnover (according to 15). GDPMan is synthesised either from mannose or from glucose, which are translocated into the cell via specific transporters. The bulk of GDPMan takes part in several glycosylation reactions, localised in the ER lumen (synthesis of MPD and LLO) and in the Golgi compartment (synthesis of the outer sugar chain of glycoproteins). In animal cells some of the glyoproteins are transported to lysosomes, and degraded enriching the intra-cellular mannose pool. In the cytoplasm GDPMan can be metabolised to GDP-Fucose (GDP-Fuc).*

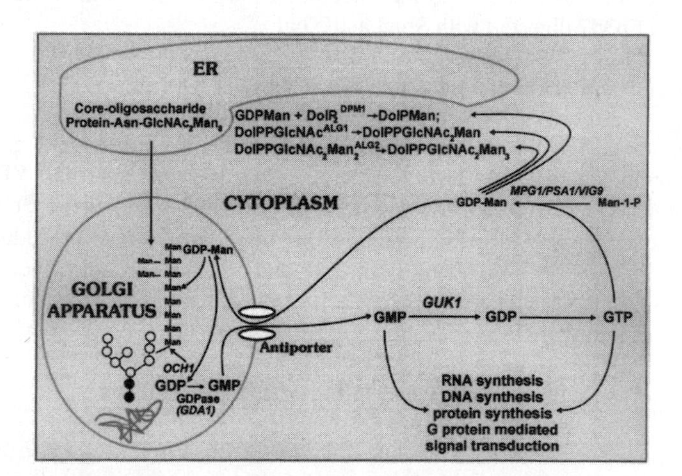

Figure 2 *Glycosylation reactions affected by the intra-cellular level of GDPMan in the yeast cell. Reactions involved in core oligosaccharide formation in the ER, i.e. synthesis of DolPMan (MPD), reaction encoded by the Dpm1-protein; elongation of the dolichol-linked oligosaccharide, catalysed by the Alg1- and Alg-2- proteins. In the Golgi compartment: addition of the outer sugar chain (α-1,6 linked mannose).*

2 METHODS

2.1. Yeast and bacterial strains and media.

S.cerevisiae strains: the MPD synthase disrupted strain ***dpm1-6*** (*MATa* , *dpm1::LEU2, leu2-3, 112 lys-801, trp1-Δ1, ura3-52*) harboring the plasmid pDM8-6, containing the temperature-sensitive *dpm1-6* (Pro $_{157}$to Leu) allele on the plasmid; the temperature-sensitive ***alg1-1*** (*MATa, ura3-52*) and ***alg2-1*** (*MATα* , *ura3-52, ade2-101*) mutant strain were used in the complementation experiment with the *S.cerevisiae MPG1* gene. The above strains and the wild type control DBY 640 (*MATa, ade2-101*). were kindly provided by Dr.Peter Orlean from University of Illinois at Urbana-Champaign.

The ***dpm1-6*** conplemented with the *DPM1* gene on the CEN-based plasmid pFL38. was constructed in our Lab.
Yeast YPD-medium or SD-medium were prepared as described previously (23).
For drug selection, luria and broth plates were supplemented with ampicilin (100mg/ml).

2.2 Cloning and sequencing of the *MPG1* gene.

The DNA encoding Mpg1 protein was prepared by PCR amplification with the following forward and reverse primers respectively: 5'-AAA TCT AGA GGA TCC CAT GCT GCC TAT GTT and 5'-TTT GAT ATC AGA GGA ACT TTA CAT GTA TCC. The PCR product, cloned into the pBluescript vector, has been sequenced on ALF authomatic sequencer (Pharmacia), either with the Pharmacia primer sequencing kit or using self-designed oligonucleotide primers. The entire 2,724 kb insert of the pBluescript vector was cloned into the integrating plasmid pFL34 and the 2m-based (high-copy) pYEp352 digested with SmaI and XbaI.

2.3 Transformation of the temperature-sensitive yeast mutants.

The yeast *dpm1-6*, the *alg1-1* and the *alg2-1* mutant strains were grown at the permissive temperature (24°C) in SD-medium without tryptophane and YPD-medium to the stationary phase (OD_{600}=1); transformed with the *MPG1* gene on the pFL34 plasmid and the 2μ-based YEp352 plasmid using the lithum acetate procedure (22) and plated on the selective media. Viability of the transformants was confirmed by the ten-fold serial dilutions on selective SD-medium at 34^0C and 37^0C. The parental strains were used as a control.

2.4 Analysis of the degree of O-glycosylation of external chitinase

Yeast strains were cultivated to OD 1-1.5 on MM with 2% glucose as a carbon source at permissive temperature; harvested by centrifugation and transfered to MM medium with 0.1% glucose, at 24^0C or 37^0C for 75 minutes. Purification of chitinase from the culture medium was accomplished by overnight incubation with chitin in the presence of protease inhibitors, followed by centrifugation and analysis of the pellet on SDS-polyacrylamide gel (23).

3 RESULTS AND DISCUSSION

3.1 Viability of the yeast mutants affected in formation of dolichol-linked saccharides is restored by overexpression of the gene encoding α-D-mannose-1-phosphate guanyltransferase

Among all strains tested, *dpm1-6* and *alg1-1* acquired the thermoresistance upon transformation with a one-copy number of *MPG1* gene and a high-copy number of *MPG1* insert at 34^0C and 37^0C, whereas for the *alg2-1* mutant the ability for growth at 34^0C and 37^0C was restored by transformation with the *MPG1* gene on the 2μ-based plasmid YEp352.

These results indicate, that in the yeast *Saccharomyces cerevisiae* the avaiability of GDPMan affects also the reactions ascribed to the endoplasmic reticulum i.e. assembly of the dolichol-linked oligosaccharide as well as MPD formation. The yeast *Saccharomyces cerevisiae dpm1-6* mutant which harbours thermosensitive allele of the *dpm1* (MPD-synthase encoding) gene on the plasmid was complemented by the genes encoding MPG- transferase from *Trichoderma reesei* (20) and the *Saccharomyces cerevisiae* (present report). Overexpression of the *MPG1* gene, in the *dpm1-6* restored synthesis of MPD *in vitro* and *in vivo*; i.e. *in vitro*: the membrane fraction from the dpm1-6 complemented with the *MPG1* gene catalyzed formation of MPD from GDPMan and DolP ; *in vivo*: synthesis of fully O-mannosylated chitinase, a process solely dependent on MPD synthesis, was restored (Fig 3). We have also compared the affinity of MPD-synthase towards GDPMan in the wild type yeast extract and in the *dpm1-6* mutant transformed with the *MPG1* gene. The data obtained indicate up to 10-fold increased affinity of MPD-synthase from the transformed *dpm1* mutant concomitant with severly impaired maximal velocity of the reaction. (Table1). The above result could be explained by the pleiotropic effect of the *dpm1* gene mutation. Our results on the Dpm1-protein of the *Aspergillus nidulans* mutant impaired in the activity of MPD-synthase as well as protein secretion, seem to support this assumption (Lenart et. al. unpublished) . On the other hand an interaction of the Dpm1 and Mpg1 protein can not be excluded at the moment.

The question arise, why MPG transferase activity complements the defect in MPD formation. The biochemical data on the synthesis of glycosyl derivatives of MPD *in vitro* indicate that the reaction is easily reversible and highly dependent on the concentration of sugar nucleotides and the product formed (24). It is thus possible that the higher concentration of the substrate GDPMan which should result from overexpressing the MPG1 gene allows the yeast Dpm1 mutant protein to function at the restrictive temperature.

The yeast *MPG1* gene restored also the viability of the yeast N-linked glycosylation mutants *alg1* and *alg2* when cultivated at non-permissive temperature. Both the Alg1- and Alg2-proteins code for mannosyltransferases responsible for the addition of the β-1,4 (Alg1-p) and α-1,6 (Alg2-p) linked mannose residues to the dolichol-linked oligosaccharide (DolPP-$GlcNAc_2$ and DolPP-$GlcNAc_2Man_2$). Supression of the *alg1* mutation by overexpression of Mpg1 protein was reported also by (9, 17). On the other hand *T.reesei mpg1* gene did not complement the temperature-sensitive *sec53-6* mutation of *S.cerevisiae* (20) which affects the preceding cytoplasmic step in GDPMan formation, the conversion of Man-6-P to Man-1-P. Thus at the present moment we

Table 1 *Overexpression of the MPG1 gene, encoding GTP:α-D-mannose-1-phosphate guanyltransferase, restores the viability of the the yeast S. cerevisiae dpm1-6 mutant impaired in Mannosylphosphodolichol (MPD) formation*

Strain	K_m *(GDPMan mM)*	V_{max} *(pmoles/mg prot/5min)*
W 303 wild type	**5.25**	**1747**
Dpm1-6/DPM1 **(control)**	**43.93**	**11179**
Dpm1-6/MPG1 (S.cerevisiae)	**2.79**	**238**
Dpm1/mpg1 Trichoderma	**0.86**	**182**

The yeast strains were grown at non-permissive temperature (37°C)

assume that the *mpg1* gene is able to suppress mutations of the proteins such as Dpm1-, Alg1- and Alg2-, involved in the steps which utilise GDPMan as a substrate.

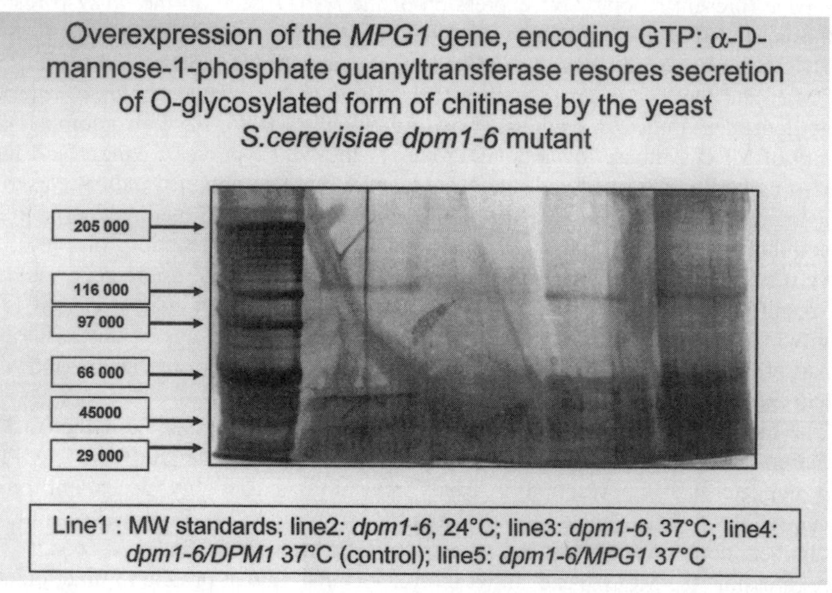
Overexpression of the MPG1 gene, encoding GTP: α-D-mannose-1-phosphate guanyltransferase resores secretion of O-glycosylated form of chitinase by the yeast S.cerevisiae dpm1-6 mutant
205 000
116 000
97 000
66 000
45000
29 000
Line1 : MW standards; line2: dpm1-6, 24°C; line3: dpm1-6, 37°C; line4: dpm1-6/DPM1 37°C (control); line5: dpm1-6/MPG1 37°C

Figure 3 *Overexpression of the MPG1 gene, encoding GTP:α-D-mannose-1-phosphate guanyltransferase, restores secretion of O-glycosylated form of chitinase by the yeast S. cerevisiae dpm1-6 mutant. Yeast strains were cultivated to OD 1-1.5 on MM with 2% glucose as a carbon source at permissive temperature; harvested by centrifugation and transfered to MM medium with 0.1% glucose, at 24°C or 37°C for 75 minutes. Purification of chitinase from the culture medium was accomplished by overnight incubation with chitin in the presence of protease inhibitors, followed by centrifugation and analysis of the pellet on SDS-polyacrylamide gel. The amount of protein loaded corresponded to 10ml culture. Protein bands were visualised by silver staining; line 1: MW standards; line 2: dpm1-6 , 24°C ; line 3: dpm1-6, 37°C; line 4:dpm1-6/DPM1 37°C (control); line 5: dpm1-6/MPG1 37°C.*

Based on the results obtained for the Dpm1p our working hypothesis is that the availability of GDPMan in the cell cytoplasm might affect the mannosyltransferases located in endoplasmic reticulum by changing their substrate affinity. Since no transporter activity for sugar nucleotides has been found in the ER membranes the question arises what is the mechanism of GDPMan translocation thorough ER membranes. We have observed the increased affinity of MPD-synthase towards GDPMan when the *MPG1* gene was overexpressed in the yeast MPD-synthase mutant. It is reasonable to consider therefore a possibility of direct interaction between the proteins involved in GDPMan and MPD formation. The cytoplasmic Mpg1-protein could interact with the Dpm1-protein in the ER-membranes and the latter may serve as a GDPMan transporter. This is further substantiated by a very recent findings on the role of dolichol in substrates translocation via ER membranes. (25).

Acknowledgement: The experimental work in the author's laboratory was financed by the grant 6PO4B 01712 from the Commitee for Scientific Research (KBN), Poland.

Abbreviations:

DolP, DolPP- dolichylphosphate, diphosphate, GDPMan- GDP-mannose, GlcNac- N-acetylglucosamine, GPI-glycosylphosphatidyl inositol, LLO- lipid linked oligosaccharide (DolPPGlcNac2Man9Glc3), MPD-dolichylphosphate mannose, MPG transferase- GDP:α-D-mannose-1-phosphate guanyltransferase, NDP-nucleotidediphosphate,
PMI-phosphomanno- isomerase, PMM-phosphomanno- mutase.

REFERENCES

1. W. Tanner and L. Lehle, Biochim. Biophys. Acta, 1987, **906**, 81.
2. C. Abeijon, P. Orlean, P.W. Robbins and C.B. Hirshberg, Proc. Natl. Acad. Sci., 1992, **86**, 6935.
3. A.Herscovics, P.Orlean, FASEB J., 1993, **7**, 540-550.
4. P. Orlean, Mol. Cell Biol., 1990, **10**, 5796-5805.
5. E.L. Kean, J. Biol. Chem., 1980, **255**, 1921-1927.
6. C. Ronin, C.B. Caseti and S. Bouchilloux, Biochim. Biophys. Acta, 1981, **674**, 48-57.
7. G. Palamarczyk, M. Maras, R. Contreras and J. Kruszewska, „Trichoderma and Glocladium", Taylor and Francis Ltd., 1998, Vol. 1, Chapter 6, p.121.
8. C. Abeijon, K. Yanagisawa, E.C. Mandon, A. Hausler, K. Moremen, C.B. Hirshberg, and P.W. Robbins, J. Cell Biol.,1993, **122**, 307.
9. Y. Shimma, M. Nishikawa, B. bin Kassim, A. Eto, J. Jigami, Mol. Gen. Genet., 1997, **256**, 469-480.
10. S.P. Ferro-Novick, C.F. and R. Schekman , J. Cell Biol., 1984, **98**, 35.
11. F. Kepes and R. Schekman, J. Biol. Chem., 1988, **263**, 9155-9161.
12. T.C. Huffaker and P.W. Robbins, Proc. Natl. Acad. Sci., 1983, **80**, 7466-7471.
13. A. Conzelmann, C. Fankhauser and C. Desponds, EMBO J., 1990, **9**, 653.

14. P. Orlean, M.J. Kuranda and C.F. Albright, Methods in Enzymol., 1991, **194**, 682-685.
15. K. Panneerselvam and H.H. Freeze, J.Clin. Invest, 1996, **94**, 1901.
16. M.A. Payton, L.S. Rheinnecker, M. Klig, M. De Tiani and E. Bowden, J. Bacteriol., 1991, **173**, 2006-2011.
17. B.K. Benton, S.D. Plump, J. Roos, W.J. Lennarz and F.R. Cross, Curr. Genet., 1996, **29**, 106.
18. H. Hashimoto, A. Sakakibara, M. Yamasaki and K. Yoda, J. Biol. Chem., 1997, **272**, 16308-16314.
19. L. Ballou, R.A. Hitzeman, M.S. Lewis and C.E. Ballou, Proc. Natl. Acad. Sci. USA, 1991, **88**, 3209.
20. J.S. Kruszewska, M. Saloheimo, M. Penttila and G. Palamarczyk, Curr. Genet., 1998, **33**, 445-450.
21. K.D. Yanagisava, C. Resnick, C. Abeijon, P.W. Robbins and C.B. Hirschberg, J. Biol. Chem., 1990, **265**, 19351-19360.
22. D. Gietz, A. St. Jean, R.A. Woods, R.H. Schiestl, Nucleic Acids Res., 1992, **20**, 1425- 1426.
23. M.J. Kuranda and P.W. Robbins, J. Biol. Chem., 1991, **266**, 19758-19767.
24. C.D. Warren and R.W. Jeanloz, Methods Enzymol., 1978, 122-137.
25. M. Sato, K. Sato, Si. Nishikawa, A. Hirata, Ji. Kato, A. Nakano, Mol. Cell Biol., 1999, **1**, 471-483.

Biochemistry of Carbohydrate Modifying Enzymes

PHYSICAL-ORGANIC PROBES OF GLYCOSIDASE MECHANISM

Michael L. Sinnott

Department of Paper Science, UMIST, POB 88, Sackville Street, Manchester M60 1QD, UK

1 INTRODUCTION

The application of the concepts and techniques of physical organic chemistry to problems of enzyme mechanisms is now half a century old, and in the last 30 years at least the two subjects of aqueous solution physical organic chemistry and mechanistic enzymology have developed symbiotically. This article will review the application of developing physical-organic probes of enzyme mechanism to glycosidase catalysis. Whilst the points made will be general, the specific examples chosen to illustrate them will be chosen where possible from work in the author's laboratories.

Koshland[1] must probably be credited with being the first to apply physical organic reasoning to glycosidase catalysis. He realised that the inversion of configuration, that appeared to be universal in bimolecular displacements at saturated carbon, implied that those glycosidases which gave the product sugar initially in the same anomeric configuration to the substrate proceeded by a double dispacement mechanism, whereas those that gave the initial sugar product inverted proceeded by a single displacement mechanism (Scheme 1).

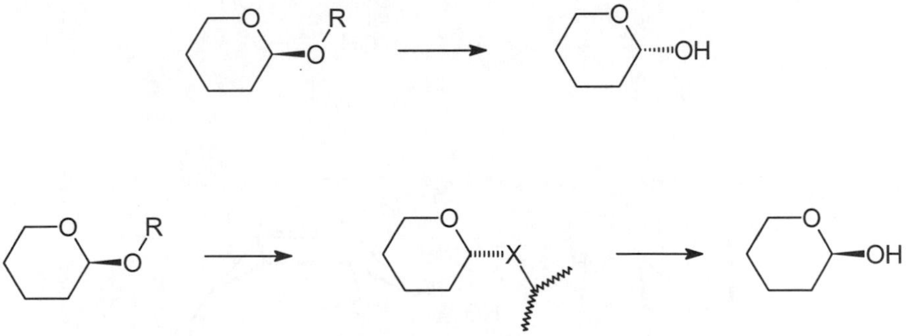

O R O OH

O R O X O OH

Scheme 1

Determination of reaction stereochemistry remains the most fundamental piece of mechanistic information about a glycosidase. Many retaining enzymes have transferase activity, and in this case circumventing the problem of mutarotation of the product sugar by using an alcohol, rather than water, as the glycosyl acceptor is simple. However, the direct demonstration of inversion of configuration when the product sugar mutarotates fast (most notably in the case of furanosidases) remains an unsolved technical problem. (Inverting glycosidases cannot have transferase activity, otherwise, by the principle of microscopic reversibility, they would have to act on both anomers of their substrates)

In the late 1950's and early 1960's attention was focused on the acid-catalysed hydrolysis of glycosides[2], and emphasis was placed on conjugation of the lone pair electrons of oxygen with the electron-deficient centre. Fully-fledged oxocarbenium ions were drawn as intermediates, and it was realised that for most effective conjugation of the oxygen lone pair with the carbenium ion centre, the lone pair should be in a pure p orbital. This means that the stable conformation of an oxocarbenium ion is one in which carbon, oxygen, and all three substituents on the $-O^{+}=C<$ moiety are coplanar. Studies of the methoxymethyl cation in superacid media, indeed, indicated a barrier to randomisation of the two methylene protons of 11.9 kcal mol^{-1}[3].

The coplanarity required in a furanosyl cation can be achieved in either the ^{3}E or E_3 envelope conformation, which is at a local energy minimum on the pseudorotational itinerary of a furanose ring anyway. The distortions of a pyranose ring involved in making C1, C2, O5, C5 and the anomeric hydrogen coplanar, though, are more complex. It was early recognised that the coplanarity could be achieved in a half-chair conformation: however, conformational studies of δ–lactones,[4] which have similar geometrical requirements to oxocarbenium ions, confirmed the *a priori* expectation that classical boat conformations, which could also accommodate the required coplanarity, could in certain circumstances be important. Permitted conformations of the glucopyranosyl cation are shown in Scheme 2.

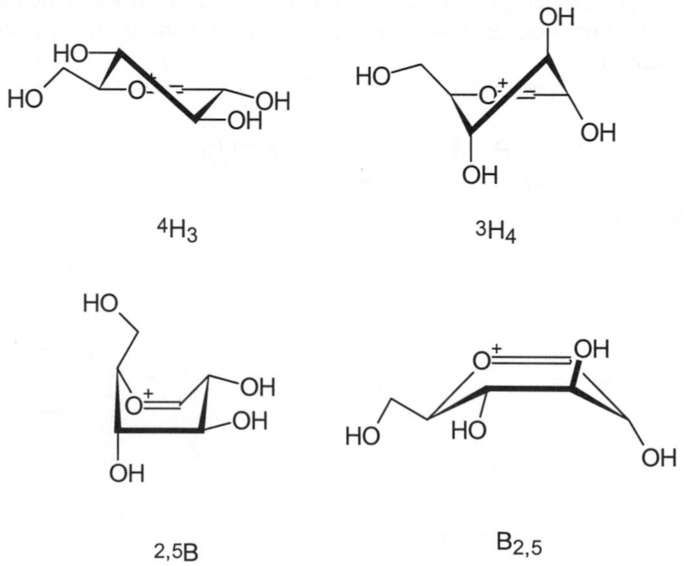

$^{4}H_{3}$ $^{3}H_{4}$

$^{2,5}B$ $B_{2,5}$

Scheme 2. Permitted conformations of the glucopyranosyl cation

Carbonium ions (later also called carbocations and carbenium ions) were popular species to invoke as intermediates in the 1960's, and when, on solution of the X-ray crystal structure of lysozyme[5], it was discovered that a $^4C_1$ chair could not be built into the –1 subsite, it was inevitable that the glycosyl-enzyme intermediate in this retentive enzyme should be written as a fully-fledged oxocarbenium ion, stabilised electrostatically by Asp 52. The observed retention of configuration was attributed simply to a shielding of the α face of this ion by the protein.

The discovery of the "Jencks clock"¶at the end of the 1970's[6] made it possible to estimate lifetimes of the more stable oxocarbenium ions in water, and it appeared from extrapolations that unstabilised oxocarbenium ions such as the methoxymethyl and glucopyranosyl cations in water had lifetimes shorter than a molecular vibration. In this medium, therefore, they were too unstable to exist, and all reactions supposedly involving them were in fact bimolecular. These data immediately called into question the physical soundness of "stabilised oxocarbenium ion" intermediates in glycosidases. In two seminal papers Craze et al.[7] and Knier and Jencks[8] established the characteristics of bimolecular nucleophilic substitutions at the methoxymethyl centre. Although indubitably bimolecular, the reactions were characterised by low sensitivity to the nucleophilicity of the attacking nucleophile and a high sensitivity to leaving group acidity. The transition state was thus "exploded", with extensive charge build-up on the central atoms, and very weak bonding to both the incoming nucleophile and the departing leaving group. Although by quantitative measures of charge buildup or rehybridisation at the reaction centre these reactions were SN1, the transition state undoubtedly contained both nucleophile and leaving group.

Bimolecular nucleophilic substitution has to be considered in terms of a More O'Ferrall-Jencks diagram in which the progress of bond-making to the nucleophile and bond breaking to the leaving group were represented on two orthogonal axes, although the inability of carbon to accommodate more than eight electrons in its valence shell means that the Northwest half of the diagram is prohibited (Figure 1b). (A similar treatment of phosphoryl transfer does not operate under this constraint, of course, so in addition to "metaphosphate", analogous to the carbenium ion in the Southeast corner, a phosphorane in the Northwest corner is a possible fully-fledged intermediate). For the simple cleavage of a single bond, with no other bonds formed or broken, it is possible to graph out the energy of the system as a function of the bond being cleaved. It is conventional, of course, to describe such a reaction in terms of the species at the maximum of the energy curve. (Figure 1a). In the same way, energy can be plotted on an axis perpendicular to the plane of Figure 1b, in the same way as contours on a map. The transition state is then the structure at the highest energy point of the "pass" leading from reactants to products. A true SN1 reaction corresponds to reaction along the Southern and Eastern edges of the diagram; an SN2 reaction on a substrate like a methyl derivative, where there is little charge buildup, to a reaction along the SW-NE diagonal. The reactions of alkoxyalkyl derivatives involve a transition state at the Southeast corner of the diagram, but not a fully-fleged oxocarbenium ion. It is therefore incorrect to write oxocarbenium ions as intermediates in glycosidase reactions, unless there is direct experimental evidence for violating physical-organic precedent.

¶ The reaction of oxocarbenium ions with water is activated and with good nucleophiles such as azide is diffusion controlled, with a rate constant of 5 x 10^9 M^{-1} s^{-1} in water at room temperature. Measurement of the ratios of product from azide to product from water gives the rate at which the oxocarbenium ion reacts with water. The results from the Jencks clock were confirmed by direct measurement based on pulse radiolysis (R. Steenken and R. A. McClelland, *J. Am. Chem. Soc.* 1989, *111*, 4967)

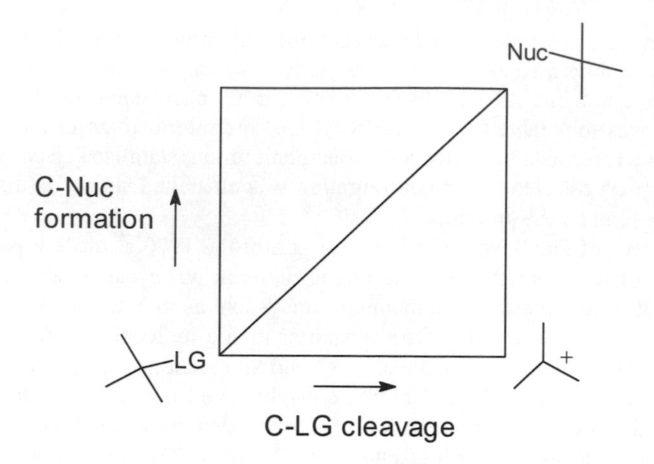

Figure 1. More O'Ferrall-Jencks diagram for substitution at saturated carbon

More recently it has become apparent that in reactions involving free oxocarbenium ions, development of C=O conjugation and C-LG cleavage are not in step. The argument was originally an extended one, based on the insensitivity of the lifetime of conjugatively stabilised carbenium ions to presumably destabilising inductively electron-withdrawing groups[9]. The attainment of planarity of the oxocarbenium ion moiety lags behind the cleavage of the C-LG bond. One of the more immediate manifestations of the effect lies in the hydrolysis of orthocarbonates[10]. Uncatalysed, spontaneous loss of p-nitrophenolate ion from $(PhO)_3CopC_6H_4NO_2$ is comparably fast to loss of p-nitrophenolate from various tetrahydropyranyl p-nitrophenolates, yet the triphenoxycarbenium ion has a lifetime of microseconds, while the best estimate of tetrahydropyranyl cation lifetime is probably around 10^{-12} sec, similar to the 2-deoxy glucosyl cation[11]. There is a barrier to the attainment of planarity that is necessary for the positive charge on the triphenyoxycarbenium ion to be delocalised over three oxygens, probably arising form steric crowding of the substrate, and this barrier is manifested both in the rate of generation of the ion and its rate of reaction with nucleophiles (Scheme 3).

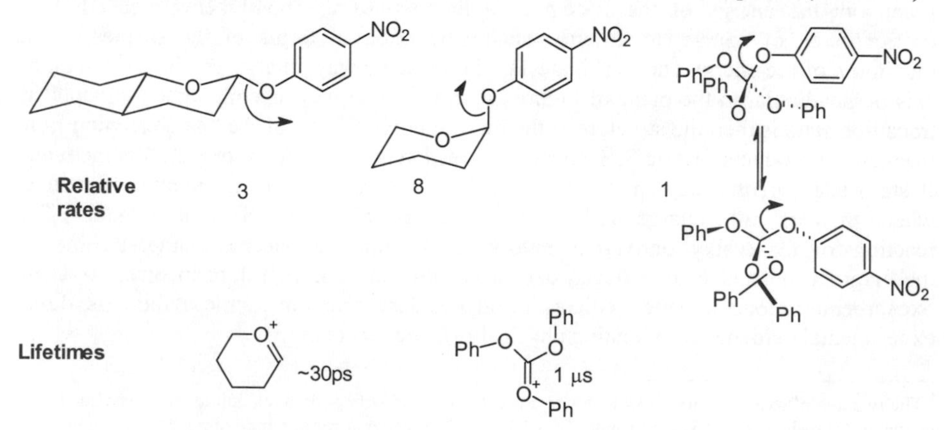

Scheme 3 Additional stabilisation of an oxocarbenium ion by two further oxygen lone pairs is not manifested at the transition state.

It is therefore necessary to consider nucleophilic substitution at an oxocarbenium ion centre in terms of a reaction coordinate in three dimensions, with an extra dimension indicating development of C-O conjugation added to the traditional More O'Ferrall-Jencks diagram. (Figure 2). (Although this axis can be described for shorthand reasons as development of C-O congugation, of course electronic motion is fast compared to nuclear motion – the Born-Oppenheimer approximation – and strictly this axis refers to the progress towards planarity of the oxocarbenium ion fragment). In a detailed study of general acid catalysis of the hydrolysis of orthocarbonates, it was shown that this extra dimension was required to explain the details of the relationships between structure-reactivity parameters[12].

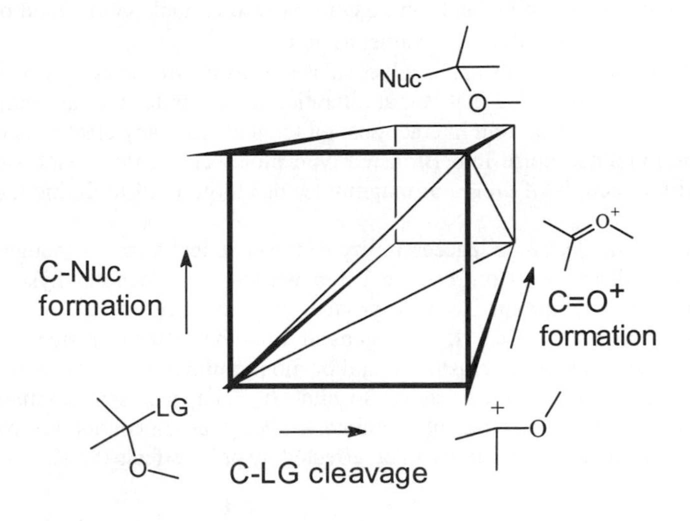

Figure 2.

USE OF STRUCTURE-REACTIVITY PARAMETERS IN CHARACTERISING GLYCOSIDASE TRANSITION STATES[13]

The basic assumption of the characterisation of transition states by structure-reactivity parameters, is that there is a linear relationship between the free energy of activation of one reaction, and the standard free energy of a related equilibrium reaction. In traditional physical organic chemistry, the reference equilibrium reaction is the acid dissociation of substituted benzoic acids, which define Hammett σ values for a substituent X as $pK_{a(PhCOOH)} - pK_{a(XC6H4COOH)}$. There are now a very large number of structure-reactivity parameters covering as wide a range of solvents and conditions as possible, and some of the reference reactions are rates rather positions of equilibrium. For reactions in aqueous solution, including enzyme reactions, though, the ionisation of benzoic acid is an irrelevance, and it is best to use aqueous pK_a values of nucleophiles and leaving groups directly wherever possible. Thus, in the seminal Knier and Jencks paper [ref] on the reactions of the *N*-methoxymethyl *N,N*-dimethylanilinium ions the leaving groups were substituted *N,N*-dimethylanilinium ions and the nucleophiles amines. It was therefore possible to use aqueous pK_a values for both variation of nucleophile and variation of leaving group. The gradient (β_{lg}) of the plot of log k against pK_a of the departing dimethyl aniline lay between -0.7 and -0.9, depending on the nucleophile. This means that the change in charge of the leaving nitrogen between ground state and the transition state is 0.7 to 0.9 of that occurring on deprotonation of dimethylanilinium ions. Strictly, to convert the β_{lg} value to a degree of C-N cleavage, the effect of change of leaving group pK_a on the position of equilibrium is also required, but a β_{eq} of approximately unity is commonly assumed in qualitative discussions.

β_{nuc} Values, the gradients of the plots of log k against pK_a of the incoming amine nucleophile, are obtained in exactly the same way, and lie in the range 0.1 to 0.2 in this system. These values indicate a degree of change development between a fifth ot a tenth of that on protonation of the amine, and can be converted to C-nucleophile bond order at the transition state by scaling with the β_{eq} value, as before.

Linear free energy relationships in glycosidase catalysis are conceptually similar, but encounter two difficulties. The first is that alteration of substrate structure introduces perturbations in the substrate-protein interactions, quite apart from any electronic effects. Such interactions introduce noise into Brønsted-type plots. Therefore a wide range of substrate reactivity (a couple of orders of magnitude) has to be used to define the trend line.

The second difficulty is the inaccessibility of the chemical steps. Although for an inverting glycosidase there is only one chemical step, with a good substrate physical steps can become kinetically important, so that product release governs k_{cat}, or enzyme-substrate binding governs k_{cat}/K_m. It is a general rule that chemical steps become kinetically accessible with poor substrates and/or non optimal pH. For a retaining glycosidase there are two chemical steps, so that k_{cat} can represent formation or decomposition of the glycosyl-enzyme intermediate. k_{cat}/K_m does not contain k_{+3}, but does contain binding parameters, and binding can be affected by polar effects (Scheme 4).

$$E + GlyX \underset{k_{-1}}{\overset{k_{+1}}{\rightleftharpoons}} E.GlyX \xrightarrow[\downarrow X]{k_{+2}} E.Gly \xrightarrow{k_{+3}} E + GlyOH$$

$$k_{cat} = \frac{k_{+2}\,k_{+3}}{k_{+2} + k_{+3}} \qquad K_m = \frac{k_{-1} + k_{+2}}{k_{+1}} \bullet \frac{k_{+3}}{k_{+2} + k_{+3}}$$

$$\frac{k_{cat}}{K_m} = \frac{k_{+1}\,k_{+2}}{k_{-1} + k_{+2}}$$

Scheme 4. Minimal kinetic mechanism for a retaining glycosidase.

In general therefore, structure-reactivity parameters for retaining glycosidases are best measured on k_{cat}/K_m, although for very unreactive substrates where the deglycosylation rate is known they can be measured on k_{cat}.

With these provisos, the transition state of a glycosidase-catalysed reaction not involving proton transfer can be characterised by measuring the effect of altering electronic demand in

(1) the leaving group,
(2) the nucleophile, and
(3) the glycone (reaction centre).

In terms of Figure 1, the β_{lg} value locates the transition state on the East-West axis and the β_{nuc} on the North-South axis. The effect of altering the electronic characterics of the glycone locates the transition state on a Northwest-Southeast diagonal. The Southwest-Northeast diagonal drawn in Figure 1 corresponds to synchronous breaking of the bond to the leaving group and formation of the bond to the nucleophile, so that there is no net change in charge at the transition state. If the effect of a substitution in the glycone on the reaction rate is half that of the effect on the equilibrium generation of the carbenium ion, then the transition state lies on a Soutwest-Northeast line that bisects the midpoint of the Southern and Western edges of the diagram. The effects at reaction centres – such as sugars - are sometimes still described by Hammett-type ρ values (e.g ρ_I or ρ^* values, the aliphatic variants), because of the difficulty of obtaining the effects on the reference equilibrium directly.

To date glycone substitutent effects are considered only in terms of the diagram of Figure 1, although in principle it should be possible to dissect out the effects of substituents close to the oxygen and close to the carbon, and thus use the extra dimension of Figure 2.

β_{lg} Values for glycosidase-catalysed hydrolysis of the corresponding glycosyl pyridinium ions are known for a number of retaining[14] and inverting[15] glycosidases that normally act on *O*-glycosides. They are in the range 0.6 – 1.0, indicative of a substantially broken bond at the transition state, although the values are in general closer to zero for the inverting enzymes, for which the pyridinium salts seem to be worse substrates. Because the enzymes cannot apply part of their normal catalytic machinery, protonation of the leaving group, the k_{cat} almost invariably represents k_{+2}. The situation is different for the NAD^+ glycohydrolases, which have a pyridine as their normal leaving group. For the calf spleen enzyme, there is no simple correlation of the rates of hydrolysis of a series of ADP-ribosyl pyridinium ions with leaving group acidity[16], but for the enzyme from the venom of the banded krait it was possible to get a β_{nuc} value of 0.43[17] for the transfer to acceptor pyridines in the transglycosylation reaction. This high value probably reflects electronic effects on binding as well as reactivity.

Alteration of the pK_a of the nucleophile is more difficult. The techniques of site-directed mutagenesis can be used to change the active site nucleophile of a retaining glycosidase to the corresponding amide. The pK_a of primary amides lies in the region of –0.5 to –1.0 [18], that of carboxylates around 4. Changing an aspartate to asparagine or glutamate to glutamine, if the reaction is governed by a similar β_{nuc} to model reactions. of 0.1 to 0.2, should change the rate of the first step of the glycosidase reaction by a factor of only ~3-10. Yet it is universally found that such a mutation results in completely inactive enzyme. It could well be that the first step does indeed take place to yield a glycosyl imidate which then loses a proton to yield an inert glycosyl-enzyme intermediate (Scheme 5).

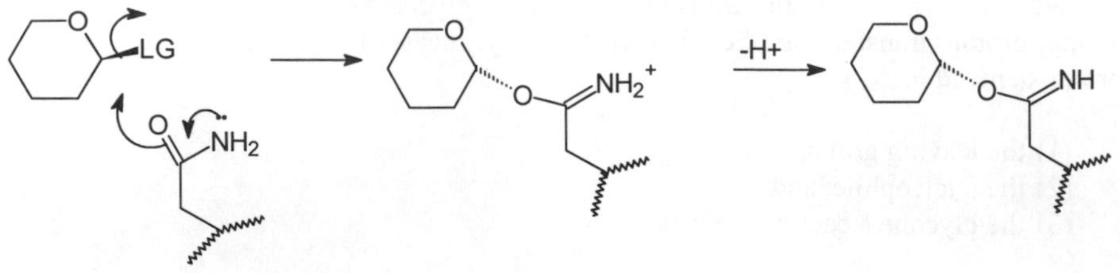

Scheme 5. Possible single turnover reaction of nucleophile Glu/Gln or Asp/Asn mutants of retaining glycosidases

An easier nucleophile to change is the acetamido group of substrates for the retaining Family 18 and 20 hexosaminidases, which use the trans 2-acetamido group of the substrate, rather than an enzyme carboxylate, as the nucleophile. There is one report of this being done, for the *N*-acetylglucosaminidase of *Aspergillus niger*[19], in which the methyl group was successively fluorinated; the ρ^* value calculated on k_{cat}/K_m was –1.4.

Systematic changes in the glycone have the disadvantage that unless they are sterically conservative, the steric effects of the substituents will outweigh and mask their electronic effects. The replacement of the OH of the natural substrate by H and F is such a change, and by comparison of the spontaneous rate of hydrolysis of a series of 2,4-dinitrophenyl deoxy and deoxyfluoro β–galactopyranosides with k_{cat}/K_m for their hydrolysis by the *lacZ* β-galactosidase of *Escherichia coli,* it was possible to show that charge development was about 80% of that in the spontaneous hydrolysis. [20] . However, specific interactions of the 2-OH with the protein played a major role, and the point for the sugar with the natural glycone did not lic on the free energy correlation.

In many glycosidases, specific interaction of the 2-OH with the protein makes the Hammett plots arising from a series of 2-substituents very noisy, if not completely invalid. An exception arises with some retaining NAD^+ - glycohydrolases,[21] A series of 2'-substituted analogues of NAD^+ were hydrolysed with a ρ_I value of –9.4 by the calf spleen enzyme, compared to a value of –7 for the spontaneous hydrolysis. Intriguingly, if the σ_I of O^-, rather than OH, is used for the point for the natural substrate, it fits on the line defined by the other substituents.

An informative substitution in a glycosidase substrate is F-for-H at position 1: the powerful inductive effect of the fluorine destabilises oxocarbenium-ion-like transition states, although this is offset in small measure by the ability of fluorine to donate electrons conjugatively. The relative importance of the two effects appears to vary with existing stability of the ion, so that α-fluorine substitution in fact stabilises unstable carbenium ions.[22] A 1-fluoro *O* glycoside would be unstable towards spontaneous hydrolysis, so the leaving group has to be fluoride for reasons of synthetic practicality. One therefore compares the rates of hydrolysis of glycosyl fluorides with those of the corresponding 1-fluoroglycosyl fluorides; in the case of retaining glycosidases the comparison has to be made on k_{cat}/K_m. A rough scaling of the effect can be made from the relative rates of spontaneous hydrolysis (at 50°, the relative rates of 1-fluoroglucosyl fluoride, α-glucosyl fluoride and β-glucosyl fluoride are 1:1000:40,000)[23], but it is now clear that the water reactions of glucosyl fluoride are bimolecular[24, 25] so that effects greater than those on spontaneous hydrolysis can be, and are, observed. The ratio is expressed logarithmically, as a free energy difference $\Delta\Delta G^\ddagger$, defined by $\Delta\Delta G^\ddagger = RT\ln\{(k_{cat}/K_m)_F/(k_{cat}/K_m)_{F2})$ which is proportional, if the usual assumption about proportionality between Brønsted-type coefficients and charge development hold, to the change in charge on the sugar between the unbound state and the first irreversible enzymic transition state.

Table 1 *Relative rates of hydrolysis of glycosyl fluorides and 1-fluoroglycosyl fluorides by various enzymes*

Enzyme	**Source**	**Parent sugar**	**ΔΔG‡ (kJ mol⁻¹)**	**Reference, Stereo-chemistry**
Glucoamylase	*Aspergillus niger*	α–glucose	21	S &K, 1991, inversion
Trehalase	Pig kidney	α–glucose	16.6	inversion
β (1→3) glucanase	*Sporotrichum dimorphosporum*	β–glucose	13	Bimali's thesis; inversion
Cel6A (CBHII)	*Trichoderma reesei*	β-cellobiose	>29	inversion
α–glucosidase	yeast	α–glucose	21.2	retention
α–glucosidase	rice	α–glucose	12.3	retention
α–galactosidase	*Pycnoporus cinnabrinus*	α–galactose	22	[26]retention
α–galactosidase	*Phanerochaete chrysosporium*	α–galactose	21	retention
β-glucosidase	Sweet almonds	β-glucose	14.5	retention
β-glucosidase A$_3$	*Aspergillus wentii*	β-glucose	14.5	retention
β–galactosidase	*Escherichia coli (lacZ)*	β-galactose	13.5	retention
β–galactosidase	*E. coli (ebg^o)*	β–galactose	10.4	[27]retention
"	*E. coli (ebg^a)*	"	6.8	"
"	*E. coli (ebg^b)*	"	7.2	"
"	*E. coli (ebg^{ab})*	"	6.3	[28] "
"	*E. coli (ebg^{abcd})*	"	6.0	"
"	*E. coli (ebg^{abcde})*	"	7.3	"
Cel7A (CBHI)	*Trichoderma reesei*	β-lactose	>33	retention

ΔΔG‡ proves to be remarkably variable from enzyme to enzyme, and even within a series of experimental evolvants of the same enzyme, as with the various *ebg* enzymes of *Escherichia coli.* Values are given in Table 1. In the case of the *ebg* enzymes, *ebg^o* is wild-type, and each superscript refers to a separate amino-acid change in the series of evolvants. Table

It is clear therefore that the details of glycosidase transition states can differ quite radically from one enzyme to another, with the amount of charge on the glycone at the transition state ranging over a factor of 5. Some enzymes, such as the cellobiohydrolases of *T. reesei* appear in fact to be characterised by more charge development at enzymic transition states than non-enzymic ones, although detailed interpretation of the non-enzymic rates is complicated by the difficulty, as yet not overcome, of determining which fluorine of the difluoride leaves first in spontaneous hydrolysis.

The behaviour of the α-galactosidase of *Phanerochaete chrysoporium* towards α-galactosyl fluoride and 1-fluoro-galactosyl fluoride is instructive. Hydrolysis of substrates with good leaving groups by this enzyme exhibits pronounced substrate inhibition, whereas hydrolysis of substrates with poor leaving groups is strictly Michaelian. All data fit a model in which substrate inhibition arises because a second molecule of substrate can bind to the galactosyl enzyme, and there turn over slowly to disaccharide glycoside slowly, if at all. The hydrolysis of α–galactosyl fluoride shows very strong substrate inhibition, whereas that of the 1-fluoro-galactosyl fluoride is strictly Michaelian. This implies that formation of the fluoro-galactosyl-enzyme intermediate is rate determining in the hydrolysis of the difluoride, but hydrolysis of the galactosyl-enzyme intermediate is rate-determining in the case of the monnofluoride, i.e. 1-fluorine substitution slows the first chemical step more than the second. This cannot be a transition state effect, since by microscopic reversibility the first and second transition states in the action of a retaining glycosidase have to be identical, at least for virtual reactions such as 1-OH exchange. Consequently, the galactosyl-enzyme intermediate in this enzyme already experiences significant charge buildup on the sugar.[1]

USE OF KINETIC ISOTOPE EFFECTS IN CHARACTERISATION OF GLYCOSIDASE TRANSITION STATES[2].

Isotopic substitution is a quite unique probe of chemical mechanism since it does not alter the potential energy surface for a reaction, unlike more drastic substitutions, such as F-for-H. Most kinetic isotope effects at ordinary temperatures arise from the effect of change in mass on the zero point energy (½h) associated with molecular vibrations. The various vibrations of a molecule are described by force constants, which are unaltered on isotopic substitution. The frequency ( ) of the vibration of heavy isotopomer will therefore be less than that of a lighter one, because the restoring force is the same, but acts on a higher mass. Consequently, heavy isotopomers possess less zero point energy than light ones. Primary kinetic isotope effects arise because it requires more energy to break a bond to a heavier isotopomer, as it has less zero point energy. Even if the vibration is weakened rather than completely eliminated, secondary effects arise because less zero point energy is given up on weakening of the bonds to heavier isotopomers. All the kinetic isotope effects that can throw light on glycosidase action are zero point effects, although other phenomena can be important with other classes of enzyme, such as quantum mechanical tunnelling in some hydrogen-transfer reactions. Informative sites of isotopic substitution in a glycoside are shown in Figure 3.

Although an isotope effect on a chemical step is a completely unambiguous mechanistic datum, in a way a substituent effect is not, deriving an isotope effect on a chemical step from a measured isotope effect is not simple. In general, isotope effects with good substrates are likely to be wholly or partly masked by physical steps not subject to an isotope effect: indeed, with the natural substrate there are sound evolutionary reasons why this should be so. Thus, a good substrate may be "sticky" ($k_{+2} > k_{-1}$), so that effects on k_{cat}/K_m are in fact effects on k_{+1}, the (usually diffusion-limited) binding of enzyme and substrate. Isotope effects on k_{cat}, even with single displacement

1. H. Brumer III, P. F. G. Sims and M. L. Sinnott, *Biochem. J.* 1999, **339**, 43.

2. M. Sinnott, C. D. Garner, E. First, and G. Davies. *Lexicon of Terms and Concepts in Mechanistic Enzymology,* p.47 (Vol. IV of *Comprehensive Biological Catalysis,* M. L. Sinnott, Ed., Academic Press, London, 1998)

enzymes, do not necessarily reflect the chemical step, since product release or conformation changes may be rate-limiting. Consequently, it is now common practice to re-measure at least some effects away from the pH optimum of the enzyme, since (at least with uncharged substrates) physical steps tend to be less pH-sensitive than chemical ones. The use of bad substrates, or of enzymes partly crippled by mutation of key residues, are other, complementary, approaches to revealing the chemical steps.

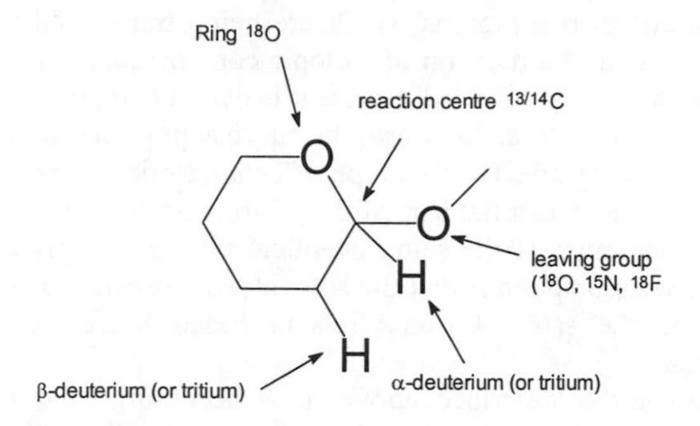

Figure 3. Mechanistically informative sites of isotopic substitution during glycoside hydrolysis

Even when fully expressed, most of the isotope effects arising from the substitutions in Figure 3 are small (k_{light}/k_{heavy} typically being 1.05-1.25 for α-deuterium (H1 as drawn), 1.03-1.10 for β-deuterium (H2 as drawn), 1.005 – 1.08 for anomeric ^{13}C, 0.985 – 0.997 for ring ^{18}O, and 1.02 – 1.05 for leaving group ^{18}O). Direct comparison of zero order reaction rates with good liquid handling techniques and an accurate, well thermostatted spectrometer can give rate-ratios with errors of 0.5 %. Competitive methods, in which isotope ratios are measured as a function of degree of reaction, can be much more accurate, but invariably give effects on k_{cat}/K_m, since one isotopomer acts as a competitive inhibitor of the transformation of the other; k_{cat} effects have to be measured by direct comparison. A further restriction on competitive methods is that the effect of hydrogen isotope substitution on binding can be significant, and comparable with expected secondary isotope effects on bond-breaking steps. Chromatographic procedures for baseline separation of hydrogen isotopomers of some molecules (e.g. $(C_6H_5)_2CO$ and $(C_6D_5)_2CO$ [29]) exist, and must be based on isotope effects on non-specific partitioning ratios.

Within the accuracy of almost all kinetic isotope effect data on enzyme reactions, it can safely be assumed that the relationship between the isotope effects arising from different non-hydrogen isotopes is intuitive – if there is a 2% ^{13}C effect (i.e. k_{12}/k_{13} =1.02), there will be a 4% ^{14}C effect. If hydrogen isotope effects arise entirely from zero point energy differences, then the Swain-Schaad relationship between deuterium and tritium effects holds:

$$k_H/k_T = (k_H/k_D)^{1.44}$$

Breakdown of this relationship has been used to calculate presumed intrinsic effects from experimental ones, but the procedure is questionable since the relationship also breaks down as a consequence of quantum mechanical tunnelling.

Primary Hydrogen Isotope Effects. The only primary hydrogen isotope effects measurable on glycosidase-catalysed reactions are those on proton transfer, obtainable by comparing rates in H_2O and D_2O. However, on putting an enzyme into D_2O, all OH and SH sites, and many NH sites will exchange, so an isotope effect at a single pH and isotopic composition is rarely informative. The "proton inventory" technique, though, in principle permits the distinction to be made between solvent isotope effects which are the sum of many small and uninterpretable effects and those which arise by significant loosening of the bonds to one or two protons which are being transferred (protons "in flight"). The rate , v, is measured as a function of isotopic composition, n. If the effect is due to one proton only, a plot of v against n is linear, if it is due to two protons, the plot is a shallow parabola and if it is due to three (or more) the curve approximates to a catenary. In order to avoid the complicating effects of isotope effects on critical ionisations, it is desirable to make the measurement on a flat part of the pH-rate profile, or, failing that, to use buffers of constant buffer ratio, of the same chemical type as the protein residues causing the ionisations. The assumption is that the solvent isotope effect on enzyme and buffer pK_a values will be the same. Carboxylic-acid based buffers are therefore appropriate for glycosidases.

Even with all the controls described above, it is becoming clear that solvent isotope effects on k_{cat}/K_m are not really interpretable, since on binding substrate many hydrogen bonds are tightened; these offset the effects on k_{+2} due to protons in flight[30]. Additionally, the different microviscosity of H_2O and D_2O will give rise to k_{cat}/K_m isotope effects for "sticky" substrates binding at the diffusion limit. Effects on k_{cat} for poor substrates for which a single chemical step is rate-determining can still be informative, though. The solvent isotope effect of 1.8 on k_{cat} for hydrolysis of p-nitrophenyl α glucopyranoside by *A. niger* glucoamylase is attributable to a single proton in flight, presumably from the nucleophilic water to the catalytic base, as the acidic leaving group does not require protonating.[31]

α-Deuterium and Tritium Isotope Effects. When the hybridisation of a carbon with a hydrogen attached changes from sp^3 to sp^2, the force constant for a bending vibration of the C-H bond weakens, and the vibration decreases in frequency. Consequently, reactions involving an sp^3 to sp^2 hybridisation change are associated with secondary hydrogen isotope effects in which the heavier isotopomer reacts slower (k_H/k_T, k_H/k_D >1). Conversely, reactions involving an sp^2 to sp^3 hybridisation change are associated with secondary hydrogen isotope effects in which the heavier isotopomer reacts faster (k_H/k_T, k_H/k_D <1). At one stage it was thought that the effect could be used to distinguish between SN1 and SN2 reactions, since the incoming nucleophile in an SN2 reaction obstructed the bending motion of the hydrogen, and so increased the effective force constant. However, it now appears that [Knier and Jencks] that the relationship between the magnitude of an α-deuterium kinetic isotope effect, and degree of oxocarbenium ion character at the transition state for a displacement at an acetal centre, only holds for a restricted range of nucleophiles of the same chemical type, if at all.

The qualitative information given by the sense of an α-deuterium effect, though, is quite unambiguous. The direct effects (k_H/k_D >1) measured on the hydrolysis of glycosyl enzyme intermediates establish unequivocally that the intermediates are covalent. If "stabilised oxocarbenium ions" [5] were involved, then the effects would be inverse (k_H/k_D <1). Table 2 sets out the effects measured to date: none are known which support the "stabilised oxocarbenium ion" model for glycosyl-enzyme intermediates.

Table 2 *α–Deuterium Kinetic Isotope Effects on the Hydrolysis of Glycosyl-Enzyme Intermediates*

Enzyme	Source	Sugar residue	αDKIE	Reference
β–galacto-sidase	*E. coli (lacZ)*	α–galacto-pyranosyl	1.25	32
β–galacto-sidase	*E. coli (ebg^a, ebg^b)*	α–galacto-pyranosyl	1.10	33
β–galacto-sidase	*E. coli (ebg^{ab})*	α–galacto-pyranosyl	1.08	34
β–glucosidase	*Stachybotris atra*	α–gluco-pyranosyl	1.11	35
β–glucosidase	*Agrobacterium faecalis*	α–gluco-pyranosyl	1.11	36
Cellulase/ xylanase Cex	*Cellulomonas fimi*	α–glucopyranosyl, cellobiosyl	1.10	37
α–galacto-sidase	*P. chrysosporium*	β–galacto-pyranosyl	1.11	29

β-Deuterium and Tritum Kinetic Isotope Effects. The effect of isotope substitution at a carbon adjacent to the reaction centre yields information both about the geometry of the sugar ring and the electron-deficiency of the reaction centre. The effect arises from electron donation from the σ- orbital of the C-H, C-D or C-T (C-L) bond into the electron-deficient p orbital of the reaction centre. The C-L bond is thereby weakened. (Figure 4).

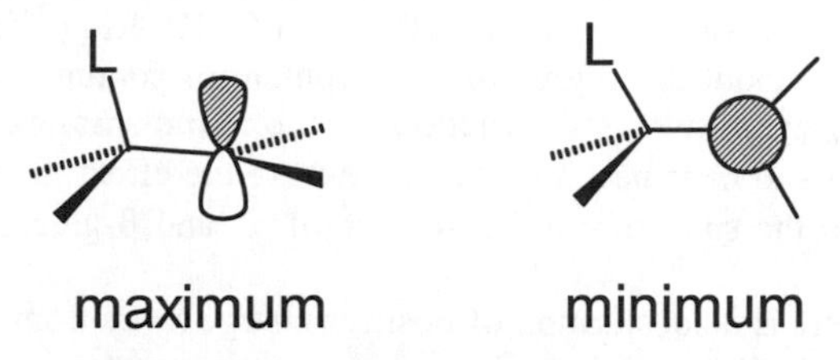

Figure 4 *Origin of β-deuterium and tritium isotope effects*

These effects are generally considered in terms of the following equation:

$$\ln(k_H/k_D)_{obs} = \cos^2\theta.\ln(k_H/k_D)_{max} + \ln(k_H/k_D)_i$$

The last term refers to a small, geometry-independent inductive effect. (Deuterium is in effect slightly more electron-donating than hydrogen, possibly because the anharmonicity of the C-L bond stretching vibration, even in the zeroth vibrational state, means that the electron-cloud of the σ bond lies closer to carbon the less the zero point energy). It is clear that a single β–hydrogen isotope effect gives one parameter – the effect – and three unknowns, $(k_H/k_D)_i$, $(k_H/k_D)_{max}$, and θ. Therefore, although a single β-hydrogen effect may be informative as part of a mapping of a transition state by multiple effects, on its own it is not very valuable.

The situation changes if there are two diastereotopic hydrons next to the reaction centre, whose effects can be measured separately, as with the glycosides of N-acetyl neuraminic acid. The inductive effect is the same for both hydrons, so an immediate qualitative conclusion about the reactive conformation can be made on the basis of the relative magnitudes of the effects. In its complex with the influenza neuraminidase, N-acetyl neuraminic acid is in the $B_{2,5}$ conformation. If the $B_{2,5}$ conformation is also the reactive conformation of the substrate, then the β-deuterium kinetic isotope effect for the *proS* hydron should be bigger than for the *proR*, This is indeed the case, although at optimal pH the effects on k_{cat}/K_m are partly masked, and those on k_{cat} fully masked, by non-covalent events which appear to be a conjoint change of the enzyme-substrate or enzyme-product complex from the initially bound chair to the boat, and the reverse[38] .

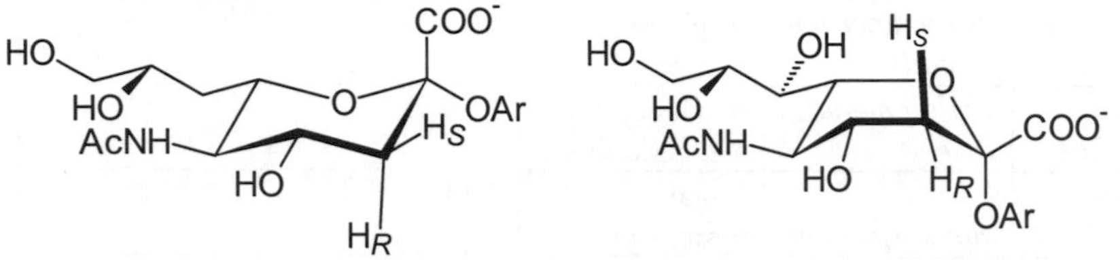

Scheme 6. Diastereotopic 3-hydrons in an N-acetyl neuraminide.

Heavy atom isotope effects are smaller than even secondary hydrogen effects, so their use has been comparatively limited, the exception being Schramm's group, which has published extensive effects on a wide range of glycohydrolases and glycotransferases. The technique largely used – radiolabelling and ^3H and ^{14}C double counting - being a competition method, permits measurement of only effects on (V/K). Ring ^{18}O effects are in principle very informative about the degree of development of conjugation, but only one such effect on an enzyme-catalysed reaction – the diphtheria-toxin-catalysed hydrolysis of NAD^+ - appears to have been measured[39], where the effect (0.988 ± 0.003) is in line with the effect on the sponatneous hydrolyses of α- and β-glucosyl fluoride.
Zhang, Bommuswamy and Sinnott

In systems where there is delocalisation of positive charge away from the reaction centre, the reaction centre ^{13}C (or ^{14}C) appears very sensitive to "push" from a nucleophile; indeed in the case of the bimolecular reaction of α-glucosyl fluoride with azide ion, an 8% ^{13}C effect is observed [Zhanget al.]. A simple physical explanation is that in such systems the motion of the reaction centre carbon is part of the reaction coordinate, and effects are maximal when the carbon is half transferred from leaving group to nucleophile, in much the same way as primary hydrogen effects are maximal when the hydrogen is transferred. Certainly, the ^{13}C effect on the yeast α-glucosidase-catalysed hydrolysis of α-glucosyl pyridinium ions is higher than on the spontaneous hydrolysis, which goes through an ion-molecule complex[40]

PROTON TRANSFER DURING GLYCOSIDASE CATALYSIS

As discussed above, hope entertained some years ago that solvent isotope effects would fairly readily enable the contribution to proton transfer during glycosidase catalysed reactions to be characterised has not been fully realised. Indications of the importance and extent of proton transfer can, though, be obtained from the dependence of rate on leaving group pK_a. Figure 5 shows the More O'Ferrall-Jencks diagram for the acid catalysed hydrolysis of an ether to give a stable carbenium ion. The magnitude of β_{lg} locates the transition state on a Northwest-Southeast diagonal.

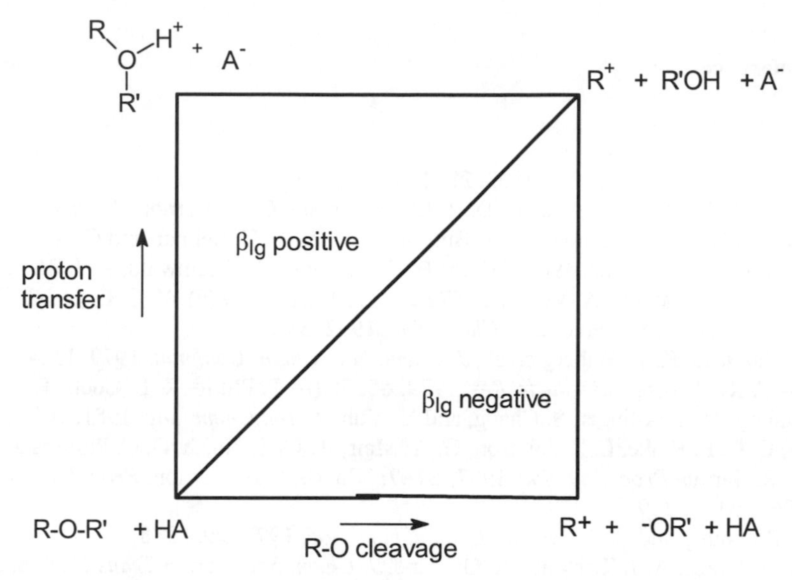

Figure 5 *More O'Ferrall-Jencks diagram for general acid catalysis of the hydrolysis of an ether via a discrete, non-delocalised carbenium ion*

Even in two dimensions, β_{lg} plots can be curved or straight, depending on the precise location of the transition state and the curvature of the free energy surface.[41] The addition of axes in additional dimensions, corresponding to bond-formation to the nucleophile and delocalisation of charge, will further complicate matters. Yet more complexity is added by the binding steps of enzymes. In free solution the general acid catalysis of the departure of a leaving group of lower pK_a than the catalysing acid is not generally seen. In an ES complex, however, the acid catalyst is inevitably present. Consequently, some curved β_{lg} plots may arise because acidic aglycones are departing without acid assistance and more basic ones are being partly protonated at the transition state. Curved β_{lg} plots may also trivially arise from a change of rate-determining step: substrates with good leaving groups may become "sticky" so that k_{cat}/K_m represents diffusion-controlled binding, or k_{cat} for substrates of a retaining glycosidase may change from representing glycosylation to deglycosylation as the leaving group gets better. Nonetheless, sophisticated attempts have been made to interpret Brønsted parameters for both glycosylation and deglycosylation of the *lac Z* β-galactosidase of *Escherichia coli* and various mutants.[42] It is unfortunate though, that the simplest interpretation of data for this enzyme is that the Mg^{2+}- saturated form at least uses electrophilic catalysis by magnesium, rather than general acid catalysis to assist aglycone departure. It is a subtle system to disentangle, since the presumed acid catalyst (Glu 461) is one of the ligands to Mg^{2+}, so the demetallated enzyme *can* use protic catalysis.

Investigations on the role of the general base in inverting glycosidases are largely confined to site-directed mutagenesis experiments which are not within the scope of this article. However, general base catalysis of the addition of alcohols to carbocations is characterised by β values from 0.23 to 0.33 depending on the alcohol and carbocation.[43] These low β values suggest that an Asp/Asn or Glu/Gln (ΔpK_a ~4) change of the catalytic base in an inverting glycosidase should have a small effect (10-20 fold reduction in k_{cat}).

References

1. D. E. Koshland, *Biol. Rev.* 1953, **28**, 416
2. (a) C. A. Bunton, T. A. Lewis, D. R. Llewellyn and C. A. Vernon, *J. Chem. Soc.* 1955, 4419 (b) C. Armour, C. A. Bunton, S. Patai, L. H. Selman and C. A. Vernon, *J. Chem. Soc.* 1961, 412 (d) B. E. C. Banks, Y. Meinwald, A. J. Rhind-Tutt, I. Sheft and C. A. Vernon *J. Chem. Soc.* 1961, 3240 (e) W. G. Overend, C. W. Rees and J. S. Sequeira, *J. Chem. Soc.* 1962, 3429
3. D. Farcasiu, K. B. Wiberg et. al., *J. Chem. Soc. Chem. Commun.* 1979. 1124
4. (a) C. R. Nelson, *Carbohydr. Res.* 1979, **68**, 55 (b) T. Philip, R. L. Cook, T. B. Malloy, N. L. Allinger, S. Chang, and Y. Yuh, *J. Am. Chem. Soc.* 1981, **103**, 2151
5. (a) C. C. F. Blake, L. N. Johnson, G. A. Mair, A. C. T. North, D. C. Phillips and V. R. Sarma, *Proc. Roy. Soc.* 1967, **B167**, 378. (b) C.A. Vernon, *Proc. Roy. Soc.* 1967, **B167**, 389.
6. P. R. Young and W. P. Jencks, *J. Am. Chem. Soc.* **1977**, *99*, 8288
7. G.-A. Craze, A. J. Kirby and R. Osborne, *J. Chem. Soc. Perkin Trans II* **1978**, 357
8. B. L. Knier and W. P. Jencks, *J. Am. Chem. Soc.*1980, **102**, 6789
9. (a) J. P. Richard, *J. Am. Chem. Soc.*1989, **111**, 1455 (b) T. L. Amyes and W. P. Jencks, *J. Am. Chem. Soc.* 1989, **111**, 7888 (c) J. P. Richard, T. L. Amyes, L. Bei and V. Stubblefield, *J. Am. Chem. Soc.* 1990, **112**, 9513 (d) T. L. Amyes and J. P. Richard, *J. Am. Chem. Soc.* 1991, **113**, 1867 (e) T. L. Amyes and J. P. Richard, *J. Chem. Soc. Chem. Commun.* 1991, 200 (f) V. Jagannadham, T. L. Amyes and J. P. Richard, *J. Am. Chem. Soc.* 1993, **115**, 8465 (g) T. L. Amyes, I. W. Stevens and J. P. Richard, *J. Org. Chem.* 1993, **58**, 6057 (h) J. P. Richard, *J. Org. Chem.* 1994, **59**, 25 (I) R. A. McClelland, F. L. Cozens, S. Steenken, T. L. Amyes and J. P. Richard, *J. Chem. Soc., Perkin Trans. II,* 1993, 1717
10. P. Kandanarachchi and M. L. Sinnott, *J. Am. Chem. Soc.* 1994, **116**, 5592
11. J. Zhu and A. J. Bennet, *J. Am. Chem. Soc.* 1998, **120**, 3887
12. P. Kandanarachchi and M. L. Sinnott, *J. Am. Chem. Soc.* 1994, **116**, 5601
13. M. Sinnott, C. D. Garner, E. First, and G. Davies. *Lexicon of Terms and Concepts in Mechanistic Enzymology,* p.58 (Vol. IV of *Comprehensive Biological Catalysis,* M. L. Sinnott, Ed., Academic Press, London, 1998)
14. (a) M. L. Sinnott and S. G. Withers, *Biochem. J.* 1974, **143**, 751 (b) G. Legler, M. L. Sinnott and S. G. Withers, *J. Chem. Soc., Perkin Trans II* 1980, 1376 (c) L. Hosie and M. L. Sinnott, *Biochem. J* 1985, **226**, 437. (d) M. A. Kelly, M. L. Sinnott and M. Herrchen, *Biochem. J.* 1987, **245**, 843
15. B. Padmaperuma and M. L. Sinnott, *Carbohydr. Res.* 1993, **250**, 79
16. F. Schiber, P. Travo and M. Pascal, *Bioorganic Chem.* 1979, **8**, 83
17. D. A. Yost and B. M. Anderson, *J. Biol. Chem* 1983, **258**, 3075
18. B. C. Challis and J. A. Challis in D. H. R. Barton and W. D. Ollis, Eds., *Comprehensive Organic Chemistry*, Pergamon, Oxford, 1979. Volume 2 (volume ed. I. O. Sutherland), pp 996, 997

19. C. S. Jones and D. J. Kosman, *J. Biol. Chem.* 1980, **255**, 11861.
20. J. D. McCarter, M. J. Adam and S. G. Withers, *Biochem. J.* 1992, **286**, 721
21. A. L. Handlon, C. Xu, H. M. Muller-Steffner, F. Schuber and N. J. Oppenheimer, *J. Am. Chem. Soc.* 1994, **116**, 12087
22. R. D. Chambers, M. Salisbury, G. Apsey and G. Moggi, *J. Chem. Soc., Chem. Commun.* 1988, 680
23. A. K. Konstantidis, and M. L. Sinnott, *Biochem. J.* 1991, **279**, 587.
24. N. S. Banait and W. P. Jencks, *J. Am. Chem. Soc.*1991**, 113,** 7951.
25. Y. Zhang, J. Bommuswamy and M. L. Sinnott, *J. Am. Chem. Soc.* 1994, **116**, 7557
26. B. Padmaperuma, Ph. D. Thesis, University of Illinois at Chicago, 1993
27. S. Krishnan, A. K. Konstantinidis and M. L. Sinnott, *Biochem. J.* 1993, **291**, 15
28. S. Krishnan, B. G. Hall, and M. L. Sinnott, *Biochem. J.* 1995, **312**, 971
29. T. Holm, *J. Am. Chem. Soc.* 1994, **116**, 8803
30. S. A. Adediran and R. F. Pratt, *Biochemistry*, 1999, **38**, 1469
31. T. Selwood and M. L. Sinnott in *Molecular Mechanisms of Bioorganic Processes*, B. T. Golding and C. Bleasdale, Eds., p 188, Royal Society of Chemistry, London, 1990.
32. M. L. Sinnott and I. J. L. Souchard, *Biochem. J.* 1973, **133**, 89
33. B. F. L. Li, D. Holdup, C. A. J. Morton and M. L. Sinnott, *Biochem. J.* 1989, **280**, 109
34. A. C. Elliott, S. Krishnan, M. L. Sinnott, J. Bommuswamy, Z. Guo, B. G. Hall and Y. Zhang, *Biochem. J.* 1992, **282**, 155
35. E. van Doorslaer, O. van Opstal, H. Kersters-Hilderson and C. K. De Bruyne, *Bioorg. Chem.* 1984, **12**, 158
36. J. B. Kempton and S. G. Withers, *Biochemistry,* 1992, **31**, 9961
37. D. Tull and S. G. Withers, *Biochemistry,* 1994, **33**, 6363
38. X. Guo, W. G. Laver, E. Vimr, and M. L. Sinnott, *J. Am. Chem. Soc.* 1994, **116**, 5572
39. P. J. Bert, S. R. Blanke, and V. L. Schramm, *J. Am. Chem. Soc.* 1997, **119**, 12079
40. X. Huang, K. S. E. Tanaka and A. J. Bennet, *J. Am. Chem. Soc.* 1997, **119**, 11147.
41. D. A. Jencks and W. P. Jencks, *J. Am. Chem. Soc.* 1977, **99**, 7948
42. J. P. Richard, *Biochemistry*, 1998, **37**, 4305
43. R. Ta-Shma and W. P. Jencks, *J. Am. Chem. Soc.* 1986, **108**, 8040.

NEW TWISTS IN ENZYMATIC GLYCOSIDE SYNTHESIS

Stephen G. Withers

Department of Chemistry
University of British Columbia
Vancouver, B.C.
CANADA
V6T 1Z1

1 INTRODUCTION

The use of oligosaccharides as therapeutic agents has not developed as rapidly as might have been expected, given the central role of glycan structures in cellular recognition processes, and the potential for interference in such processes as a route to new therapies. One of the major factors hindering their development has been the difficulty of synthesising oligosaccharides on the large scales (multiple kilograms) needed both for clinical trials and eventual production. Classical organic synthesis has made astounding advances in the past decades, enabling almost any oligosaccharide conceivable to be made. However, this process is extremely time-consuming, and only works well on the small scale. Further, the cost of such synthesis on the large scale would be prohibitive for all but the simplest oligosaccharides. Attention has therefore long turned to the use of enzymatic syntheses for large scale production.

In many ways the most obvious enzymatic approach involves the use of Nature's own synthetic enzymes, the glycosyl transferases. These enzymes use a nucleotide phosphosugar (e.g. UDP glucose) as their glycosyl donor, and transfer the sugar with complete control of stereochemistry and regiochemistry to a specific hydroxyl group on their acceptor sugar. They would therefore appear to be perfectly suited for the application. However, there are several problems, including the very high cost of the donor sugar and the relative paucity of transferases available in a suitably pure, stable, water-soluble form. These problems are slowly being overcome in several ways, and indeed synthesis has been achieved on multi-kilogram scales in several industrial settings. Improvements include cheaper sources of the nucleotide phosphosugars, which are becoming available from recombinant cell lines, and also clever cofactor recycling schemes that have been developed to generate the donors at lower cost and also avoid the problems of product (UDP) inhibition. Further, a large number of glycosyl transferases have now been cloned, and greater success in expressing these in suitable forms has now been enjoyed, particularly with the move towards bacterial enzymes rather than their more recalcitrant mammalian counterparts. Nonetheless, the use of glycosyl transferases is still not trivial and alternative approaches are being sought.

Glycosidases used in the "reverse" mode provide an alternative. The use of glycosidases for the synthesis of oligosaccharides has a long history, dating back over 85 years.[1] However their use in this role has been severely limited by several problems, the major one of which is that hydrolysis, rather than synthesis, is of course the preferred process in aqueous solution. Yields therefore tend to be poor. A second problem can be that of control of the regiochemistry (1,3 versus 1,4 etc.) of bond formation. Several approaches have been followed in attempting to solve these problems, and these are discussed briefly within this article. More detailed accounts can be found in some useful recent reviews,[2-4] as well as in a number of earlier overviews.[5-7] The major part of the article, however, concerns a discussion of a new method of enzymatic synthesis involving mutant glycosidases.

1.1 Glycosidase Mechanisms

Hydrolysis occurs with one of two possible stereochemical outcomes, inversion or retention of anomeric configuration, all enzymes within a family hydrolysing their substrate with the same stereochemical outcome. These different outcomes demand different mechanisms, proposals for which were made by Koshland in 1953.[8]

Subsequent mechanistic and structural studies have largely substantiated these proposals and have identified the key active site residues as a pair of carboxylic acids in almost all cases: see the following reviews for background information.[9-14] Glycosidases hydrolysing with inversion of anomeric configuration employ a mechanism involving a direct displacement of the leaving group in an acid/base catalysed process proceeding via an oxocarbenium ion-like transition state as shown in Figure 1. The two carboxylic acids are appropriately spaced (10 – 11 Å) to perform their role as general acid/base catalysts.

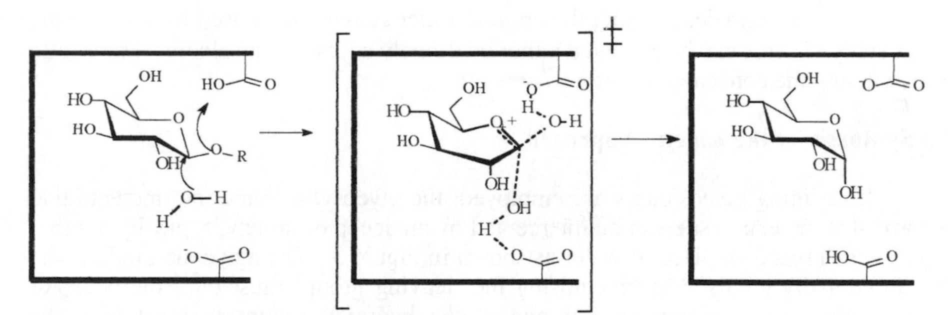

Figure 1 *Mechanism of an inverting glucosidase*

The vast majority of glycosidases that hydrolyse with net retention of anomeric configuration also have an active site containing a pair of carboxylic acids, but in this case only approximately 5 Å apart. This smaller separation between the two groups reflects their different roles in the two step double-displacement mechanism. One residue functions as a general acid catalyst, protonating the glycosidic oxygen, while the other functions as a nucleophile, attacking at the anomeric centre to form a covalent glycosyl-enzyme intermediate as shown in Figure 2.

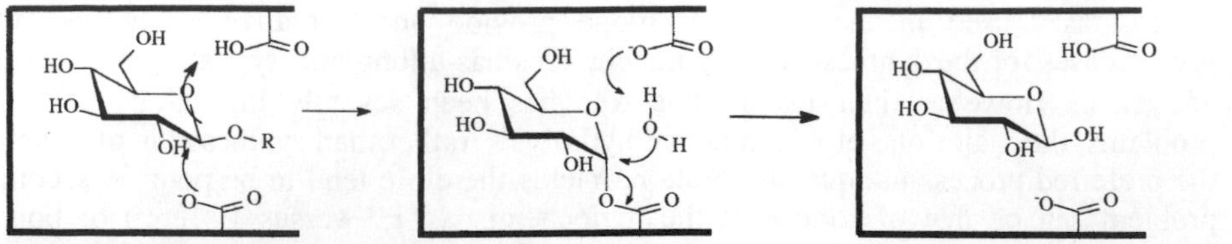

Figure 2 *Mechanism of a retaining β-glucosidase*

In a second step, water attacks this intermediate in a general base-catalysed process to yield the product of retained anomeric configuration. Both steps again proceed via oxocarbenium ion-like transition states (not shown).

1.2 Synthesis by the "Thermodynamic" Approach

By running the reaction in the presence of very high concentrations (typically molar) of sugar the condensation reaction becomes, to at least some extent, favourable. Organic solvents or high salt concentrations can help in displacing this equilibrium, as long as the enzyme is compatible with, and the sugars soluble in, this medium. The advantage of this approach is its simplicity and cheapness, but yields are typically poor (<15% in most cases), with minimal control of regiochemistry. Coupling of two sugars in this way is therefore not generally feasible since it produces intractable mixtures, though there are occasional exceptions involving two units of the same sugar, where useful (20 %) yields of a disaccharide have been obtained.[15,16] This approach can, however, provide an attractive route to the synthesis of glycosides of simple hydrophilic (ideally liquid) alcohols, reaction being performed in the presence of a very high concentration of the alcohol. This approach has been used to provide good (up to 61%) yields when the reaction was performed at an elevated temperature (50 ^{0}C) and time was spent in determining the optimal water activity to be used for the enzyme under study.[17] This can be achieved either by directly adjusting the alcohol/water ratio or by varying the concentration of a co-solvent.

1.3 Synthesis by the Kinetic Approach

If retaining glycosidases are employed, the glycosyl-enzyme intermediate that is formed at the active site can be intercepted by an acceptor moiety, typically another sugar, which reacts in place of water as shown in Figure 3. The aglycone binding site that is normally involved in recognising the 'leaving group' must bind the acceptor sugar in the correct orientation for one of the hydroxyl groups to react with the glycosyl-enzyme in a general base-catalysed process. Interactions between this second sugar and the protein help to stabilise the transition state for glycosyl transfer, thereby lowering the activation barrier relative to that for hydrolysis in the exactly the same way that such interactions serve to stabilise the transition state for disaccharide hydrolysis.

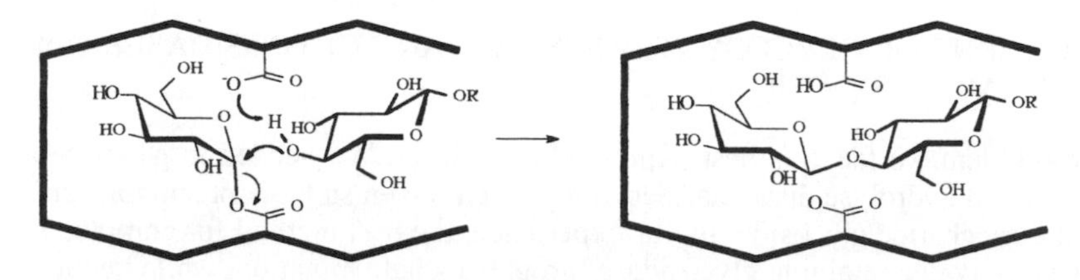

Figure 3 *Mechanism of transglycosylation*

Use of this approach in synthesis therefore involves reaction of an activated donor sugar with a suitable acceptor sugar in the presence of the appropriate retaining glycosidase. In order for this to work the transglycosylation process must obviously occur in preference to hydrolysis. Some control over this can be effected through the concentration of the acceptor sugar used, since occupancy of the acceptor site requires sugar concentrations higher than the dissociation constant for that site. In addition, the product disaccharide must function as a significantly poorer glycosyl donor than the activated donor substrate employed. Control over this is effected by using the most reactive (highest k_{cat}/K_m) donor possible since the ratio of the rates of consumption of donor versus product (k_{rel}) depends upon their relative k_{cat}/K_m values as follows.

$$k_{rel} = \frac{(k_{cat}/K_m)_P \times [P]}{(k_{cat}/K_m)_D \times [D]}$$

Due to their generally high k_{cat}/K_m values nitrophenyl glycosides and glycosyl fluorides are particularly attractive donors in this regard. In addition, the departed aglycone (fluoride or p-nitrophenolate) is itself a very poor acceptor thus will not take part in further reaction and can be very easily separated from products.

If these factors are controlled well, then respectable yields in the range of 10 – 60 % can, in some cases, be obtained. However this generally requires exquisite timing of product harvesting, and/or the use of some trick to "pull" the equilibrium of the reaction in the synthetic direction, usually developed separately for each enzyme. Approaches include the use of co-solvents to decrease the water activity (REF), or continuous removal of the product. Techniques employed for this include preferential product absorption onto activated charcoal chromatography columns or use of a second "coupling" enzyme, for example a glycosyl transferase, that reacts only with the product of interest and ideally converts it directly to the desired product.[18, 19]

While these approaches can be made to work in some cases, none of these is a truly general approach: each one had to be developed on a case-by-case basis. There therefore existed a need for a more general approach to solve this problem of hydrolysis. Our approach to this problem involved the use of protein engineering methods to generate a mutant that could synthesise disaccharides but could not hydrolyse these products once formed. The approach developed is described in the following section.

2 DEVELOPMENT OF GLYCOSYNTHASES, MUTANT GLYCOSIDASES FOR OLIGOSACCHARIDE SYNTHESIS.

The problem we faced is best expressed as follows. We needed to generate a mutant that cannot hydrolyse disaccharides, but yet can, given suitable precursors, still synthesise a disaccharide glycoside. In our experience, the best method for completely eliminating activity in a retaining glycosidase through a single mutation, while keeping the structure intact, involves mutation of the catalytic nucleophile to a non-nucleophilic residue such as alanine. If this mutation and purification are carried out with considerable care to minimise the amount of translational misincorporation and of contamination from wild type enzyme or other active mutants during purification, then such Ala mutants of retaining glycosidases are essentially devoid of activity. Maximal observed rates are at least 10^5 times lower than those of their wild type enzymes,[20] thus should not cleave products *if* they can be made to first from disaccharides. Hope that the mutant could function as a synthetic enzyme derives from the fact that the mutant folds correctly, as evidenced by CD measurements, and even X-ray crystallographic analyses in several cases (Brayer, G., Rose, D. and Withers. S. G., Unpublished results). The mutant therefore has an active site capable of binding both the glycone and aglycone moieties, and this active site is optimised for stabilization of the oxocarbenium ion-like transition state involved in glycosyl transfer (although lacking the negative charge from the carboxylate). In addition the active site contains an acid/base residue suitably poised to assist in its normal role. Indeed, as has been shown by direct ^{13}C-NMR titrations in a suitably labeled sample of a retaining xylanase from *Bacillus circulans,*[21] the removal of the charged nucleophilic residue should lower the pKa of the acid/base residue, placing it in the deprotonated form necessary for it to function as the general base. In full, therefore, the enzyme is set up to carry out the transglycosylation step, but cannot do so because it has not formed, and cannot form, a reactive α-glycosyl-enzyme intermediate.

Our answer to this problem was to *provide* the mutant with a glycosyl donor that was a chemical mimic of this reactive α-glycosyl-enzyme intermediate incorporating a leaving group that is sufficiently small to fit into the space in the active site generated by removal of the catalytic nucleophile. Our reagent of choice was an α-glycosyl fluoride since such compounds are reactive glycosyl donors with a very small substituent/leaving group, fluoride, and they are very readily synthesised. Incubation of the mutant with this donor in the presence of a suitable glycosyl acceptor should lead to product formation as shown in Figure 4.

We first developed this approach with the β-glucosidase/galactosidase from *Agrobacterium sp.* (Abg), a Family 1 β-glycosidase with broad specificity that has been studied extensively in our group.[20,22-30] The identity of the active site nucleophile was known[28] and indeed mutants at this position had been generated

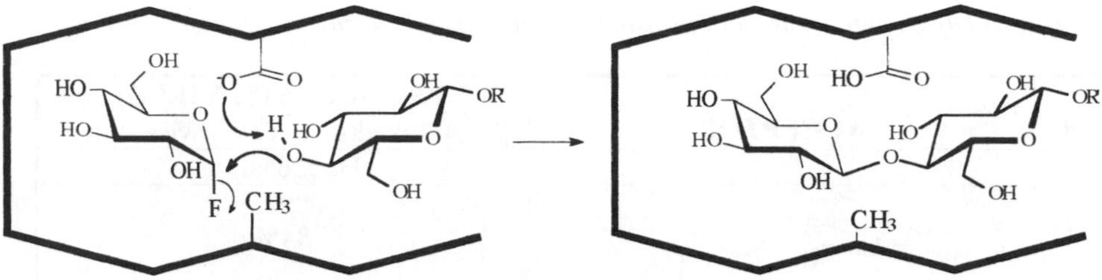

Figure 4 *Glycosynthase Mechanism*

previously.[20] As donors we focussed upon α-galactosyl fluoride and α-glucosyl fluoride since the highest rates were anticipated in those cases, given the substrate preferences of this enzyme. A wide range of different sugars was tested as acceptors to probe the preferences of this enzyme, and to explore its synthetic potential. Since HF is liberated on a stoicheometric basis the reaction requires high buffer concentrations to keep the medium neutral. Those chosen were either 150 mM sodium phosphate, or 150 mM ammonium bicarbonate. This approach has been published.[31]

As an illustration of this approach (2 mL scale) the Glu358Ala mutant of Abg was incubated with α-galactosyl fluoride (40 mM) as donor and p-nitrophenyl β-cellobioside (30 mM) as acceptor in the presence of AbgGlu358Ala (0.01 mg) in 150 mM ammonium bicarbonate buffer. After 8 hours TLC analysis indicated that reaction was complete and the reaction was worked up by ultrafiltration to remove the enzyme (which can be re-used). Lyophilisation, then HPLC purification, yielded 34 mg (92% isolated yield) of Gal-β-1,4-Glu-β-1,4-Glu-β-p-nitrophenyl.

The enzyme was shown to be capable of transferring galactose residues to a range of different acceptors, as shown in Table 1. The products formed were themselves poor acceptors for the enzyme, as also is galactose, since the enzyme will not transfer to an axial 4-hydroxyl, and only slowly to the other positions. Reaction therefore terminated after a single transfer. As can be seen, the enzyme will transfer to a range of sugars including glucosides, cellobiosides, xylosides and mannosides. Interestingly, in all cases except when xylosides are the acceptors, transfer occurs to the 4-position. This is in contrast to the case with the wild type enzyme working in the "normal" transglycosylations mode, when a mixture of 3- and 4-linked products is typically observed.[32] This difference in behaviour is probably a consequence of the fact that synthesis via the Glycosynthase is strictly irreversible, the product mixture observed being truly kinetically controlled. Such is not necessarily the case for the wild type enzyme since it can rearrange the initially formed products by re-forming the glycosyl-enzyme intermediate using the product as donor. Thus reaction will tend towards a thermodynamically controlled mixture, though of course the truly thermodynamic product mixture is that resulting from hydrolysis. The observation of only 1,3-linked products when the acceptor is a xyloside is of interest, and is consistent with a strong preference for 1,3-linked products with the wild type enzyme under similar conditions. It is also reminiscent of similar results on other enzymes from Clan GHA.[33]

Table 1. *Glycosynthase Reactions using α-Galactosyl Fluoride as Donor*

#	ACCEPTOR	**PRODUCTS (% YIELD)** β-1,4 linked(unless otherwise stated)
		Disaccharide
1		**84%**
2		**81%** (β-1,3 linked)
3		**64%**
4		**66%**
		Trisaccharide
5		**92%**
6		**88%**

α-Glucosyl fluoride also acts as an excellent donor, however the products, being glucosides, are also excellent acceptors, resulting in longer oligomer products. The size of the oligosaccharide product can be controlled to a reasonable extent through control of the number of equivalents of donor employed relative to acceptor, though such control is not perfect. Examples of such transfers are shown in Table 2, the glycosyl acceptor specificity observed being similar to that seen when α-galactosyl fluoride was the donor. The length of the oligosaccharides formed is largely determined by the ratio of donor to acceptor sugar employed. Indeed, oligosaccharides of four to five sugars in length can be made in this way using this mutant enzyme, thereby allowing the relatively facile synthesis of useful chromogenic substrates for cellulases, such as the nitrophenyl and methylumbelliferyl cello-oligosaccharides. In addition this technology has been used recently to synthesise a series of spin-labeled

Table 2. *Glycosynthase Reactions using α-Glucosyl Fluoride as Donor*

#	ACCEPTOR	PRODUCTS (% YIELD) β-1,4 linked			
		Disacch.	Trisacch	Tetrasac	**Total Yield†**
7		48%	34%	-	**82%**
8		38%	24%	10%	**72%**
9		41%	29%	6%	**76%**
10		38%	42%	4%	**84%**
11a b		-	75%* 23%*	8%* 54%*	**83%*** **77%***
12a b		44% 12%*	29% 66%*	- 8%*	**73%** **86%***
13		12%	51%	3%	**66%** (β-1,3 linked)
14		31%	42%	6%	**79%**
15a b			79% 8%*	13% 64%*	**92%** **72%***
16			64%	21%	**85%**
17			59%	12%	**71%**

All reactions with 1-1.4 equiv. or (*) with 2.2-3.0 equiv. of glycosyl donor.

cello-oligosaccharides for use in NMR experiments to determine the orientation of binding of oligosaccharides to a cellulose binding module.[34]

A final, interesting example, which illustrates the power of this approach, is that of the synthesis of some 2-fluoro-cellooligosaccharide-based cellulase inactivators. Incubation of AbgGlu358Ala with 2,4-dinitrophenyl 2-deoxy-2-fluoro β-glucoside plus 1.5 equivalents of α–glucosyl fluoride yielded an intentional mixture of 2,4-dinitrophenyl 2-deoxy-2-fluoro β-cellobioside and -trioside (84% total yield) that was readily separated by HPLC. These compounds have already found use in structural studies of glycosyl-enzyme complexes.[35-38] It is interesting to note that this reaction could not be carried out by standard glycosidase transglycosylation since the 2-fluorosugar would have inactivated the enzyme, a process that is not possible with the mutant.

This approach therefore provides very good yields of product in a very simple reaction starting with easily accessible reagents and has good potential as a method for synthesis of oligosaccharides, particularly on the large scale. The method has therefore been patented and the mutant enzymes given the name "Glycosynthases" in order to differentiate them from the natural glycosyl transferases, but to reflect their origins as glycosidases. The approach indeed appears to be reasonably general, as it has been applied to several different retaining glycosidases, both by our own groups, and by others (see other papers from this conference). Indeed a second example has recently been published.[39] Our current efforts are focussed upon extending this technology to the synthesis of a greater range of oligosaccharides. This is being attempted in two ways. One approach involves the generation of the equivalent mutation in other retaining glycosidases that exhibit a strong propensity to transglycosylate. The other involves subjecting our existing Glycosynthases to random mutagenesis, preferably through so-called 'directed evolution', in conjunction with efficient, high-capacity screens. Given the success of others in changing specificities of enzymes, even glycosidases, using such approaches[40] it seems reasonable that a range of Glycosynthases of differing synthetic capability could be generated in this way.

Acknowledgments

I thank my colleagues and collaborators for their assistance in this work. In particular I would like to thank Tony Warren, Lloyd Mackenzie, Karen Rupitz and Qingping Wang for their invaluable contributions. In addition I thank the Protein Engineering Network of Centres of Excellence of Canada and the Natural Sciences and Engineering Research Council of Canada for financial assistance.

References

1 E. Bouquelot, *Ann. Chim.*, 1913, **29**, 145.
2 D. H. G. Crout and G. Vic, *Curr. Opin. Chem. Biol.*, 1998, **2**, 98.
3 G. M. Watt, P. A. Lowden, and S. L. Flitsch, *Curr. Opin. Struct. Biol.*, 1997, **7**, 652.
4 V. Kren and J. Thiem, *Chem. Rev.*, 1997, **26**, 463.

5 J. Thiem, *FEMS Microbiol. Rev.*, 1995, **16**, 193.
6 C. H. Wong, R. L. Halcomb, Y. Ichikawa, and T. Kajimoto, *Angew. Chem. Int. Ed. Engl.*, 1995, **34**, 521.
7 K. G. I. Nilsson, *Tibtech*, 1988, **6**, 256.
8 D. E. Koshland, *Biol. Rev.*, 1953, **28**, 416.
9 M. L. Sinnott, *Chem. Rev.*, 1990, **90**, 1171.
10 J. D. McCarter and S. G. Withers, *Curr. Opin. Struct. Biol.*, 1994, **4**, 885.
11 G. Davies and B. Henrissat, *Structure*, 1995, **3**, 853.
12 G. Davies, M. L. Sinnott, and S. G. Withers, in '*Comprehensive Biological Catalysis*, Academic Press Inc., ed. M. L. Sinnott, Vol 1, 119, 1998.
13 H. Ly and S. G. Withers, *Ann. Rev. Biochem.*, 1999, **In Press**.
14 D. L. Zechel and S. G. Withers, , New York, 1999.
15 S. Suwasono and R. A. Rastall, *Biotechnol. Lett.*, 1996, **18**, 851.
16 K. Ajisaka, I. Matsuo, M. Isomura, H. Fujimoto, M. Shirakabe, and M. Okawa, *Carbohydr. Res.*, 1995, **270**, 123.
17 G. Vic and D. H. G. Crout, *Carbohydr. Res.*, 1995, **279**, 315.
18 Y. Ito and J. C. Paulson, *J. Amer. Chem. Soc.*, 1993, **115**, 7862.
19 V. Ķren and J. Thiem, *Angew. Chem. Int. Ed.*, 1995, **34**, 893.
20 Q. Wang, R. W. Graham, D. Trimbur, R. A. J. Warren, and S. G. Withers, *J. Am. Chem. Soc.*, 1994, **116**, 11594.
21 L. P. McIntosh, G. Hand, P. E. Johnson, M. D. Joshi, M. Korner, L. A. Plesniak, L. Ziser, W. W. Wakarchuk, and S. G. Withers, *Biochemistry*, 1996, **35**, 9958.
22 J. C. Gebler, D. E. Trimbur, A. J. Warren, R. Aebersold, M. Namchuk, and S. G. Withers, *Biochemistry*, 1995, **34**, 14547.
23 M. N. Namchuk and S. G. Withers, *Biochemistry*, 1995, **34**, 16194.
24 Q. Wang and S. G. Withers, *J. Am. Chem. Soc.*, 1995, **117**, 10137.
25 Q. Wang, D. Trimbur, R. Graham, R. A. Warren, and S. G. Withers, *Biochemistry*, 1995, **34**, 14554.
26 J. B. Kempton and S. G. Withers, *Biochemistry*, 1992, **31**, 9961.
27 I. P. Street, J. B. Kempton, and S. G. Withers, *Biochemistry*, 1992, **31**, 9970.
28 S. G. Withers, R. A. J. Warren, I. P. Street, K. Rupitz, J. B. Kempton, and R. Aebersold, *J. Am. Chem. Soc.*, 1990, **112**, 5887.
29 S. G. Withers, K. Rupitz, D. Trimbur, and R. A. Warren, *Biochemistry*, 1992, **31**, 9979.
30 D. E. Trimbur, R. A. Warren, and S. G. Withers, *J. Biol. Chem.*, 1992, **267**, 10248.
31 L. F. Mackenzie, Q. Wang, R. A. J. Warren, and S. G. Withers, *J. Am. Chem. Soc.*, 1998, **120**, 5583.
32 H. Prade, L. F. Mackenzie, and S. G. Withers, *Carbohydr. Res.*, 1998, **305**, 371.

33 R. Lopez and A. Fernandez-Mayoralas, *J. Org. Chem.*, 1994, **59**, 737.

34 P. E. Johnson, E. Brun, L. Mackenzie, S. G. Withers, and L. P. McIntosh, *J. Mol. Biol.*, 1999, **287**, 609.

35 A. White, D. Tull, K. Johns, S. G. Withers, and D. R. Rose, *Nat Struct Biol*, 1996, **3**, 149.

36 G. Sulzenbacher, L. F. Mackenzie, K. Wilson, S. G. Withers, C. Dupont, and G. J. Davies, *Biochemistry*, 1999, **38**, 4826.

37 L. F. Mackenzie, G. Sulzenbacher, C. Divne, T. A. Jones, H. F. Woldike, M. Schulein, S. G. Withers, and G. J. Davies, *Biochem. J.*, 1998, **335**, 409.

38 G. J. Davies, L. F. Mackenzie, A. Varrot, M. Dauter, A. M. Brzozowski, M. Schulein, and S. G. Withers, *Biochemistry*, 1998, **37**, 11707.

39 C. Malet and A. Planas, *FEBS Letts.*, 1998, **440**, 208.

40 J. H. Zhang, G. Dawes, and W. P. Stemmer, *Proc. Natl. Acad. Sci. U. S. A.*, 1997, **94**, 4504.

DIFFERENCES IN CATALYTIC PROPERTIES OF ACETYLXYLAN ESTERASES AND NON-HEMICELLULOLYTIC ACETYLESTERASES

P. Biely[1*], G. L. Côté[2], L. Kremnický[1], and R.V.Greene[2]

[1]Institute of Chemistry, Slovak Academy of Sciences, 84238 Bratislava, Slovakia;
[2]Biopolymer Research Unit, National Center for Agricultural Utilization Research, Agricultural Research Service, United States Department of Agriculture, Peoria, Illinois 61604, USA;

1 ABSTRACT

The substrate specificities of acetylxylanesterases (AcXEs) from *Schizophyllum commune*, *Trichoderma reesei* and *Streptomyces lividans* were compared with the specificities of wheat germ lipase, orange peel esterase and *Candida cylindracea* lipase. Investigated substrates included aryl acylates, acetylated methyl glycosides and acetylxylan. The latter three enzymes were unable to deacetylate xylan to any significant degree. Only AcXE from *Streptomyces lividans* did not hydrolyze aryl acylates. With the exception of *C. cylindracea* lipase, the relative activities of the enzymes on aryl acylates were 4-nitrophenyl acetate > propionate > butyrate. The AcXEs showed preference for deacetylation at positions 2 and 3 of methyl 2,3,4-tri-*O*-acetyl β-D-xylopyranoside and methyl 2,3,4,6-tetra-*O*-acetyl β-D-glucopyranoside. This regiospecificity corresponded to the function of the AcXEs in acetylxylan degradation and was found to be complementary to that exhibited by the non-hemicellulolytic enzymes. This complementation offers new possibilities for chemoenzymic synthetic carbohydrate chemistry. Based on the kinetics of deacetylation of diacetates of methyl β-D-xylopyranoside, it has been hypothesized that the mechanism of deacetylation by AcXEs at positions 2 and 3, when the neighboring positions (O-3 and O-2) are not acetylated, involves enzyme-catalyzed formation of a five-membered transition state from which the acetyl group is subsequently released.

2 INTRODUCTION

The presence of acetyl ester groups in hemicellulose is believed to have two physiological roles: i) protection of polysaccharide against microbial degradation by limiting access of hydrolases to the backbone[1]; ii) alteration of physico-chemical properties of cell walls due to changes in polysaccharide solubility[2]. Certain microorganisms have evolved enzymes that remove acetyl groups from polysaccharides and enzymes which remove acetyl groups from partially acetylated D-glucurono-D-xylan are called acetylxylan esterases (AcXEs)[1-5]. AcXEs render xylopyranosyl residues of the xylan main chain more accessible to cleavage by endo-β-1,4-xylanases (EC 3.2.1.8)[6].

Since the discovery of AcXEs in 1985[3], a number of enzymes exhibiting this substrate specificity have been described and are now recognized to be common components of microbial hemicellulolytic and cellulolytic enzyme systems[7-10]. Comparison of their complete or partial amino acid sequences led to the recognition that there are several types or groups of AcXEs. According to Henrissat and Coutinho[11] (3rd Carbohydr. Bioeng. Workshop, unpublished results), seven groups of proteins that exhibit AcXE activity exist among the ten carbohydrate esterase (CE) families. Substrate specificity and catalytic properties of three AcXEs, belonging to three of the seven AcXE groups, were investigated.

The first group includes AcXEs from *Streptomyces lividans*[8], *Streptomyces thermoviolaceus*[12] and the NodB-like domain of *Cellulomonas fimi*[13] (Table 1). The distinctive feature of these AcXEs is high specificity for the polymeric substrate and lack of activity on low molecular mass substrates, such as 4-nitrophenyl acetate or 4-methylumbelliferyl acetate[22]. The AcXE sequence from *S. lividans* suggests that other enzymes from this group might possess xylan-binding domains as well[8]. The enzymes in this group seem to represent the real substrate-specific AcXEs.

Table 1 *Types of acetylxylan esterases examined*

AcXE group (CE* family)[11]	Genus or species	M. w. (kDa)	Activity on low M.w. substrates	% Homology with the first family member**	Ref.
I (4)	*Streptomyces lividans*	34	No		14, 8
	Streptomyces thermoviolaceus	34.3	Unknown	80	12
	NodB-like domain of xylanase D	23	No	54	13
II (5)	*Trichoderma reesei*	34	Yes		7, 15
	Penicillium purpurogenum AcXEII	23	Yes	67	9, 16
III (1)	*Schizophyllum commune*	31	Yes		17, 18
	Penicillium purpurogenum AcXEI	23	Yes	41	9
	Aspergillus sp.	?	?	65-67	19-21

*Carbydrate esterase family. ***Concerns the whole sequence or just a part of NH_2-terminal.

The second group of AcXEs contains AcXE from *Trichoderma reesei* and AcXE II[7,15] from *Penicillium purpurogenum*[9,16] (Table 1). These enzymes exhibit activity on 4-nitrophenyl-, 4-methylumbelliferyl- and naphthyl-acetate. Although the *T. reesei* AcXE was reported to be inactive on acetyl galactoglucomannan[23], it deacetylated per-*O*-acetyl methyl β-D-mannopyranoside[24]. This is in contrast to AcXE from *S. lividans*.

The third group investigated includes AcXE from *Schizophyllum commune*[17,18]. A comparison of the 50 NH_2-terminal amino acid sequence of this enzyme (unpublished data[24]) with gene sequences and partial amino acid sequences of other proteins indicates that this group of AcXEs belongs to carbohydrate esterase family 1. Family 1 also includes *Aspergillus* AcXEs[19-21] and AcXEI from *P. purpurogenum*[9,11]. A common feature of these enzymes appears to be relatively low substrate specificity. Based on the properties of AcXE from *S. commune*[18] one can expect the enzymes of this group to deacetylate acetylated sugars and low molecular mass substrates quite efficiently. *S.*

commune AcXE was shown to catalyze very selective 2,3-deacetylation of 2,3,4,6-tetra-*O*-acetyl methyl β-D-mannopyranoside[18].

In the present article, we describe results in which certain catalytic properties of selected AcXEs were compared with those of non-hemicellulolytic acetylesterases or lipases frequently used in the regioselective deacetylation of carbohydrates or in acetylation of carbohydrates in the presence of organic solvents[25-27].

3 MATERIALS AND METHODS

3.1 Enzymes

AcXE from *S. lividans* was from a genetically modified strain which overproduces AcXE together with XynB[8]. Purified enzyme was a gift from Drs. Claude Dupont and Dieter Kluepfel (Institute Armand-Frappier, Laval, Quebec, Canada). AcXE from *T. reesei*[7], used in our previous study[24], was kindly provided by Dr. Maija Tenkanen (VTT, Espoo, Finland). AcXE from *S. commune* was purified from cellulose-spent medium as described[18]. Wheat germ lipase, type I, L-3001 (EC 3.1.1.3), acetylesterase from orange peel (EC 3.1.1.6) and *Candida cylindracea* lipase, type VII, L-1754 (EC 3.1.1.3) were purchased from Sigma. All three commercial enzymes were dialyzed before use.

3.2 Enzyme assays and other tests

Acetyl esterase activity was determined on 4-nitrophenyl acetate using the procedure of Johnson et al.[28], which was modified as described[18]. One unit is defined as the amount of enzyme releasing 1 μmol of 4-nitrophenol per minute. A similar procedure was used to determine enzymatic activity on 4-nitrophenyl-propionate and -butyrate.

Deacetylation of sugar acetates was compared using two substrates: methyl 2,3,4-tri-*O*-acetyl β-D-xylopyranoside (2,3,4-tri-*O*-Ac-Me-β-Xyl*p*) and methyl 2,3,4,6-tetra-*O*-acetyl-β-D-glucopyranoside (2,3,4,6-tetra-*O*-Ac-Me-β-Glc*p*). Saturated solutions of acetylated glycosides (14.5 and 10.4 mM, respectively) in 0.1 M phosphate buffer (pH 6.0) were incubated with enzyme at 40°C. Deacetylation was monitored by GLC of trimethylsilyl derivatives[18]. The rate of the first deacetylation step was calculated on the basis of the disappearance of fully acetylated substrate. Subsequent deacetylation steps were calculated utilizing the production rate of derivatives with lower acetate content.

The ability of the enzymes to deacetylate beechwood acetylxylan was compared on the time basis required to visualize precipitation in a 1% polysaccharide solution. Solutions were buffered with 0.1 M phosphate (pH 6.0). Assays were conducted at 40°C and were calibrated to one AcXE unit, or, in the case of *S. lividans* AcXE, to an equivalent amount of enzyme which, at a similar rate, deacetylated 2,3,4-tri-*O*-Ac-Me-β-Xyl*p* and 2,3,4,6-tetra-*O*-Ac-Me-β-Glc*p*. In other words, the action of AcXEs on acetylxylan was compared using amounts of enzyme that catalyzed similar deacetylation rates of acetylated methyl glycosides.

Previous data describing the action of AcXEs on diacetates of methyl β-D-xylopyranoside[18,22,24] were compared with the action of wheat germ lipase. The rates of deacetylation were determined and products were analyzed by GLC after trimethylsilylation of lyophilized reaction mixtures[18].

4 RESULTS

Comparative data pertaining to catalytic properties of AcXEs and non-hemicellulolytic acetylesterases or lipases are shown in Table 2. Activities on various substrates were compared by normalization to activity on 4-nitrophenyl-acetate, with the exception of AcXE from *S. lividans* which did not hydrolyze acylated nitrophenols. This latter enzyme was tested on acetylxylan using a similar level of activity as exhibited by the other two AcXEs on acetylated glycosides.

Table 2 *Substrate specificities of AcXEs and other esterases*

Enzyme	Relative activity on 4-NPh-Ac,-Prop,-But	Initial rate of deacetylation of tri-Ac-Me-β-Xyl*p* (mM/min/U/ml)	Initial rate of deacetylation of tetra-Ac-Me-β-Glc*p* (mM/min/U/ml)	Precipitation of acetylxylan solution [h] at 1 mU/ml
S. lividans AcXE*	No activity	3.5	5.4	0.016 (~ 1 min)
T. reesei AcXE	1 : 1.1 : 0.2	5.6	5.6	1.6
S. commune AcXE	1 : 0.44 : 0.05	3.6	6.6	25
Wheat germ lipase	1 : 0.96 : 0.13	1.8	0.56	no precipitation
Orange peel AcE	1 : 0.81 : 0.00	0.87	0.07	no precipitation
C. cylindracea lipase	1 : 4 : 5	0.43	0.86	no precipitation

*Enzyme was used at a concentration which caused a similar rate of deacetylation of acetylated glycosides as other AcXE.

Candida cylindracea lipase was the only case where the relative activity on acylated 4-nitrophenols increased with increasing acyl chain length. All other enzymes, including wheat germ lipase, behaved more as carbohydrate esterases than as lipases. Sugar acetates were deacetylated 3-10 times faster by AcXEs relative to acetylesterases/lipases in reactions where activity units on 4-nitrophenyl-acetate were held constant. There were smaller rate differences in the first deacetylation of the xyloside and glucoside by AcXEs than with other enzymes. Acetylxylan was deacetylated by all three AcXEs. However, the affinity towards polymeric substrate varied with different AcXEs. When enzyme activity based on the first deacetylation step towards sugar glycosides was equal, *T. reesei* AcXE caused precipitation of acetylxylan solution (due to deacetylation) 15-times faster and *S. lividans* AcXE 100-times faster than did AcXE from *S. commune*. The non-hemicellulolytic esterases or lipases did not catalyze measurable deacetylation of acetylated xylan.

Differences were observed with respect to the order in which the acetyl groups are removed from 2,3,4-tri-*O*-Ac-Me-β-Xyl*p* and 2,3,4,6-tetra-*O*-Ac-Me-β-Glc*p*. Deacetylation time course plots of 2,3,4-tri-*O*-Ac-Me-β-Xyl*p* by four enzymes are shown in Figure 1. The *S. lividans* AcXE reaction did not generate measurable mono-deacetylated intermediates. This enzyme yielded the 4-*O*-acetyl derivative as the major product. This is because the enzyme catalyzes the second deacetylation much faster than the first, once either position 2 or 3 is deacetylated. The persistence of low levels of 2,3-

diacetate results from an occasional, rare deacetylation of the substrate at position 4, creating a product which serves as a poor substrate for further deacetylation. AcXE from *S. commune* showed a preference for deacetylation at position 3, resulting in the transient accumulation of 2,4-di-*O*-Ac-Me-β-Xyl*p*. This product was further deacetylated to give a mixture of 2- and 4-monoacetates. *T. reesei* AcXE was intermediate to the *S. lividans* and *S. commune* enzymes in terms of specific hydrolysis rates of variously acetylated substrates. Wheat germ lipase and orange peel acetylesterase both exhibited very different deacetylation regioselectivity of 2,3,4-tri-*O*-Ac-Me-β-Xyl*p* (Figure 1). They had a marked preference to first deacetylate position 4 and then position 3. The AcXEs behaved similarly towards 2,3,4,6-tetra-*O*-Ac-Me-β-D-Glc*p*. With this glycoside, wheat germ lipase and orange peel acetylesterase first deacetylated position 2 and then positions 6 and 4. *C. cylindracea* lipase deacetylated position 6 first and then position 4. The differences between AcXEs and non-hemicellulolytic esterases in deacetylation regioselectivity of

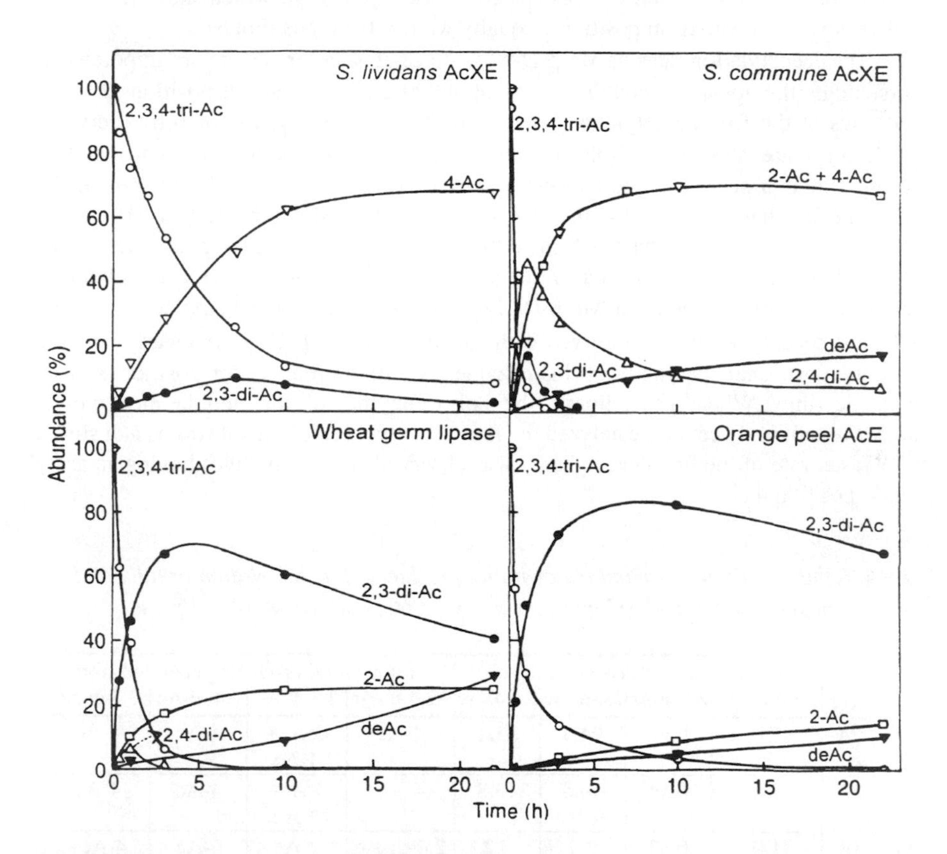

Figure 1 *Time course of deacetylation of 2,3,4-tri-*O*-Ac-Me-β-Xyl*p *by AcXEs of* S. lividans *and* S. commune *and wheat germ lipase and orange peel acetylesterase (the abbreviations attached to individual curves mark the substrate and products of deacetylation)*

Table 3 *Comparison of the action of AcXEs and non-hemicellulolytic esterases/lipases on acetylated methyl glycosides*

Enzyme	Order of the removal of the acetyl groups in	
	2,3,4-tri-*O*-Ac-Me-β-Xyl*p*	2,3,4,6-tetra-*O*-Ac-Me-β-Glc*p*
AcXE *S. lividans**	2-Ac=3-Ac>4-Ac	2-Ac=3-Ac>4-Ac=6-Ac
AcXE *T. reesei*	2-Ac=3-Ac>4-Ac	2-Ac=3-Ac>4-Ac=6-Ac
AcXE *S. commune*	3-Ac>2-Ac=4-Ac	3-Ac>2-Ac>4-Ac=6-Ac
Wheat germ lipase	4-Ac>3-Ac>2-Ac	2-Ac>6-Ac>4-Ac
Orange peel AcE	4-Ac>3-Ac	2-Ac>6-Ac>4-Ac
C. cylindracea lipase	4-Ac>3-Ac	6-Ac>4-Ac

acetylated methyl glycosides are summarized in Table 3. The apparent "double deacetylation" of sugar acetates is designated by the equal sign, which indicates that an acetyl group is removed from position 2 equally well as from position 3.

The deacetylation rates of Me-β-D-Xyl*p* diacetates and triacetate are important for understanding the apparent "double deacetylation" of carbohydrates at positions 2 and 3. Initial rates of the first deacetylation relative to that of deacetylation of fully acetylated Me-β-D-Xyl*p* are shown in Table 4. Derivatives that contain acetyl groups at both positions 2 and 3 served as the slowest substrates for AcXEs from *S. lividans* and *T. reesei*. The 2,4-diacetate and 3,4-diacetate were deacetylated by the *S. lividans* enzyme 100- and 30-times faster than the former two derivatives. A similar phenomenon was observed with AcXE from *T. reesei*, which deacetylated 2,4- and 3,4-diacetates about 10-fold faster than fully acetylated Me-β-D-Xyl*p*. AcXE from *S. commune* was different. The fastest deacetylation was observed with 3,4-di-*O*-Ac-Me-β-D-Xyl*p*, clearly denoting the unique positional specificity of this enzyme; i.e., its preference for position 3 of the pyranoside ring. Wheat germ lipase, the only enzyme of the non hemicellulolytic esterases tested on diacetates, catalyzed the first deacetylation of all substrates at a similar rate. A faster rate of the first deacetylation was observed with derivatives having an acetyl group at position 4.

Table 4 *Relative initial rates and major initial products of deacetylation of fully and doubly acetylated methyl β-D-xylopyranoside by AcXEs and wheat germ lipase*

Enzyme	Ref.	Relative rate of deacetylation of				Major initial products generated from			
		2,3,4-tri-Ac	2,3-di-Ac	2,4-di-Ac	3,4-di-Ac	2,3,4-tri-Ac	2,3-di-Ac	2,4-di-Ac	3,4-di-Ac
AcXE *S. lividans*	24	1.0	0.92	**94.4**	**32.4**	4-Ac	deAc+ 4-Ac	4-Ac	4-Ac
AcXE *T. reesei*	22	1.0	0.54	**10.8**	**10.8**	4-Ac	4-Ac+ 2-Ac	4-Ac	4-Ac
AcXE *S. commune*	18	1.0	0.83	0.4	**5.2**	2,4-di-Ac	2-Ac	4-Ac	4-Ac+ 2-Ac
Lipase Wheat germ		1.0	0.54	0.6	1.0	2,3-di-Ac	3-Ac	4-Ac+ 2-Ac	2-Ac

5 DISCUSSION

All three AcXEs exhibited higher affinities towards acetylated carbohydrates and lower affinities towards synthetic low molecular mass substrates than did non-hemicellulolytic esterases or lipases. Non-hemicellulolytic esterases/lipases did not deacetylate acetylxylan to a degree that would lead to precipitation of the polysaccharide from solution. The AcXEs also exhibited different regioselectivity in the deacetylation of carbohydrates compared to non-hemicellulolytic enzymes. Deacetylation of 2,3,4-tri-*O*-Ac-Me-β-Xyl*p* and 2,3,4,6-tetra-*O*-Ac-Me-β-Glc*p* by the AcXEs principally occurred at positions 2 and 3. This regioselectivity corresponds well to enzymatic function in acetylxylan degradation, where xylopyranosyl residues bear acetyl groups at these two positions. Furthermore, this regioselectivity is complementary to that catalyzed by other types of esterases and lipases used in carbohydrate chemistry and represents a new tool for chemoenzymic synthesis.

Relative to AcXEs from *T. reesei* and *S. commune*, AcXE from *S. lividans* does not attack low molecular mass substrates and shows unusually high affinity for polymeric substrate, based on the first deacetylation rate of fully acetylated methyl glycosides. *T. reesei* AcXE, in this regard, exhibits properties which are intermediate to the two other AcXEs. The high affinity of *S. lividans* AcXE for acetylxylan correlates well with the striking preference of this enzyme for deacetylation at position 2 or 3 in sugar acetates when the neighboring position 2 or 3 is not substituted. Sugar derivatives in which both positions 2 and 3 are acetylated are poor substrates for *S. lividans* AcXE. These results suggest that *S. lividans* AcXE, and possibly also *T. reesei* AcXE, catalyze deacetylation of position 2 or 3 when only one of these positions is acetylated, via a five-membered transitory state (Figure 2) from which the acetyl group is finally released. Such intermediates are believed to be involved in the spontaneous migration of acetyl groups along the pyranoid rings of carbohydrates[29]. Indeed, it has been clearly demonstrated in aqueous systems that intermediates of this type are involved in the spontaneous interconversion of diacetates of Me-β-Xylp[18]. Syntheses of acetylated sugar derivatives, designed to further study the postulated deacetylation mechanism depicted in Figure 2, are underway in our laboratories.

As this study documents, there are significant differences in substrate specificity and catalytic versatility among the investigated AcXEs. It is worth reiterating that these enzymes belong to three different carbohydrate esterase families[11]. It is probable that many of the findings regarding the catalytic properties of these three AcXEs will be generally applicable to other members of their individual families.

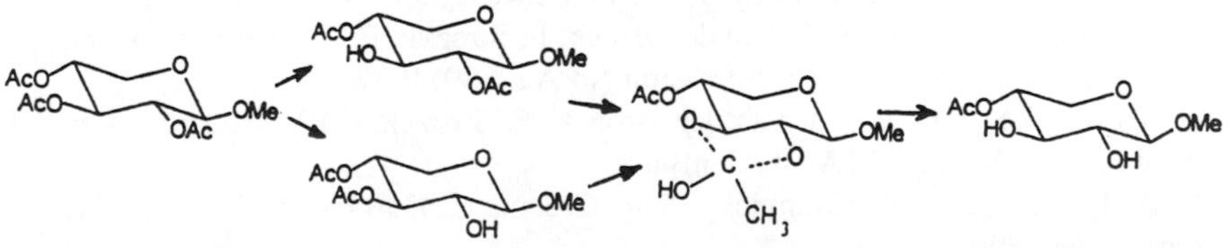

Figure 2 *The five-membered intermediate proposed to be common for deacetylation of positions 2 and 3 by AcXE from S. lividans.*

Acknowledgements

This work was supported by the United States Department of Agriculture, Agricultural Research Service (Agreement No. 58-3620-6-F133) and by a grant from the Slovak Grant Agency VEGA (2/4147/1998).

References

1. G. Williamson, P.A. Kroon and C.B. Faulds, *Microbiology*, 1998, **144**, 2011.
2. P. Biely, C.R. MacKenzie and H. Schneider, *Can. J. Microbiol.*, 1988, **34**, 767.
3. P. Biely, J. Puls and H. Schneider, *FEBS Lett.*, 1985, **186**, 80.
4. K. Poutanen, M. Ratto, J. Puls and L. Viikari, *J. Biotechnol.*, 1987, **6**, 49.
5. L.P. Christov and B.A. Prior, *Enzyme Microb. Technol.* 1993, **15**, 460.
6. P. Biely, C.R. MacKenzie, J. Puls and H. Schneider, *Bio/Technology*, 1986, **4**, 731.
7. M. Sundberg and K. Poutanen, *Biotechnol. Appl. Biochem.* 1991, **13**, 1.
8. C. Dupont, N. Daigneault, F. Shareck, R. Morosoli and D. Kluepfel, *Biochem. J.*, 1996, **319**, 881.
9. L. Egana, R. Gutierrez, V. Caputo, A. Peirano, J. Steiner and J. Eyzaguirre, *Biotechnol. Appl. Biochem.*, 1996, **24**, 33.
10. B.P. Darlymple, D.H. Cybinski, I. Layton, C.S. McSweeney, G./P. Xue, Y.J. Swadling and J.B. Lowry, *Microbiology*, 1997, **143**, 2605.
11. B. Henrissat and P. Coutinho, Proc. 3rd Carbohydr. Bioengin. Workshop, Newcastle upon Tyne, in press.
12. H. Tsujibo, T. Ohtsuki, T. Iio, I. Yamazaki, K. Miyamoto, M. Sugiyama and Y. Inamori, *Appl. Environ. Microbiol.*, 1997, **63**, 661.
13. J.I. Laurie, J.H. Clarke, A. Ciruela, C.B. Faulds, G. Williamson, H.J. Gilbert, J.E. Rixon, J. Millward-Sadler and G.P. Hazlewood, *FEMS Microbiol. Lett.*, 1997, **148**, 261.
14. F. Shareck, P. Biely, R. Morosoli and D. Kluepfel, *Gene*, 1995, **153**, 105.
15. E. Margoles-Clark, M. Tenkanen, H. Söderlund and M. Pentillä, *Eur. J. Biochem.*, 1996, **237**, 553.
16. R. Gutierrez, E. Cederlund, L. Hjelmqvist, A. Peirano, F. Herrera, D. Ghosh, W. Duax, H. Jörnvall and J. Eyzaguirre, *FEBS Lett.*, 1998, **423**, 35.
17. N. Halgasova, E. Kutejova and J. Timko, *Biochem. J.*, 1994, **298**, 751.
18. P. Biely, G.L. Cote, L. Kremnicky, D. Weisleder, and R.V. Greene, *Biochim. Biophys. Acta*, 1996, **1298**, 209.
19. J. Linden, M. Samara, S. Decker, E. Johnson, M. Boyer, M. Pecs, W. Adney and M. Himmel, *Appl. Biochem. Biotechnol.*, 1994, **45/46**, 383.
20. L.H. de Graff, J. Visser, H.C. van den Brock, F. Strozyk, F.J.M. Kormelink and J.C.P. Boonman, GenBank Updates, Accession No A22880, 1992.
21. T. Koseki, S.S. Furuse, H.C.van den Broeck, F. Strozyk, F.J.M. Kormelink and J.C. Boonman, 1992, DDBJ/EMBL/GenBank.
22. P. Biely, G.L. Cote, L. Kremnicky, R.V. Greene, C. Dupont and D. Kluepfel, *FEBS Lett.* , 1996, **396**, 257.
23. M. Tenkanen, *Biotechnol. Appl. Biochem.*, 1998, **27**, 19.
24. P. Biely, G.L. Cote, L. Kremnicky, R.V. Greene and M. Tenkanen, *FEBS Lett.*, 1997, **420**, 121.

25. H.M. Sweers and C.-H. Wong, *J. Am. Chem. Soc.*, 1986, **108**, 6421.
26. D.G. Druckhammer, W.J. Hennen, R.L. Pederson, C.F. Barbas III., C.M. Gautheron, T. Krach and C.-H. Wong, *Synthesis*, 1991, **1991**, 499.
27. R. Lopez, C. Perez, A. Fernandez-Mayoralas and S. Conde, *J. Carbohydr. Chem.*, 1993, **12**, 165.
28. K.G. Johnson, J.D. Fontana and C.R. MacKenzie, *Methods Enzymol.*, 1988, **160**, 551.
29. K.-F. Hsiao, H.-K. Lin, D.-L. Leu, S.-H. Wu ad K.-T. Wang, *Bioorg. Med. Chem. Lett.*, 1994, **4**, 1629.

CARBOHYDRASES AS TOOLS FOR UNDERSTANDING HOW COLD ENZYMES ARE ADAPTED TO FUNCTION AT LOW TEMPERATURES

D. R. Humphry, S. P. Cummings and G. W. Black

School of Sciences,
University of Sunderland,
Sunderland SR1 3SD UK

1. INTRODUCTION

The properties of enzymes produced by extremophiles; *i.e.* organisms capable of growth at extreme temperature, pH, pressure or salinity; are commonly used to gain further insight into the structure-function and structure-stability relationships of enzymes.[1] Although much is known about enzymes from mesophiles and thermophiles,[2] it is amazing that so little is known about the enzymes produced by psychrophiles, especially since up to 70% of the globe is subjected to temperatures below 5°C.

The enzymes produced by psychrophiles are known as cold enzymes and, to date, only a handful of genes encoding cold enzymes have been cloned and the gene products characterised. To date only four lipase genes,[3,4] one α-amylase,[5] one subtilisin gene,[6] two β-galactosidase genes,[7,8] one pyruvate kinase gene,[9] and one aspartate carbamoyltransferase gene[10] have been cloned and characterised. Recently, the structure of one of these enzymes, the α-amylase, has been solved.[11] These studies have shown that cold enzymes display higher activity at low temperatures when compared with their mesophilic counterparts. For example, the α-amylase from the psychrotolerant bacterium, *Alteromonas haloplanctis* has a turnover number (k_{cat}) 5.5 times high than that for the corresponding enzyme from the mesophile *Bacillus amyloliquefaciens*, at 15°C.[6]

However, according to the following equation: $k = (\kappa K_B/h)e^{-\Delta G^*/RT}$ (where k, rate constant; κ, transmission coefficient; K_B, Boltzmann constant; h, Planck constant; ΔG*, activation free energy; R, gas constant; and T, temperature); the rate of a chemical reaction is exponentially dependent upon temperature. Therefore, any decrease in temperature causes an exponential decrease in the reaction rate, the magnitude of which is dependent upon the activation free energy. Accordingly, biochemical reactions display a reaction rate 2 to 3 times lower when the temperature is decreased by 10°C (*i.e.* Q_{10} = 2 to 3). As a consequence, the activity of a mesophilic enzyme is between 16 to 80 times lower when the reaction temperature is shifted from 37°C to 0°C ($\approx 2^4$ to 3^4). However, the generation times of psychrophilic bacteria near 0°C are of the same order as those of mesophilic bacteria at 37°C. It follows therefore, that cold enzymes synthesized by psychrophiles must be adapted to function at low temperatures. If they were not, then their reaction rates at low temperatures would be similar to those of their mesophilic counterparts, and they are clearly not. So how are cold enzymes adapted?

It is known that cold enzymes have a reduced activation free energy value in comparison to their mesophilic counterparts[12] (Figure 1). It has, therefore, been postulated that cold enzymes have an increased structural flexibility which enhances the conformational complementarity between enzyme and substrate allowing more efficient interaction and hence a lowering of the amount of free energy needed to produce the transition state enzyme-substrate complex.[13] The very fact that cold enzymes are usually very thermolabile supports the idea that these enzymes have a looser, and therefore possibly a more flexible structure.

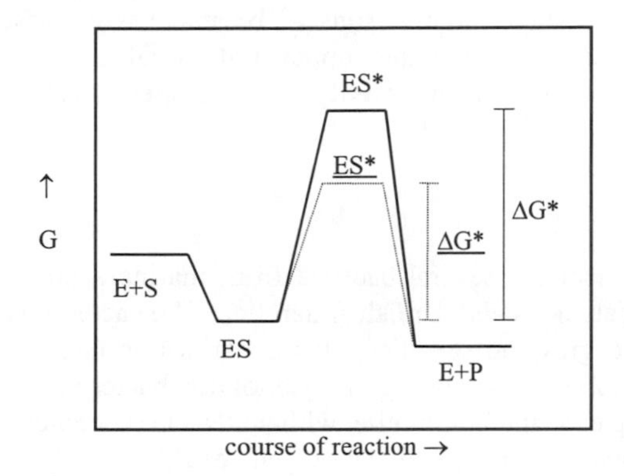

Figure 1. *E, enzyme; S, substrate; ES, enzyme-substrate complex; ES*, transition state mesophilic enzyme-substrate complex; <u>ES*</u>, transition state cold enzyme-substrate complex; P, product; ΔG*, activation free energy for mesophilic enzymes; <u>ΔG*</u>, activation free energy for cold enzymes; solid line, values for mesophilic enzymes; dashed line, values for cold enzymes.*

The comparison of the structures of several cold enzymes with their mesophilic homologues, whether via crystallographic means or computer modelling, reveals the following: the α-amylase from *A. haloplanctis* lacks surface salt bridges;[5] the cold trypsins from salmon and bovine contain less hydrogen bonds between the two β-barrel domains;[14] and the psychrophilic subtilisins and β-lactamases possess less aromatic interactions.[6,15] All the above indicate that cold enzymes are less compact than mesophilic enzymes. This is supported by circular diachroism spectra of psychrophilic and mesophilic α-amylases, which suggest that the conformation of the cold enzyme was less compact.[5] However, there is still no proof that cold enzymes are more flexible and furthermore, that this putative flexibility engenders higher enzymatic activities at lower temperatures.

An essential requirement for investigators attempting to elucidate the mechanisms by which cold enzymes function at low temperatures is the availability of data from mesophilic homologues. Often such data is not available. However, there are numerous examples of mesophilic xylanases and cellulases. Currently the sequences of hundreds of xylanase- and cellulase-encoding genes are known and the three-dimensional structures of tens of xylanases and cellulases have been solved.[16] However, to date, no genes encoding cold xylanases or cellulases have been sequenced or the structures of their encoded enzymes solved. In addition, many mechanistic analyses of mesophilic xylanases and cellulases have been reported. For example, White and colleagues[17] have determined the three dimensional structure of the 2-deoxy-2-fluorocellobiosyl-enzyme intermediate formed on the xylanase, Cex, from *Cellulomonas fimi*. They showed that the interactions which are proposed to be formed with the sugar 2-hydroxyl and the enzyme contribute enormously to transition state stabilisation as indicated by increased enzyme efficiency. We therefore believe that similar analyses of cold xylanases and cellulases will allow direct comparisons to be made with mesophilic xylanases and cellulases, providing us with a unique opportunity to advance our understanding of the factors which govern cold enzyme activity at low temperatures.

2. RESULTS

We have recently isolated several bacteria from marine sediment and littoral detritus collected from Adelaide Island, British Antarctica. The bacterial isolations were selected for their ability to grow on specific polysaccharides as their sole carbon and energy source. We are currently analysing a xylanolytic bacterium (isolate *A2i*) which is aerobic, Gram-negative and non-motile, with a DNA G+C content of 63%.

Phylogenetic analysis of the 16S rDNA gene of *A2i* indicates that it is a new species of *Flavobacterium* (Figure 2). *A2i* showed an optimum growth rate at 15°C, a significant rate of growth at 0°C and no growth at 25°C or higher (Figure 3) and can therefore be classified as a psychrophile according to the definition of Morita.[18] Growth on various substrates and the induction of xylanase activity is shown in Table 1. Interestingly, *A2i* showed no growth on cellulose. Recently, analysis of culture supernatants of *A2i* and, for comparison, the xylanolytic mesophile *Pseudomonas fluorescens* subspecies *cellulosa* (*Pfc*), have shown that a 37°C-unit* of xylanase activity from *A2i* culture supernatant is approximately twice as active against soluble xylan at temperatures lower than 23°C when compared to a 37°C-unit* of xylanase activity from *Pfc* (Figure 4). In fact optimum xylanase activity from *Pfc* was seen at 60°C, compared to 37°C for *A2i*. We have also shown that the thermal half-life of the xylanase activity from *A2i* is significantly shorter than that for *Pfc* (Table 2). It would therefore appear that the xylanase(s) produced by *A2i* conform to the 'rules' normally associated with cold enzymes, *i.e.* a higher activity at low temperatures, compared to their mesophilic counterparts, and an increased thermal lability.

* A 37°C-unit of xylanase activity is equivalent to 1 μmole of reducing sugar released per minute at 37°C

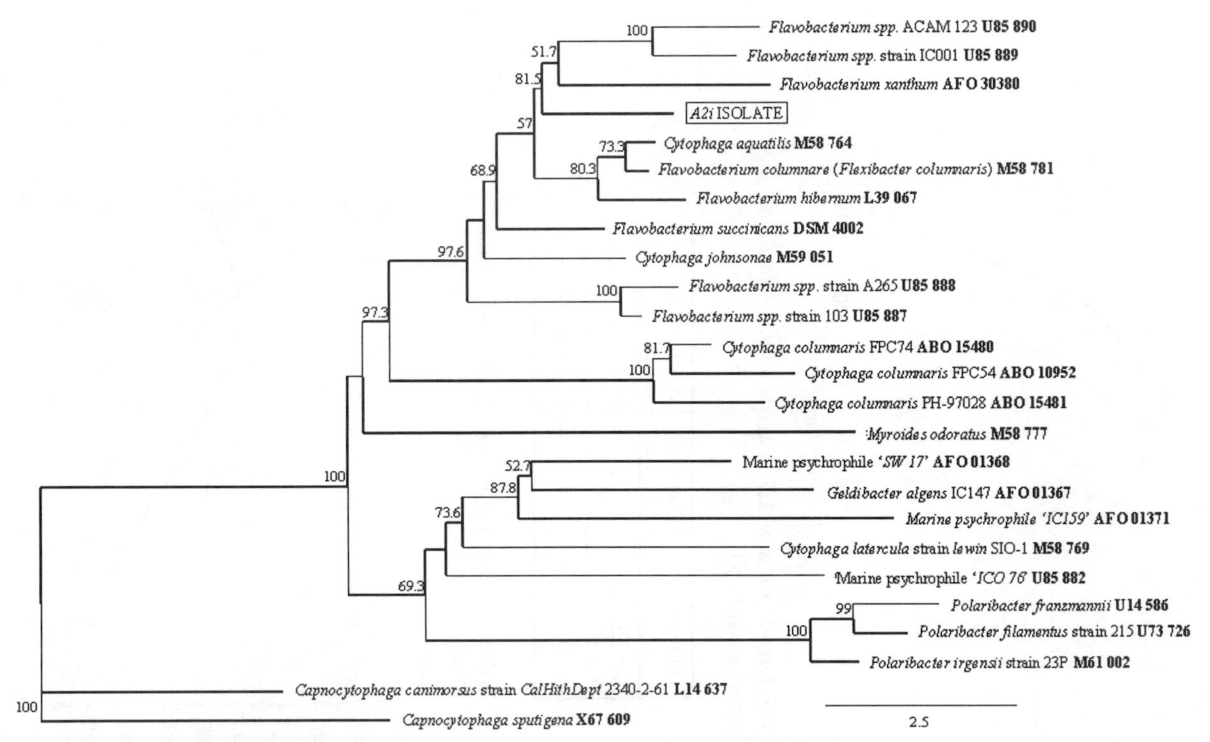

Figure 2. *An unrooted phylogenetic dendogram produced using the Seqboot, DNAdist, Neighbour and Consense programs of the PHYLIP software package. The Treeview program[19] was used to add a scalebar to the dendogram (the scale bar is equivalent to 2.5 nucleotide substitutions per 100 nucleotides). Genbank accession numbers were used where available. The dendogram is based on the 16S rDNA sequence of the A2i isolate compared to those from 24 other organisms taken from the BLAST database and the Ribosomal Database Project II Websites.*

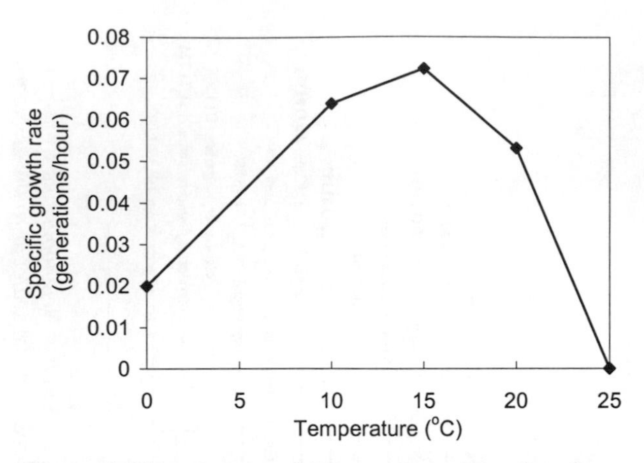

Figure 3. *Growth rates of A2i on xylan minimal media at different temperatures.*

Substrate	Growth on substrate	Xylanase induction
glucose	+	-
xylose	+	+
xylan	+	+
CMC	-	N/A

Table 1. *Growth of A2i on different substrates. CMC, carboxymethyl cellulose; N/A, not applicable*

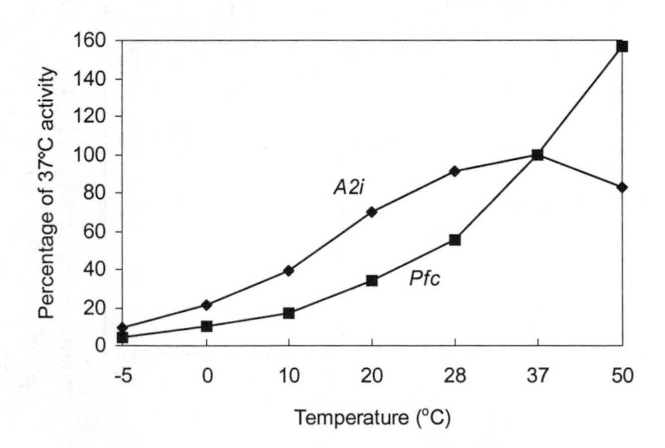

Figure 4. *Xylanase activity of A2i and Pfc supernatants expressed as percentages of enzyme activity at 37°C.*

Temp. (°C)	Half-life (min.)	
	A2i	*Pfc*
20	3564	TLAD
37	20.5	5760
50	1.5	90

Table 2. *Thermostability half-lives at various temperatures. TLAD, too long to accurately determine.*

3. DISCUSSION

The cloning and hyperexpression of the cold xylanase gene(s) from *A2i in Escherichia coli* will facilitate the purification, mechanistic analysis, and ultimately, the determination of their three-dimensional structure. Comparison with mesophilic xylanases may allow us to understand how these cold enzymes are adapted to function at low temperatures.

It will also be interesting to (i) see if the cold xylanases of *A2i* possess cellulose binding domains since *A2i* appears to be non-cellulolytic; and (ii) determine if cold xylanases from *A2i* are non-aggregating, since they are synthesized by an aerobe; or aggregating, since they function in a low temperature environment, which is a low energy environment, and therefore, similar to anaerobic environments.[20]

4. REFERENCES

1. V. V. Mozhaev, I. V. Berezin and K. Martinek , *CRC Crit. Rev. Biochem.*, 1988, **23**, 235.
2. L. Menéndez-Arias and P. Argos, *J. Mol. Biol.*, 1989, **206**, 397
3. G. Feller, M. Thiry, J.L. Arpigny and C. Gerday C, *Gene*, 1991, **102**, 111.
4. D-W. Choo, T. Kurihara, T. Suzuki, K. Soda and N. Esaki, *Appl. Environ. Microbiol.*, 1998, **64**, 486.
5. G. Feller, T. Lonhienne, C. Deroanne, C. Libioulle, J.V. Beeumen and C Gerday, *J. Biol. Chem.*, 1992, **267**, 5217.
6. S. Davail, G. Feller, E. Narinx, C. Gerday, *J. Biol. Chem.*, 1994, **269**, 17448.
7. D. E. Trimbur, K. R. Gutshall, P. Prema, J. E. Brenchley, *Appl. Environ. Microbiol.*, 1994, **60**, 4544.
8. K. R. Gutshall, D. E. Trimbur, J. J. Kasmir and J. E. Brenchley, *J. Bacteriol.*, 1995, **177**, 1981.
9. K. Tanka, H. Sakai, T. Ohta and H. Matsuzawa, *Biosci., Biotechnol. Biochem.*, 1995, **59**, 1536.
10. Y. Xu, Y. Zhang, Z. Liang, M.V. de Casteele, C. Legrain and N. Glansdorff, *Microbiol.*, 1998, **144**, 1435.
11. N. Aghajati, G. Feller, C. Gerday and R. Haser, *Protein Sci.*, 1998, **7**, 564.
12. P.S. Low, J.L. Bada and G. N. Somero, *Proc. Natl. Acad. Sci. USA*, 1973, **70**, 430.
13. G. Feller and C. Gerday, *Cell. Mol. Life Sci.*, 1997, **53**, 830.

14. A.O. Smalås, E.S. Heimstad, A. Hordviik, N.P. Willassen and R. Male, *Proteins*, 1994, **20**, 149.
15. G. Feller, Z. Zekhnini, J. Lamotte-Brasseur and C. Gerday, *Eur. J. Biochem.*, 1997, **244**, 186.
16. http://afmb.cnrs-mrs.fr/~pedro/CAZY/ghf.html
17. A. White, D. Tull, K. Johns, N.R. Gilkes, S.G. Withers and D.R. Rose, *Nature Struct. Biol.*, 1996, **3**, 149.
18. R.Y. Morita, *Bacteriol. Rev.*, 1975, **39**, 144.
19. R.D.M. Page, *Comput. Appl. Biosci.*, 1996, **12**, 357.
20. S.P. Cummings and G.W. Black, *Extremophiles*, 1999, in press.

STRUCTURE AND FUNCTION RELATIONSHIP OF FAMILY 11 XYLANASES

K. A. McAllister, L. Marrone and A. J. Clarke

Department of Microbiology
University of Guelph
Guelph, Ontario N1G 2W1
Canada

1 INTRODUCTION

Xylanases (EC 3.2.1.8; 1,4-β-D-xylan xylanohydrolase) catalyze the random hydrolysis of the xylan backbone of heteroxylans, thereby exposing cellulose fibrils to attack by the cellulolytic enzymes. Based on amino acid sequence comparisons together with hydrophobic cluster analysis, the xylanases have been organized into two of the families of glycosidases, Families 10 and 11. In all, 65 discrete enzymes or catalytic domains comprise these Families and all are xylanases, 30 belonging to Family 11. Nevertheless, differences in the organization of the catalytic domains, their physicochemical properties, and their kinetic behaviours are observed even within a Family of enzymes. This chapter reviews the literature pertaining to the structure and function relationship of the Family 11 xylanases and presents recent findings on the role of a non-conserved residue in conferring substrate specificity.

2 DOMAIN STRUCTURE OF XYLANASES

The molecular architectures of representative xylanases are depicted in Figure 1. These range from the simplest structure involving a single catalytic domain and associated signal peptide, as exemplified by *Bacillus circulans* Xyn A (1), to the most complex arrangement of five discrete domains, as in *Thermoanaerobacterium saccharolyticum* Xyn A (2). The most striking examples of multiple catalytic domains, sometimes with associated multi-functionality, are the xylanases of rumen microbes. The microorganisms *Ruminococcus flavefaciens* 17 (3), *Fibrobacter succinogenes* S85 (4), and *Neocallismatix patriciarum* (5) each produce a xylanase with three domains, two of which are catalytically active. The third domain has yet to be assigned a function.

2.1 Catalytic Domains

Xylanase C (Xyn C) produced by the anaerobic, ruminant bacterium *F. succinogenes* S85 (the type strain) has a molecular mass of 63,850 Da and is comprised

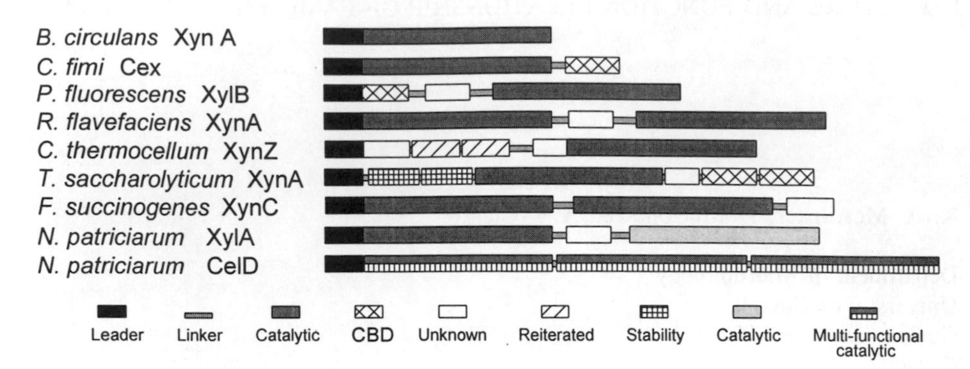

Figure 1 *Molecular architecture of representative xylanases*

of three distinct domains, A, B and C, which are separated by serine-rich linker peptides (Figure 1). Deletion analysis of the Xyn C gene indicated that domains A and B function catalytically as xylanases (4). Both domains are members of Family 11 glycosidases and they share 54% identity (Figure 2). The catalytic domains I and II of *N. patriciarum* xylanase are essentially homologous (3,5), and both also belong to Family 11 glycosidases. In contrast, the catalytic domains of *R. flavefaciens* Xyn A have very low identity of 19%, and in fact, domain A is assigned to Family 10 while domain C is homologous to Family 11 enzymes (3). Cel D is also produced by *N. patriciarum* and it has three multi-functional catalytic domains. Each catalytic domain demonstrates substantial activity against cellulose, xylan, and lichenan and appears to bind crystalline cellulose. This suggests endoglucanase, cellobiohydrolase, xylanase activities and possibly a cellulose-binding domain (CBD) exists within the same domain emphasizing the true multi-functionality of this enzyme (6).

The significance of linking multiple catalytic domains is not clear. It may increase the efficiency of xylan hydrolysis to xylose by sequential activity of the domains, to generate larger oligosaccharides which function as substrate for the other. It could also confer a competitive advantage in that it may minimize release of hydrolysis products to competing microbes or to generate hydrolysis products that would not be generated as easily by one domain (3).

2.2 Cellulose Binding Domains

CBDs are classified into ten families with a growing number of newly characterized CBDs which are unclassified. Xylanase CBDs are usually produced as discrete non-catalytic domains separated from the catalytic domain by a linker sequence. *Cellulomonas fimi* Xyl D, *C. flavigena, Ps. fluorescens* sb. *cellulosa* Xyl A, B, C, D possess CBDs with high affinity for Avicel (7-10). *P. fluorescens* sb. *cellulosa* Xyl A and C, with CBDs removed, demonstrated a reduced activity against a complex mixture of cellulose:heteroxylan (11). The CBD appears to mediate the attachment of the enzyme to the cellulose of the plant cell wall, facilitating digestion of associated xylan which may impede the access to the energy-giving cellulose (12).

```
                         *      ↓                               ↓
Fs(a)   106   VSGTPSQLVEYYVIDNTLANMPGSWIGNERKGTITVDGGTYIVYRNTRT-GPA   (P35811)
Fs(b)   119   YGWTVDPLVEYYIVDDW-FNKPGANLLGQRKGEFTVDGDTYEIWQNTRVQQPS   (P35811)
Bc       69   YGWTRSPLIEYYVVDSWGTYRP-TGTY---KGTVKSDGGTYDIYTTTRYNAPS   (P09850)
Bp       84   YGWTQQPLAEYYIVDSWGTYRP-TGAY---KGSFYADGGTYDIYETTRVNQPS   (P00694)
Bs       69   YGWTRSPLIEYYVVDSWGTYRP-TGTT---KGTVKSDGGTYDIYTTTRYNAPS   (P18429)
Ca       85   YGWTQSPLVEYYIVDSWGTWRPPGGTS---KGTITVDGGIYDIYETTRINQPS   (P17137)
Ss       76   YGWTSNPLVEYYIVDNWGSYRP-TGEY---RGTVHSDGGTYDIYKTTRYNAPS   (X81045)
Sl(C)    76   YGWTSNPLVEYYIVDNWGSYRP-TGTY---KGTVTSDGGTYDIYQTTRYNAPS   (P26220)
Sl(B)    78   YGWTSNPLVEYYIVDNWGTYRP-TGEY---KGTVTSDGGTYDIYKTTRVNKPS   (P26515)
Rf       86   YGWTRNPLMEYYIVEGWGDWRPPGNDG-EVKGTVSANGNYYDIRKTMRYNQPS   (Q53317)
Np(a)   103   RGVQGVPLVEYYIIEDWVDWVPDAQ-G---RMVTI-DGAQYKIFQMDHT-GPT   (P29127)
Np(b)   378   RGLNGVPLVEYYIIEDWVDWVPDAQ-G---KMVTI-DGAQYKIFQMDHT-GPT   (P29127)
Sc       78   YGWTRSSLIEYYIVESYGSYDPSSAAS--HKGSVTCNGATYDILSTWRYNAPS   (P35809)
An       70   YGWVNYPQAEYYIVEDYGDYNPCSSAT--SLGTVYSDGSTYQVCTDTRTNEPS   (P55330)
At       70   YGWVNYPQAEYYIVEDYGDYNPCSSAT--SLGTVYSDGSTYQVCTDTRTNEPS   (P55331)
Th       78   YGWSRNPLIEYYIVENFGTYNPSTGAT--KLGEVTSDGSVYDIYRTQRVNQPS   (P48793)
Tr(II)   77   YGWSRNPLIEYYIVENFGTYNPSTGAT--KLGEVTSDGSVYDIYRTQRVNQPS   (P36217)
```

Figure 2 *Amino acid sequence alignment of Family 11 xylanases. The arrows denote the homologous residues to Trp135 and Trp161 of* F. succinogenes *XynC-B, while the asterisk identifies the catalytic anion/nucleophile. Residues in bold, homology among at least 9 of the 17 sequences; hatched residues, total conservation; (accession numbers).*

2.3 Linker Sequences

Short linker sequences rich in proline or hydroxylamino acids exist in both cellulases and xylanases. These linkers function as flexible, extended hinge regions between functional domains (13). In cellulases, they provide the flexible link between CBD and catalytic domains to allow cleavage of more than one bond without desorption (12). Xylanases are believed to be served by linkers in a similar way as the linker provides the flexibility necessary for the catalytic domain to fully access soluble heteroxylans (11). Linker sequences have been identified in every xylanase that exhibits multiple domains but they do not affect function of catalytic domains or CBDs (7).

2.4 Reiterated Repeat and Thermostability Sequences

In addition to the domains of known and putative functions, there exists in many xylanases regions of reiterated repeats of amino acids of unknown function that appear not to directly effect catalysis. It is speculated the reiterated domains may play a role in secretion or alternatively, may represent a docking sequence which interacts with the other enzymes to form a plant-cell wall multi-enzyme complex (8,14,15). Additional function attributed to domains within xylanases is thermostability. Although the effect on thermostability is not absolute, there is a noticeable influence. A deletion of 1.75 kb from the N-terminal end of Xyn A of *T. saccharolyticum* reduced thermostability as demonstrated by a decrease in optimum temperature from 75-65°C (2). A N-terminal truncated derivative of *T. maritima* Xyn A exhibited a similar decrease in thermostability (16). A domain exhibiting high similarity to *T. maritima* N-terminal domain, in the mesophile *R. flavefaciens* Xyn D, conferred stability on the enzyme suggesting the domain functions less to specifically increase thermostability and more for general stability (16).

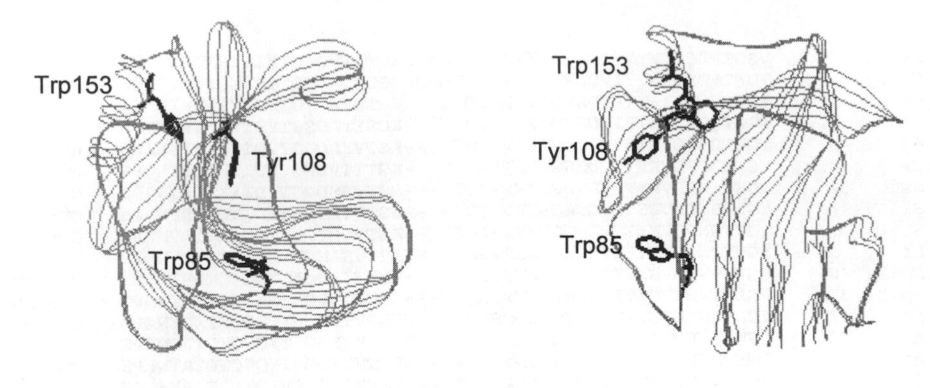

Figure 3 *Two views of the three-dimensional crystal structure of the Family 11* B. circulans *xylanase (PDB accession: BCX). The side chains of the homologous residues to Trp135, Trp161, and Trp202 of* F. succinogenes *XynC-B are depicted.*

3 THREE-DIMENSIONAL STRUCTURE OF XYLANASES

The three-dimensional structure of a number of xylanases have been solved, in some cases to very high resolution. However, these structures are limited to either the single domain Family 11 xylanases and the catalytic domains only of Family 10 xylanases. Apparently, the presence of flexible linkers between domains poses problems with the crystallization of intact multi-domain enzymes (17,18).

The crystalline structures for the Family 11 *T. reesei* Xyn I and II (19,20), *B. circulans* xylanase (1), and *B. pumilus* IPO xylanase (21) are quite similar, differing mainly in the number of β strands comprising the β-sheets. Figure 3 depicts a ribbon diagram of the three-dimensional structure of the *B. circulans* xylanase, representative of Family 11 enzymes (1). These xylanases are generally composed of two β-sheets and an α-helix which interact to form the deep cleft of the active site. The configuration has been likened to a right hand, where the two β-sheets form the fingers, the twisted β-sheet and α-helix form the palm, and the extended loop forms the thumb (1,19). The α-helix is packed against the hydrophobic face of the β-sheets. The cleft exhibits conformational change when co-crystallized with xylose (19).

4 CATALYTIC MECHANISM OF XYLANASES

All characterized xylanases catalyze the hydrolysis of xylan with the retention of anomeric configuration in the product which would thus involve a double-displacement mechanism (reviewed in references 22-24). The general features of this mechanism require, i) an acid catalyst to protonate the glycosidic linkage to be hydrolysed which then serves as general base to abstract a proton from water, and ii) an anion/nucleophile which stabilizes the cationic oxocarbenium ion transition states which both precede and follow the formation of a covalent adduct. The essential amino acids involved in double displacement hydrolysis are the carboxylates of glutamic and aspartic acid residues. For a carboxylic acid to function as an acid catalyst at the pH ranges optimal for this class of enzymes, it must exhibit an elevated p*K*a.

Differential chemical modification studies employing a carboxyl-specific carbodiimide served to identify Glu87 as the putative catalytic nucleophile of *Schizophyllum commune* xylanase A (25). This assignment was confirmed by Wither's group using the mechanism-based inhibitor, dinitrophenolate 2-deoxy-2-fluoroxylanoside. The homologous residues Glu78 and Glu78 in *B. circulans* (26) and *B. subtilis* (27) xylanases were found to be covalently bound by the inhibitor. These glutamates represent one of the two completely conserved acidic residues in the Family 11 xylanases. Examination of the known three-dimensional structures of these enzymes reveals that the other conserved residue is positioned directly opposite on the other side of the active site cleft. The distance between the two acidic residues is approximately 5Å, the distance expected between the two catalytic residues of a retaining enzymes (24). Replacement of Glu78 and Glu172, and Glu93 and Glu183 in *B. circulans* xylanase (1) and *B. pumilus* IPO Xyn A (28), respectively, by site-directed mutagenesis of their respective genes confirmed their essential nature for substrate hydrolysis. These data thus indicate that the homologous residues to Glu172 and Glu183 act as the acid catalysts.

Fourier transform infrared spectroscopic studies coupled with site-directed mutagenesis served to confirm the assignment of the two functions of the catalytic Glu residues in the *B. circulans* xylanase (29). The pK_a of the carboxyl groups of the residues homologous to Glu172 must be elevated for them to cycle between two ionization states as required of the acid catalyst in a double-displacement mechanism. This observed property has been attributed to electrostatic effects between them and the carboxyl side chain of the stabilizing anion/nucleophile (29,30), combined with conformational changes at the active site induced by the binding of ligand (19,30).

A unique situation has been noted involving the *T. reesei* Xyn I and *Aspergillus niger* Xyn I as both enzymes exhibit an uncharacteristically low pH optimum of 3.0 (20,31). This appears to be caused by the replacement of a highly conserved Asn residue with an Asp, and in both cases this residue is in the proximity of the acid catalytic Glu residue. It is postulated that a low pH is required to retain the replacement Asp residue in its protonated state to minimize ionic interactions between it and the acid catalyst which would otherwise influence catalysis (31). This phenomenon demonstrates the profound influence that the replacement of a non-catalytically essential residue may have on the catalytic mechanism of an enzyme.

5 BINDING OF SUBSTRATE TO XYLANASES

5.1 Chemical Modification Studies

Preliminary chemical modification studies using *N*-bromosuccinimide (NBS) alluded to the involvement of Trp residues in hydrolysis by the xylanases from *Chainia* sp., *Bacillus* sp. (32), *Streptomyces* sp. T7 (33), *F. succinogenes* (34), and *S. commune* (35). Upon oxidation of *S. commune* xylanase A with NBS in the absence of substrate, the absorbance at 280 nm dropped characteristically, and then proceeded to rise with a corresponding drop in enzyme activity. This phenomenon was shown to be due to oxidation of Tyr residues, which is known to increase absorbance at 280 nm and made estimating the number of oxidized Trp very difficult. The increased absorbance at 280 nm was prevented by protection with substrate, suggesting a Tyr residue may be essential. Following tryptic digestion of the oxidized enzyme and amino acid sequencing of the substrate-protected peptides lacking absorbance at 350 nm, Tyr97 was identified as a potentially essential binding residue in *S. commume* xylanase A (35). Fourth derivative spectroscopy confirmed the oxidation of four of eight Trp residues by NBS and the ligand protection of one residue. It is was conjectured the Trp may undergo stacking interactions with the sugar rings of ligands (35).

5.2 X-Ray Crystallography

Co-crystallization of a catalytically-inactive derivative of *B. circulans* Xyn A with xylotetraose revealed a number of interactions involving aromatic residues, although only two of the four sugar residues could be mapped by X-ray crystallography (1). A stacking interaction was observed between Trp9 and one of the xylose residues. The homologous Trp18 residue of Xyn II was also observed to participate in binding of substrate to this *T. reesei* xylanase (19). In addition, three Tyr residues in both enzymes are positioned within the binding-site cleft to interact with substrate (1,19). Tyr166 and Tyr69 of the *B. circulans* enzyme appear to form H-bonds to the two observed xylose residues, with the latter also binding to the nucleophilic Glu78 residue, thus functioning to orient both substrate and this catalytic residue. Tyr80 is postulated to interact with a longer substrate, as well as form an H-bond to Glu172 thereby positioning the acid/base catalytic residue (1). Replacement of these two residues by site-directed mutagenesis resulted in dramatic and total loss of activity, respectively. Kinetic analysis of the Tyr80Phe xylanase A derivative demonstrated the expected increase in K_m for xylan substrates, and decrease in catalytic activity (1).

Interestingly, Tyr88 of *B. circulans* Xyn A, the homologous residue to Tyr97 of *S. commune* xylanase A, did not appear to be involved with ligand binding, but this may be due to the length of the substrate used. Using xylo-oligosaccharides ranging in degrees of polymerization of between two and seven, a kinetic analysis indicated that the active-site cleft of the *S. commune* xylanase is comprised of at least seven substrate-binding sub-sites, designated -IV, -III, -II, -I, I, II, III (36). It was later shown that substrate bound predominantly to subsites -II, -III, -IV (37), and Tyr97 may constitute one of these. In the crystallographic studies of *B. circulans* Xyn A, xylotetraose spanned the catalytic site occupying sub-sites -II, -I, I, II (1), suggesting that the Tyr 97 equivalent of this enzyme, Tyr88, may comprise one of the sub-sites further removed, such as -III or -IV (35).

The active-site cleft of the *T. reesei* XYN II has been proposed to be comprised of five subsites, labelled -2,-1,+1,+2,+3 (19). Trp18 is located at sub-site -2 while Tyr179 and Tyr96 help to form subsites +2 and +3, respectively. In addition, Tyr77 and Tyr171 are thought to be positioned such that they too may H-bond to substrate. Refinement of the binding sites was recently facilitated by X-ray crystallography of the xylanase inhibited by the mechanism-based inhibitor 2',3'-epoxypropanyl-β-D-xyloside. This reagent was observed to be covalently bound to the catalytic nucleophile Glu86 and the xylosyl residue occupied subsite -2, packing against Trp18 (38). Additional H-bonding interactions to Tyr177 and Tyr171 were noted, confirming previous observations (19). Changes seen upon binding of the inhibitor corresponded to alterations at the base of the thumb-like region, believed to form a hinge which regulates the opening and closing of the thumb. These data led to the proposal that upon entry of the correct substrate into the active site, the xylose moiety is packed against Trp18 in sub-site -2, initiating the conformational changes, the "closing" of the thumb, which is necessary for optimal catalytic activity (38).

5.3 Role of Non-Conserved Residues in Conferring Binding Specificity

As noted above (section 2.1), Xyn C from *F. succinogenes* S85 contains two catalytic domains, A and B, both of which are homologous to Family 11 xylanases (4). Domains A and B+C have been separately cloned and expressed in *Escherichia coli,* thereby permitting a comparison of the kinetic properties of each catalytic domain relative to the complete enzyme (Table 1). Domain B of Xyn C (XynC-B) has a broader substrate range and releases more xylose compared to domain A (XynC-A), although the latter has better Michaelis-Menten paramenters (K_M, V_{max} and k_{cat}) (34). XynC-A demonstrated a high affinity for xylopentaose, hydrolyzing it to equimolar amounts of xylobiose and xylotriose which were subsequently hydrolysed to equal amounts of xylose and xylobiose, suggesting a uni-molecular reaction mechanism. In contrast, XynC-B displayed a lower affinity and k_{cat} for this substrate and released mainly xylobiose and xylose, characteristic of a multi-molecular reaction mechanism which would combine transglycosylation and shifted binding with hydrolysis.

Although the two catalytic domains of Xyn C differ in their substrate specificities, they still share significant amino acid homology (42.9 % identity, 71.3% similarity), including conservation of residues around the putative catalytic Glu residues and a number of other residues involved in binding of substrate (Figure 2). This enzyme thus provides an excellent model to investigate the role of non-conserved amino acid residues in conferring substrate specificity.

5.3.1. Identification of Essential Binding Residues by Chemical Modifications. In a previous study by Zhu *et al.* (34), each of the four Trp residues in XynC-A were found to be oxidizable by *N*-bromosuccinimide (NBS), resulting in complete inactivation of the enzyme. Of these four Trp residues, three homologous residues exist in XynC-B but NBS oxidation of this catalytic domain resulted in modification of only two of the seven Trp residues present. Given the role of Trp residues in binding of substrate to many carbohydrases, these observations warranted further investigation.

Table 1 *Comparison of kinetic parameters for hydrolysis of xylan by* F. succinogenes *xylanases*

Enzyme	K_M (mg·ml^{-1})	k_{cat} (s^{-1})	k_{cat}/K_M (s·ml·mg^{-1})
Xyn C[1]	2.09	6.40	3.06
XynC-A[1]	1.83	345	188
XynC-BC[1]	2.38	61.2	25.7
XynC-B	1.34	111	82.8
(Trp135Thr)XynC-B	2.02	113	55.9
(Trp135Ala)XynC-B	1.20	109	90.8

1. Data obtained from reference 34.

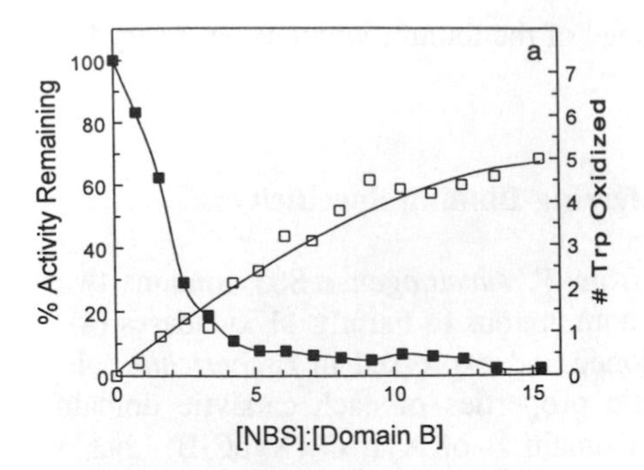

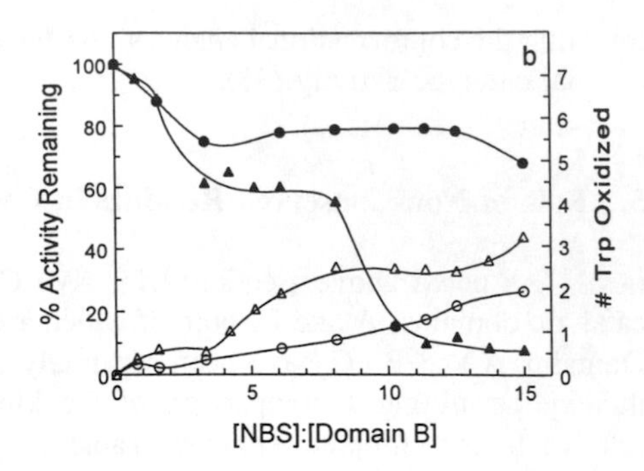

Figure 4. *NBS oxidation of XynC-B in the (a) absence (■ □), and (b) presence of 10 mM xylose (▲ △) or 16 mM xylohexaose (● O). Closed symbols, activity; open symbols, number of oxidized Trp determined by fourth-derivative absorbance spectroscopy (35).*

In our hands, NBS oxidation under controlled conditions resulted in the modification of five of the seven Trp residues present in XynC-B (Figure 4a). Hydrolytic activity of the enzyme was lost rapidly upon initiation of oxidation, as a molar ratio of about two NBS molecules per molar equivalent of protein was sufficient to inhibit 50 % of enzyme activity, and the addition of five molar equivalents of NBS resulted in less than 10% activity. Pre-incubation of XynC-B with the competitive inhibitor D-xylose resulted in the apparent protection of two Trp residues from oxidation (Figure 4b). Xylose-protection of the enzyme also resulted in a maintenance of activity, with 60 % activity still evident after addition of 8-9 molar equivalents of NBS. This protection was enhanced by the inclusion of xylohexaose in reaction mixtures. Under these conditions, however, a further Trp residue was protected from NBS oxidation.

The identification of the three protected Trp residues was achieved by differential labelling and peptide mapping of NBS oxidized preparations of the xylanase employing a combination of both electrospray mass spectroscopic (ESMS) analysis and N-terminal sequencing. Thus, the peptides containing Trp135, Trp161 and Trp202 isolated from xylohexaose-protected XynC-B were each characterized as having UV absorbance properties of an intact Trp. On the other hand, the λ_{max} of these peptides prepared from the fully oxidized enzyme derivative (*viz.* no substrate protection) had shifted to approximately 250 nm from 280 nm, consistent with the formation of oxindolealanine.

The three Trp residues identified above are conserved to varying degrees within the Family 11 xylanases. Trp202 (Trp153 in *B. circulans* Xyn A) is fully conserved within the Family and is located on the only α-helix present in the enzymes, at the interface between it and the back of the β-sheet which forms the active site cleft (Figure 3). Clearly, this residue could not possibly participate in direct interactions with substrate. It is conceivable that this Trp residue is exposed to reagent in the resting state of the enzyme, but conformational changes that accompany binding of substrate result in its "protection" from oxidation. It is also conceivable that given the fact that this residue is totally conserved in the Family 11 xylanases, it plays a critical role in the maintenance of structural integrity and function and its oxidation may cause loss of catalytic activity by hindering required conformational changes.

Whereas Trp 161 is not strictly conserved, there is a high conservation of aromatic residues at this position (Figure 2). The majority of the homologues to Trp 161 within the Family are Tyr residues (found in about half of the characterized xylanases) and indeed, Trp161 is represented by Tyr150, Tyr108, and Tyr118 in the *F. succinogenes* XynC-A, *B. circulans* Xyn A, and *T. reesei* Xyn II enzymes. From the three-dimensional structures of the latter two enzymes (1,19), Trp 161 would be located on β-strand 8 of β-sheet B which is packed against the α-helix and oriented in such a way as to readily allow interaction with substrate (Figure 3). The replacement of a Tyr residue with Trp in XynC-B is not likely to greatly influence substrate specificities and thereby contribute to the differences in activity apparent between this domain and XynC-A (47). However, the situation pertaining to Trp135 of XynC-B is much more interesting. As depicted in Figure 2, the majority of the residues occupying the position of Trp135 are other Trp residues, but such is not the case with XynC-A where Trp135 is replaced with a Thr residue. Moreover, this difference would appear to be quite significant as Trp135 is located in a highly conserved cluster of amino acids which includes the catalytic nucleophile Glu residue at position 128 of XynC-B.

5.3.2 Genetic Engineering of F. succinogenes *XynC-B.* Given the significant difference in chemical nature between the two residues that occupy positions 122 and 135 of XynC-A and XynC-B, respectively, a XynC-B derivative was engineered by site-directed mutagenesis to replace Trp135 with either Thr or Ala. Th constructs were prepared using the PCR-based mutagenesis protocol of Papworth *et al.*(39) (Stratagene, LaJolla, CA) and expressed as a fusion protein with glutathione S-transferase (40) (Amersham Pharmacia Biotech, Baie d'Urfé, Que). The fusion proteins were isolated by affinity chromatography on glutathione-Sepharose 4B, digested with thrombin and the XynC-B derviatives were purified by anion-exchange chromatography on MonoQ.

The kinetics of reactions catalyzed by the Trp135Thr XynC-B were compared to those obtained with wild-type Xyn C, XynC-B, XynC-BC, and XynC-A. As seen in Table 1, the replacement of this single amino acid to that present in XynC-A conferred changes in the binding properties of the derivative such that it more closely resembled XynC-A, while no change to k_{cat} was detected. In contrast, the more conservative replacement of Trp135 with Ala resulted in negligible change of either K_M or k_{cat}. With each XynC-B derivative, however, the product distribution of xylopentaose hydrolysis did not change as xylotriose and xylobiose were released as the major reaction products.

Hence, it would appear that Trp135 does indeed contribute to the different substrate profile of XynC-B compared to XynC-A, underscoring the impact that an individual "non-essential" residue may have on the activity of an enzyme.

References

1. W. W. Wakarchuk, R. L. Campbell, W. L. Sung, J. Davoodi and M. Yaguchi, *Protein Science*, 1994, **3**, 467.
2. Y.E. Lee, S.E. Lowe, B. Henrissat and J.G. Zeikus, *J. Bacteriol.*, 1993, **175**, 5890.
3. J. X. Zhang and H. J. Flint, *Mol. Microbiol.,* 1992, **6**, 1013.
4. F. W. Paradis, H. Zhu, P. J. Krell, J. P. Phillips and C. F. Forsberg, *J. Bacteriol.*, 1993, **175**, 7666.
5. H. J. Gilbert, G. P. Hazlewood, J. I. Laurie, C. G. Orpin and G. P. Xue, *Mol. Microbiol.*, 1992, **6**, 2065.
6. G. P. Xue, K. S. Gobius and C. G. Orpin, *J. Gen. Microbiol.*, 1992, **138**, 2397.
7. L. M. A. Ferreira, A. J. Durrant, J. Hall, G. P. Hazlewood and H. J. Gilbert, *Biochem. J.*, 1990 **269**, 261.
8. L. E. Kellett, D. M. Poole, L. M. A. Ferreira, A. J. Durrant, G. P. Hazlewood and

H. J. Gilbert, *Biochem. J.*, 1990, **272**, 369.
9. S. J. Millward-Sadler, D. M. Poole, B. Henrissat, G. P. Hazlewood, J. H. Clarke and H. J. Gilbert, *Mol. Microbiol.*, 1994, **11**, 375.
10. J. T. Pembroke, M. McMahon and B. Sweeney, *Biotech. Lett.*, 1995, **17**, 331.
11. G. W. Black, J. E. Rixon, J. H. Clarke, G. P. Hazlewood, M. K. Theodorou, P. Morris and H. J. Gilbert, *Biochem. J.*, 1996, **319**, 515.
12. P. Béguin and J. P. Aubert, *FEMS Microbiol. Rev.*, 1994, **13**, 25.
13. N. R. Gilkes, B. Henrissat, D. G. Kilburn, R. C. Miller Jr. and R. A. J. Warren, *Microbiol. Rev.*, 1991, **55**, 303.
14. O. Grépinet, M. C. Chebrou P. Béguin, *J. Bacteriol.,* 1988, **170**, 4582.
15. J. Hall, G. P. Hazlewood, P. J. Barker and H. J. Gilbert, *Gene*, 1988, **69**, 29.
16. C. Winterhalter, P. Heinrich, A. Candusso, G. Wich and W. Liebl, *Mol. Microbiol.*, 1995, **15**, 431.
17. S. Bedarkar, N. R. Gilkes, D. G. Kilburn, E. Kwan, D. R. Rose, R. C. Miller, A. J. Warren and S. G. Withers, *J. Mol. Biol.*, 1992, **228**, 693.
18. R. W. Pickersgill, J. A. Jenkins, M. Scott and I. Connerton, *J. Mol. Biol.*, 1993, **229**, 246.
19. A. Törrönen, A. Harkki and J. Rouvinen, *EMBO. J.*, 1994, **13**, 2493.
20. A. Törrönen and J. Rouvinen, *Biochemistry*, 1995, **34**, 847.
21. H. Okada, "Recent Advances in Biotechnology and Applied Microbiology-Proceedings of 8th International Conference on Global Impacts". pp. 427-437
22. A. J. Clarke, "Biodegradation of Cellulose", Technomic Publishing Co., Lancaster, PA, 1997.
23. M. L. Sinnott, *Chem. Rev.*, 1990, **90**, 1171.
24. J. D. McCarter and S. G. Withers, *Curr. Opin. Struct. Biol.*, 1994, **4**, 885.
25. M. R. Bray and A. J. Clarke, *Eur. J. Biochem.*, 1994, **219**, 821.
26. S. G. Withers and R. Aebersold, *Protein Science*, 1995, **4**, 361.
27. S. Miao, L. Ziser, R. Aebersold and S. G. Withers, *Biochemistry*, 1994, **33**, 7027.
28. E. P. Ko, H. Akatsuka, H. Moriyama, A. Shinmyo, Y. Hata, Y. Katsube, I. Urabe and H. Okada, *Biochem. J.*, 1992, **288**, 117.
29. J. Davoodi, W. W. Wakarchuk, R. L. Campbell, P. R. Carey and W. K. Surewicz, *Eur. J. Biochem.*, 1995, **232**, 839.
30. L. P. McIntosh, G. Hand, P. E. Johnson, M. D. Joshi, M. Körner, L. A. Plesniak, L. Ziser, W. W. Wakarchuk and S. G. Withers, *Biochemistry*, 1996, **35**, 9958.
31. U. Krengel and B. W. Dijkstra, *J. Mol. Biol.*, 1996, **263**, 70.
32. V. Deshpande, J. Hinge and M. Rao, *Biochim. Biophys. Acta.*, 1990, **1041**, 172.
33. S. S. Keskar, M. C. Srinivasan and V. V. Deshpande, *Biochem. J.,* 1989, **261**, 49.
34. H. Zhu, F. W. Paradis, P. J. Krell, J. P. Phillips and C. W. Forsberg, *J. Bacteriol.,* 1994, **176**, 3885.
35. M. R. Bray and A. J. Clarke, *Biochemistry*, 1995, **34,** 2006.
36. M. R. Bray and A. J. Clarke, "Xylan and Xylanases" (J. Visser, G. Beldman, M. A. Kusters-van Someren, A. J. G. Voragen, eds), Elsevier, Amsterdam, 1992, p. 423.
37. M.R. Bray, PhD Thesis, University of Guelph, 1993.
38. R. Havukainen, A. Törrönen, T. Laitinen and J. Rouvinen, *Biochemistry*, 1996, **35**, 9617.
39. C. Papworth, J. Bramon and D. A. Wright, *Strategies,* 1996, **9**, 3.
40. D. B. Smith and K. S. Johnson, *Gene*, 1998, **67**, 31.

STRUCTURE-FUNCTION RELATIONSHIPS IN POLYGALACTURONASES: A SITE-DIRECTED MUTAGENESIS APPROACH

J.A.E. Benen, H.C.M. Kester, S. Armand, P. Sanchez-Torres, L. Parenicova, S. Pages and J. Visser

Section Molecular Genetics of Industrial Microorganisms,
Wageningen Agricultural University, Dreyenlaan 2,
6703 HA Wageningen, The Netherlands.

1. INTRODUCTION

Pectin is one of the major plant cell wall carbohydrates present in the middle lamella and the primary cell wall where it functions as a glue to hold cells together. Pectin is also one of the most complex carbohydrates. It consists of smooth regions of α-1,4 linked D-galacturonic acid (Gal*p*A) interspersed by streches in which α-1,2 L-rhamnose alternate with Gal*p*A, the hairy regions or rhamnogalacturonan I. To the rhamnose, side chains of arabinan or galactan can be attached. In naturally occuring pectin generally 70 % of the Gal*p*A residues are methylesterified. Within the rhamnogalacturonan I region stretches of homogalacturonan high in xylose substition at O2 or O3 are found as well. These regions are called xylogalacturonan.

For the complete decomposition of pectin saprophytic and phytopathogenic microorganism have a large repertoir of pectinolytic enzymes at their disposal. These enzymes comprise side chain degrading activities like galactanases, β-galactosidases, arabinanases and arabinofuranosidases, methyl- and acetyl esterases and main chain depolymerases. The latter group consists of two major classes of enzymes: the lyases and the hydrolases. In each of those classes, enzymes specific for the smooth regions, *viz.* pectate and pectin lyases and polygalacturonases, and hairy regions, *viz.* rhamnogalacturonan lyases and rhamnogalacturonases, are found. Also enzymes have been found that are able to degrade the highly xylose substituted xylogalacturonan[1].

Thus far twenty-one *Aspergillus niger* pectinase genes have been identified and an expression system has been developed which allows the individual overexpression of each member of such a gene family[2]. This has allowed the detailed analysis of the structures of the genes as well as the biochemical properties of the enzymes. The recent elucidation of 3D structures for pectate and pectin lyases, rhamnogalacturonase and polygalacturonases and a pectin methylesterase, has also opened the way to the understanding of the substrate specificities of the pectinases as defined by their actual subsite architecture. In this contribution we will give a detailed update of our studies of the endopolygalacturonases from *A. niger*.

2. RESULTS AND DISCUSSION

2.1. Genetic Organization and Relatedness of *A. niger* Polygalacturonase Genes

The first *A. niger* polygalacturonase (PG) genes *pga*I and *pga*II were cloned via reverse genetics by Bussink *et al.*[3,4]. Bussink *et al.*[5], in addition to cloning *pga*C, also demonstrated the presence of an endoPG gene family in *A. niger*. Pa⊆enicova *et al.* cloned the other members, *pga*A[6], *pga*B[6], *pga*D[7] and *pga*E[2]. A schematic representation of the *pga*-genes is give in Fig. 1. Although the number of introns varies from one to three, their positions are conserved. Similarly, potential N-glycosylation sites are conserved as well. For PGII it has been shown that this site is indeed glycosylated[8]. Determination of the N-terminal residues of PGI, PGII and PGE revealed that the enzymes were synthesized as prepro-enzymes. The pre-sequence directs the enzymes to the secretory pathway where the pre-sequence is cleaved by a signal peptidase. Next the pro-sequence is cleaved by a protease in analogy to KEX2. Except for PGII, which contains a monobasic cleavage site (Arg), all other PGs contain a dibasic cleavage site (Lys-Arg or Lys-Lys).

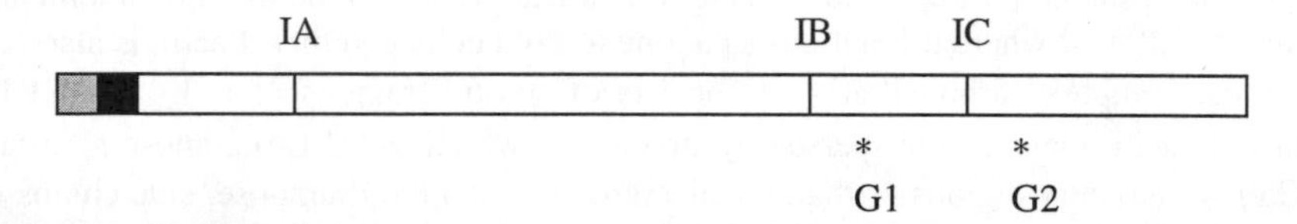

Figure 1. *Schematic representation of endopolygalacturonase genes from* A. niger. The grey box indicates the pre-sequence. The black box indicates the pro-sequence. IA, IB and IC signify introns. The asterisks indicate potential glycosylation sites G1 and G2. Site G2 is only present in the *pga*I gene. Site G1 is present in all six *pga*-genes, I, II, A, B, C and E. Intron IA is present in all six *pga*-genes except for *pga*II. Intron IB is present in all *pga*-genes except for *pga*A. Intron IC is only present in *pga*C and *pga*E.

At the primary structure level, sequence identities among *A. niger* PGs can vary considerably. Recently, Wubben *et al.*[9] constructed a phylogenetic tree based on the amino acid sequences of all known fungal endopolygalacturonases using a plant gene as an out-species. This analysis shows that the *A. niger* PGs cluster into four different groups. The constitutively expressed PGB, clusters in group I together with other PGs for which constitutive expression was reported. The highly active enzymes PGI, II and A, cluster in group II. PGC and PGE, both poorly active on the model substrate polygalacturonic acid, are found together in group IV whereas PGD belongs to yet another class. Both endo- and exopolygalacturonases together with the related rhamno-galacturonases and several plant pollen enzymes were placed in family 28 of the general classification system of glycosyl hydrolases[10].

2.2. Kinetic Properties of *A. niger* EndoPGs

One of the intriguing questions we are trying to solve with our studies is why the fungus has seven endoPGs at its disposal. In addition to the detailed analysis of the regulation of the expression of the individual genes, which has already revealed the constitutive expression of *pga*A and *pga*B[6], we consider the analysis of the kinetic properties, including the mode of action on defined substrates, of utmost importance to understand the individual roles of the enzymes during pectin degradation. To allow best comparison of the kinetic properties of the enzymes, the simplest possible substrates were used. These included polygalacturonic acid (completely demethylated pectin), (reduced) oligogalacturonides of defined length (n = 2 - 8) and lemon pectins of various degree of methylation.

In order to find out whether our kinetic studies might be complicated by side-reactions like transglycosylation reactions, known to occur with retaining enzymes, we studied the mechanism of hydrolysis of PGI, PGII and exoPG by NMR[11]. All three enzymes, and therefore family 28 of hydrolases, act by inversion of configuration and thus side-reactions are very unlikely to occur.

2.2.1. A. niger *EndoPGs Acting on Polymer Substrates.* In a standard assay, using 2.5 mg/ml polygalacturonic acid as the substrate at the pH optimum of the enzymes, the specific activities of the endoPGs differed strongly[2,6,12]. Whereas for PGC and PGE specific activities were calculated as 25 U/mg and 40 U/mg, respectively, for the other PGs those were: PGI, 800 U/mg; PGII, 2000 U/mg; PGA, 1200 U/mg and PGB, 1800 U/mg. For PGB this activity was determined in the presence of 200 mM NaCl. In the standard 50 mM sodium acetate buffer, the specific activity for PGB at its pH optimum is 400 U/mg. Unlike the other PGs, which were most active at pH 4.0-4.2, PGB was most active at pH 5.0.

Determination of the kinetic parameters V_{max} and K_m revealed that for all enzymes, except for PGE, K_m was below 0.5 mg/ml polygalacturonic acid. Therefore, the specific activities approximate V_{max} for those enzymes. For PGE V_{max} and K_m were calculated as 80 U/mg and 2.5 mg/ml polygalacturonic acid. The observations that both PGC and PGE have a low V_{max} and that for PGE K_m is high, indicates that polygalacturonic acid is not the natural substrate for those enzymes. Whether these enzymes might require (a) methylated galacturonic acid function(s) for effective binding and catalysis was investigated using pectins of various degree of esterification (DE)[6,12]. These data are compiled in Table 1.

In a standard assay all enzymes except PGA and PGB appeared most active on the unmethylated substrate. Thus, methylation is not the required function for PGC and PGE. For PGII the strongest decrease in activity was observed upon increasing DE indicating that indeed unmethylated polygalacturonic acid is the natural substrate for this enzyme. PGA and PGB appeared to be more active on 22 % and 45% methylated substrate in a standard assay which is again an indication for the differences in substrate specificity among the PGs. The preference of partially methylated substrates by PGA and PGB is in accordance with the constitutive expression of these enzymes.

Table 1. *Effect of degree of methylation of pectin on the specific activity of endopolygalacturonases from* A. niger. Conditions: 0.25 % (mass/vol) of pectin in 50 mM sodium acetate pH 4.2, 30 °C. *) 200 mM NaCl added.

Endopolygalacturonase	Degree of esterification (%)					
	0	7	22	45	60	75
	Relative specific activity					
I	100	97	87	43	18	3
II	100	68	50	24	9	2
A	100	125	135	52	7	6
B	100	150	168	131	62	27
B*	100	98	92	45	18	16
C	100	102	86	36	16	5
E	100	103	71	38	16	4

2.2.2. Mode of action of A. niger *PGs.* To gain greater insight into the substrate specificity of the enzymes the number of subsites and the binding energies associated with these substites were studied[2,6,12].

In a first approach to analyse the substrate specificity the product progression during polygalacturonic acid hydrolysis was studied. A typical endolytic enzyme should display a transient increase of oligogalacturonates with higher degree of polymerization (DP) at the beginning of the reaction, which are then gradually converted to smaller oligomers. This type of behaviour was indeed observed for PGII, PGB and PGE. However, for PGI, PGA and PGC the transient accumulation of higher oligomers was small and coincided with a strong increase of galacturonate monomers.

Further analysis of this phenomenon using purified oligogalacturonates of defined DP revealed that this type of product distribution is caused by multiple attack on a single chain, also known as processivity. For PGI this behavior occurred when DP > 5[12], for PGC when DP > 6[12] and for PGA when DP > 7[6].

The mode of action analysis comprised the determination of the rate of hydrolysis of each oligogalacturonate at a known concentration and the determination of the bond cleavage frequencies (BCFs) for each oligomer. Using reduced oligomers, where the reduced end serves as a tag, it was established that hydrolysis takes place from the reducing end for all six PGs[2,6,12]. Differences in BCF become apparent from DP 5 onward, when apparently the number of productive binding modes increases. Those differences in BCF are a consequence of the differences in affinity for the substrate at the individual subsites.

Due to the fact that PGE had rather poor affinity for the oligogalacturonates a subsite map could be established[2] according to the method outlined by Suganuma *et al.* [13]. For PGI, PGII and PGC this method could not be applied since the affinities were too high (K_m < 50 μM)[12]. In fact, the K_m values can not be determined using current technology. This precluded the calculation of subsite maps according to Thoma *et al.*[14].

However, provisionary subsite maps could be obtained for PGI and PGII which account for the observed differences in BCFs[12].

2.3. Site directed mutagenesis of PGII

At the onset of our site directed mutagenesis studies (SDM) no 3D-structures for family 28 glycosyl hydrolases were known. We therefore initially concentrated on charged conserved residues among endoPGs as obtained from nucleotide database analyses. The earliest report on a residue implied in catalysis by endoPGs identified via chemical modification involved a His[15]. Indeed, one His residue, H223 in PGII, appeared conserved among all endoPGs. In addition to this His, the following residues were conserved throughout the endoPGs: D180, D201, D202, R256 and K258. We sub-jected all these residues to SDM and studied the effect of the mutation on catalysis and BCF[16]. The results of these studies are compiled in Tables 2 and 3, respectively. During the course of our studies the 3D-structure of rhamnogalacturonase (RHG), a family 28 member, was published[17]. The RHGs show several residues conserved among the endoPGs. These residues are the counterparts of D180, D201 and K258, whereas D202 appears conservatively changed into E202. Recently, the 3D-structure for *Erwinia carotovora* endoPG[18] was reported while in collaboration with Yovka van Santen and Bauke Dijkstra from the Groningen University, The Netherlands, we solved the 3D-structure for the *A. niger* PGII[19].

The 3D-structures of endoPGs show that the residues mutated are all intimately grouped, at the bottom of a cleft which can accommodate the substrate. In view of the dramatic effects on catalysis upon mutation of either of the residues, it can be concluded that they all are involved in catalysis. Furthermore, since RHGs and endoPGs both require a galacturonic acid (Gal*p*A) unit in subsite –1, it can be concluded that residues D180, D201, D202 and K258, should be located at subsite –1 or near the scissile bond. Before discussing the individual mutant enzymes some general remarks on the effect of the mutations on the BCFs have to be made. First, only productive modes, covering subsites -1 and +1, are monitored, thus any ΔG change at subsite -1 and +1 affects binding in productive modes equally and will therefore not, or only slightly, change the BCFs. The observed differences in BCFs are a result of a changed effectiveness of hydrolysis of a certain positional isomer. The difference in effectiveness may be the result of invoking subsite +2 to compensate for the loss of binding energy at subsite –1 or +1 and therefore mutations at -1 and +1 will increase the release of dimers when the residue mutated directly interacts with the substrate

Table 2 *pH optimum and kinetic parameters of mutant PGII.* Kinetic parameters were determined using 50 mM NaAc, pH 4.2. pH optima (pH opt) were determined in McIlvaine buffers. Polygalacturonic acid served as a substrate at 30 °C.

Enzyme	pH opt	Kinetic parameters	
		$V_{max\ app}$ U/mg	$K_{m\ app}$ mg/ml
Wild type	4.2	2000	<0.15
R256Q	3.8	278	1.7
K258N	3.8	16.2	2.8
H223A	4.1	10.0	0.15
H223C	3.8	21.5	0.80
H223Q	3.9	0.36	1.1
H223S	4.1	1.7	1.5
D180E	4.2	0.24	0.3
D180N	3.9	1.4	1.5
D201E	4.2	0.04	0.3
D201N	3.9	0.19	0.3
D202E	4.2	12.7	0.7
D202N	4.1	0.3	1.5
D180E/D201E	4.2	0.04	<0.15

2.3.1. SDM of R256 and K258. Of the residues mutated, mutant R256Q retains the highest activity and displays one of the highest K_m values. The shift of the BCFs in favour of dimer formation and its position in the 3D structure are compatible with a location at subsite +1. Indeed, the positive charge may be involved in interaction with the charged substrate. This is also very likely for K258 since upon mutation of this residue (K258N), negatively charged divalent ions like citrate and phosphate completely abolished activity. Furthermore, enzyme K258N displayed the highest K_m and showed a much more severe effect on catalysis than R256Q. The role of K258 might therefore be to interact with the carboxylate part of the substrate bound at subsite –1.

2.3.2. SDM of H223. As mentioned above, a His residue was the first residue implied in catalysis. SDM of this residue indeed dramatically reduced catalysis. The effect on catalysis as well as on the BCFs appeared dependent on the side chain engineered. Enzyme H223C which still has some protonating/deprotonating capacity showed the highest activity. However, this rather large side chain also had the most severe effect on the BCFs, a very strong increase of dimer formation. Much smaller effects on BCFs were observed when a small side chain like Ala or Ser was engineered. These results suggest that although H223 has a prominent effect on catalysis, its role is most probably indirect.

Table 3. *Bond cleavage frequencies for wild type and mutant PGII.* 0.5 mM oligogalacturonates were incubated with (mutant) PGII in 0.5 ml 50 mM Na-acetate pH 4.2. At regular intervals 50 μl aliquots were withdrawn and the reaction products were analysed and quantitated by HPAEC-PAD. The boldface typescript indicates the reducing end. Bond cleavage frequencies are given in percentages. Gn signifies $(GalpA)_n$.

Enzyme	Gn											
	5			G	-	G	-	G	-	G	-	**G**
Wild type									33		67	
R256Q									50		50	
K256N									52		48	
H223A									27		73	
H223C									80		20	
H223Q									48		52	
H223S									28		72	
D180E									50		50	
D180N									36		64	
D201E									39		61	
D201N									32		68	
D202E									52		48	
D202N									30		70	
D180E/D201E									35		65	
	6	G	-	G	-	G	-	G	-	G	-	**G**
Wild type							8		57		35	
R256Q							15		74		11	
K256N							18		70		12	
H223A							10		65		25	
H223C							14		84		2	
H223Q							6		72		22	
H223S							16		64		20	
D180E							9		73		18	
D180N							10		68		22	
D201E							9		76		15	
D201N							11		65		24	
D202E							9		81		10	
D202N							11		62		27	
D180E/D201E							11		63		26	

2.3.3. SDM of D180, D201 and D202. Nowadays it is well established that in all glycosyl hydrolases studied in this respect two acidic residues are immediately responsible for catalysis. One of them serving as the nucleophile and the other as the base. Furthermore, within one family of hydrolases those residues are strictly conserved. Thus, despite SDM of D202E/N having among the strongest effects on catalysis, the presence of a Glu at this position in RHG argues against a role as nucleophile or base. This leaves us with D180 and D201 to fulfil the roles of nucleophile and base. However, SDM of D180E/N, D201E/N and D180E/D201E did not completely abolish activity. The situation is further complicated by the fact that neither of the three acidic residues, D180, D201 and D202 (spaced at 4-5 X) is located at a distance compatible with the generally accepted notion that for inverting hydrolases the nucleophile and base should be spaced some 9-9.5 X[20]. We therefore have to conclude that in family 28 of hydrolases the active site configuration has diverged from the architecture found in other inverting glycosyl hydrolases. A direct role for D202 in catalysis can still not be excluded.

Currently, we are working on the elucidation of the 3D-structure of the PGII-substrate complex. In addition to this, our SDM studies are now directed to residues located at subsites other than –1 and +1 to understand the role of the residues involved in defining substrate specificity.

Acknowledgement

This work was financially supported by EC grants AIR2-CT94-1345 and ERBBIO-CT96-0685 to J.V., by TMR grant BIO4-98-5007 to P. S.-T. and by EMBO grant ALTF-386-1998 to S.P.

References

1. H.C.M. Kester, J.A.E. Benen and J. Visser, *Biotech. Appl Biochem.*, 1999, In the press.
2. L. Parenicová, J.A.E. Benen,. H.C.M. Kester and J. Visser, *Eur. J. Bioc*hem., 1998, **251**, 72.
3. H.J.D. Bussink, H.C.M. Kester and J. Visser, *FEBS Le*tters, 1990, **273**, 127.
4. H.J.D. Bussink, K.B. Brouwer, L.H. de Graaff, H.C.M. Kester and J.Visser, *Curr.Genet.*, 1991, **20**, 301.
5. H.J.D. Bussink, F.P. Buxton, B.A. Fraaye, L.H. de Graaff and J. Visser, *Eur. J. Biochem.*, 1992, **208**, 83.
6. L. Pa⊆enicová, J.A.E. Benen, H.C.M. Kester and J. Visser. *Eur. J. Biochem.* Submitted
7. L. Pa⊆enicová and J. Visser, unpublished
8. Y. Yang, C. Bergmann, J. Benen and R. Orlando, *Rapid Comm. in Mass Spectr.*, 1997, **11**, 1257.
9. J.P. Wubben, W. Mulder, A. Ten-Have, J.A.L. van Kan and J. Visser, *Appl. Envir. Microbiol.*, 1999, **65**, 1596.
10. B. Henrissat, *Biochem. J.*, 1991, **280**, 309.
11. P. Biely, J.A.E. Benen, K. Heinrichová, H.C.M. Kester and J. Visser, *FEBS Lett.*, 1996, **382**, 249.
12. J.A.E. Benen, H.C.M. Kester and J. Visser, *Eur. J. Biochem.*, 1999, **159**, 577.
13. T. Suganuma, R. Matsuno, M. Ohnishi and K. Hiromi, *J. Biochem.*, 1978, **84**, 293.
14. J.A. Thoma, G.V.K. Rao, C. Brothers and J. Spradlin, *J. Biol. Chem.*, 1971, **246**, 5621.
15. L. Rexová-Benková and M. Mracková, *Biochim. Biophys. Acta,* 1978, **523**, 162.
16. S. Armand, M.J.M. Wagemaker, H.C.M. Kester, P. Sanchez-Torres, Y. van Santen, B.W. Dijkstra, J. Visser and J.A.E. Benen, *J.Biol. Chem.* submitted
17. T.N. Petersen, S. Kauppinen and S. Larsen, *Structure*, 1997, **5**, 533.
18. R. Pickersgill, D. Smith, K. Worboys and J. Jenkins, *J. Biol. Chem.,* 1998, **273**, 24660.
19. Y. van Santen and B.W. Dijkstra, unpublished
20. J. McCarter and S.G. Withers, *Curr. Opin. Struct. Biol*, 1994, **4**, 885.

THE CONCEPT OF α-AMYLASE FAMILY

T. Kuriki

Biochemical Research Laboratory
Ezaki Glico Co., Ltd.
4-6-5 Utajima, Nishiyodogawa-ku
Osaka 555-8502, Japan

1 INTRODUCTION

The most important reason that enzymes are essential for life is that enzymes recognize each of the reactions which should be catalyzed by them, much more precisely than other catalysts. The specificities of enzymes are composed of two factors; the substrate-specificity and the reaction-specificity. The term "amylase" can be generally defined as the enzymes which hydrolyzes *O*-glycosyl linkage of starch. Starch is the most popular polysaccharide as a food ingredient. It is a mixture of the amylose which is composed of glucose-polymer linked essentially only α-1,4 linkages and the amylopectin which is composed of α-1,4-linked glucose-polymer branched by α-1,6 linkages.

We proposed a general concept for an enzyme family, α-amylase family including most of amylases and related enzymes in 1992, based on the structural similarity and the common catalytic mechanisms[1]. The pioneering work by Svensson's group[2], which reported the similarity in the secondary structure among amylases and their related enzymes, stimulated our work. However, the study on neopullulanase was the key to open the door for the concept. We describe here two crucial points which are essential for the concept of α-amylase family: the common catalytic mechanisms and the structural similarities among amylases and related enzymes.

2 AMBIGUITY OF CLASSIFICATION OF AMLASES AND RELATED ENZYMES

Amylases and related enzymes which specifically catalyze the hydrolysis or synthesis of the glucosidic linkages of starch are represented by four types as follows:

1. Hydrolysis of α-1,4-glucosidic linkages; α-amylase (EC 3.2.1.1)
2. Hydrolysis of α-1,6-glucosidic linkages; pullulanase (EC 3.2.1.41) or isoamylase (EC 3.2.1.68)
3. Transglycosylation to form α-1,4-glucosidic linkages; cyclodextrin glucanotransferase

(CGTase, EC 2.4.1.19)

4. Transglycosylation to form α-1,6-glucosidic linkages; 1,4-α-D-glucan:1,4-α-D-glucan 6-α-D-(1,4-α-D-glucano)-transferase (branching enzyme, EC 2.4.1.18)

Bisubstrate reactions catalyzed by nucleotide-diphospho-sugar glycosyltransferases such as α-1,4-glucosidic linkage extension by starch synthase (EC 2.4.1.21) are irrelevant to the main subject here. Both these four reactions and the classification of these enzymes have been clearly distinguished. Each of the four reactions are representatively catalyzed by four individual types of enzymes, described above. However, some exception have been reported: α-amylases weekly catalyze α-1,4 transglycosylation in addition to the main reaction, α-1,4 hydrolysis[3]. The reverse is true. CGTases feebly do catalyze α-1,4 hydrolysis in addition to the main reaction, α-1,4 transglycosylation[4]. We may, therefore, reasonably conclude that the boundary between "glucanohydrolases" and "glucanotransferases" is not necessarily clear. Furthermore, some α-amylases have been reported to catalyze α-1,6 hydrolysis[5,6]. Some pullulanases from thermophile have recently been reported to hydrolyze not only α-1,6- but also α-1,4-glucosidic linkages[7-9]. Thus, we have known the existence of enzymes on the boundary between α-amylase and pullulanase/isoamylase which have been classified as the enzymes specific to α-1,4- and α-1,6-glucosidic linkages, respectively. Therefore, it is now widely accepted that the relationship between enzymes and their reaction-/substrate-specificity is not necessarily fixed each other. Nevertheless, these facts have not been attracted the attention and been thought as the trifling exceptions or the trivial side-reactions.

3 THE STRUCTURAL SIMILARITY

Many primary sequences of amylases and the related enzymes from various origins have been reported. The rapid increase of the number of available sequences has made it possible to compare the sequences each other. Friedberg[10] first suggested the homology in the α-amylases from some different origins. Nakajima et al.[11] cleary pointed out the existence of four highly conserved regions and related the regions with the catalytic and the substrate-binding sites in eleven different α-amylases. Binder et al.[12] indicated that a CGTase also had the same four highly conserved regions as α-amylases. Amemura et al.[13] suggested that isoamylase and pullulanase also had the similar conserved regions.

We found a new type of pullulan-hydrolyzing enzyme, neopullulanase (EC 3.2.1.135)[14], and proved that the enzyme catalyzes hydrolysis of α-1,4-, and α-1,6-glucosidic linkages[15], as well as transglycosylation to form α-1,4-, and α-1,6-glucosidic linkages[1]. The introduction of several replacements of the amino acid residues that constitute the active center of the neopullulanase presented the evidence that one active center of the enzyme participated in all four reactions described above[16]. We demonstrated that neopullulanase also has the above-mentioned four highly conserved regions (Figure 1). The four highly conserved regions covered all of the three catalytic residues and substrate-binding residues that bind glucosyl residues adjacent to the scissile linkage in the substrates by the enzyme, according to the substrate-binding model of Taka-amylase A, the α-amylase from *Aspergillus oryzae*, proposed by Matsuura et al.[17]. Therefore, we hypothesized that α-amylase, pullulanase/isoamylase, CGTase, and branching enzyme catalyze α-1,4 and α-1,6 hydrolysis and α-1,4 and α-1,6 transglycosylation, respectively, in the same mechanisms[18].

Enzyme	Origin	Region1	Region2	Region3	Region4
			O	O	O
α-Amylase	*Aspergillus oryzae*	117DVVANH	202GLRIDTVKH	230EVLD	292FVENHD
CGTase	*Bacillus macerans*	135DFAPNH	225GIRFDAVKH	258EWFL	324FIDNHD
Pullulanase	*Klebsiella aerogenes*	600DVVYNH	671GFRFDLMGY	704EGWD	827YVSKHD
Isoamylase	*Pseudomonas amyloderamosa*	292DVVYNH	371GFRFDLASV	435EPWA	505FIDVHD
Branching enzyme	*Escherichia coli*	335DWVPGH	401ALRVDAVAS	458EEST	521LPLSHD
Neopullulanase	*Bacillus stearothermophilus*	242DAVFNH	324GWRLDVANE	357EIWH	419LLGSHD
α-Amylase-pullulanase	*Clostridium thermohydrosulfulcum*	488DGVFNH	594GWRLDVANE	627ENWN	699LLGSHD
α-Glucosidase	*Saccharomyces carlsbergenesis*	106DLVINH	210GFRIDTAGL	276EVAH	344YIENHD
Clyclodextrinase	*Thermoanaerobacter ethanolicus*	238DAVFNH	321GWRLDVANE	354EVWH	416LIGSHD
Oligo-1,6-glucosidase	*Bacillus cereus*	98DLVVNH	195GFRMDVINF	255EMPG	324YWNNHD
Dextran glucosidase	*Streptococcus mutans*	98DLVVNH	190GFRMDVIDM	236ETWG	308FWNNHD
Amylomaltase	*Streptococcus pneumoniae*	224DMWAND	291IVRIDHFRG	332EELG	391YTGTHD
Glycogen debranching enzyme	Human	298DVVYNH	504GVRLDNCHS	534ELFT	603MDITHD

Figure 1 Enzymes belong to α-amylase family and the four highly conserved regions[45]. Invariable three catalytic sites are indicated by open circle.

Branching enzymes exclusively catalyze α-1,6 transglycosylation and do neither hydrolysis nor α-1,4 transglycosylation[19]. Homology in the primary structure between glycogen branching enzyme and amylases was first reported by Preiss' group[20]. Baba et al.[21] subsequently indicated that branching enzymes contain the four highly conserved regions.

The three-dimensional structure of Taka-amylase A was first determined by X-ray crystallographic analysis among the amylases and the related enzymes[17]. The structure is based on (β/α)8-barrel as its main domain followed by anti-parallel β-sandwich. It took seven years until the same topological motif was reported for other amylase; CGTase[22,23]. The three-dimensional structure of isoamylase was recently published[24], indicating that the structure was composed of the similar (β/α)8-barrel as its main domain, as expected[2,18].

4 COMMON CATALYTIC MECHANISM

Based on the facts described above, the hypothesis that most of amylases and their related enzymes posses similar structure leaves no room for doubt. As mentioned before, Svensson's group reported the similarity in the secondary structure among amylases and their related enzymes and use a term "α-amylase superfamily members", since these enzymes have similar structures as "proteins"[2]. However, we thought that the structural similarity was not sufficient to gather these enzymes into one family and that we should clarify the common catalytic mechanism of them. Because, (β/α)8-barrel structure was first reported for triose phosphate isomerase[25], and was therefore called TIM-barrel. Triose phosphate isomerase has no relation to amylases and the related enzymes with respect to its action pattern and function. Thus, there exist lots of enzymes which do not relate to their function at all, but have similar structures. It is quite reasonable to think that the number of the folding pattern of a polypeptide as a protein stably exists in water is limited and that the protein folding is not necessarily different each other during the evolutionary events; divergence or convergent evolution from ancestral proteins. Indeed, Chothia[26] speculated the basic topological motifs of proteins exist in nature are most likely limited around one thousand. We should be careful about the fact that the structural similarity of protein does not necessarily relate to the similarity of the function of the protein. Therefore, we think it is a mistake to describe that the proteins that do not have functions as enzymes, such as amino acid transport-related proteins and 4F2 heavy-chain cell surface antigens, are belong to α-amylase family[27].

Based on the background described above, we proved that the four reactions, α-1,4 and α-1,6 hydrolysis and α-1,4 and α-1,6 transglycosylation, could be catalyzed by the same mechanism[16]. The amino acid residues located in the active center of the neopullulanase were tentatively identified according to a molecular model of Taka-amylase A[17] and homology analysis of the amino acid sequence of neopullulanase, Taka-amylase A, and other amylolytic enzymes[18]. When one of the amino acid residues corresponding to the catalytic sites of Taka-amylase A; Glu-230, Asp-206, and Asp-297, was replaced by their amide-form (Gln or Asn) or oppositely charged amino acid residues (His), neopullulanase activities toward α-1,4 and α-1,6-glucosidic linkages simultaneously disappeared. When the amino acid residues corresponding to the substrate-binding sites that bind glucosyl residues adjacent to the scissile glucosidic linkage were changed, the specificities of the mutated neopullulanase toward α-1,4 and α-1,6-glucosidic linkages were obviously altered from that of the wild-type enzyme[16]. As described before, neopullulanase strongly catalyzed α-1,4 and α-1,6 transglycosylation as well as α-1,4 and α-1,6 hydrolysis. Hence, one active center of neopullulanase participates in all the four reactions. We may lead from these results to the conclusion that the four reactions can be catalyzed in the same mechanism[1].

5 CONCEPT AND DEFINITION OF α-AMYLASE FAMILY

Based on the structural similarity and the common catalytic mechanisms among most of amylases and the related enzymes, we consequently proposed a general idea for one enzyme family, α-amylase family[1]. We define α-amylase family for the enzymes that satisfy the following four requirements:

i) They act on α-glucosidic linkages.
ii) They hydrolyze α-glucosidic linkages to produce α-anomeric mono- and oligo-saccharides or form α-glucosidic linkages by the transglycosylations.
iii) They have four highly conserved regions in their primary sequence which contain all catalytic and most of important substrate-binding sites.
iv) They have Asp, Glu, and Asp residues as catalytic sites corresponding to Asp-206, Glu-230, and Asp-297 of Taka-amylase A. These three residues have been indicated as the catalytic sites of Taka-amylase A[28] and α-amylases from *Bacillus subtilis*[29], *Bacillus stearothermophilus*[30], and barley[31], as well as neopullulanase[16], cyclodextrinase[32], CGTase[22, 33], amylopullulanase[8], and branching enzyme[34, 35] by using site-directed mutagenesis. These results are consistent with the X-ray crystallographic analyses of α-amylases[17,36-39], CGTases[22,23,40], an oligo-1,6-glucosidase[41], and an isoamylase[24].

In order to explain the concept of α-amylase family, we schematically represented the relation of specificities for the target linkage and reaction of the enzymes typically belonging to α-amylase family as in Figure 2[1,42]. Most α-amylases which typically catalyze hydrolysis of α-1,4-glucosidic linkages should be located in the lefthand-upper corner. CGTase which catalyzes transglycosylation to form α-1,4-glucosidic linkages should be in the bottom-left corner. Pullulanase and isoamylase which exclusively hydrolyze α-1,6-glucosidic linkages should be in the righthand-upper corner. Branching enzyme which exclusively catalyze transglycosylation to form α-1,6-glucosidic linkages should be in the bottom-right corner. A type of α-amylase, bacterial saccharifying α-amylase from *B. subtilis* which significantly catalyzes α-1,4 transglycosylation[3] could be in the position just below the lefthand-upper corner. Amylopullulanases which hydrolyze not only α-1,6-glucosidic linkages of pullulan but also α-1,4-glucosidic linkages of starch[7-9] should be in the middle position between α-amylase and pullulanase. Neopullulanase catalyzes all four reactions, thus, the enzyme should be located in the center position. The enzyme can be a bridge connecting α-amylase, pullulanase/isoamylase, CGTase, and branching enzyme.

6 ENZYMES BELONG TO α-AMYLASE FAMILY

We indicated α-amylase, pullulanase, isoamylase, CGTase, branching enzyme, α-glucosidase, oligo-1,6-glucosidase, amylopullulanase, and neopullulanase as the enzymes typically belong to α-amylase family[1]. Subsequently, Svensson[43] added cyclodextrinase, dextran glucosidase, amylomaltase, and glycogen debranching enzyme to the family. As described before, Janacek et al.[27] demonstrated that amino acid transport-related proteins and 4F2 heavy-chain cell surface antigens also belong to α-amylase family. Although their idea is very intriguing from the viewpoint of possible evolutionary events in the α-

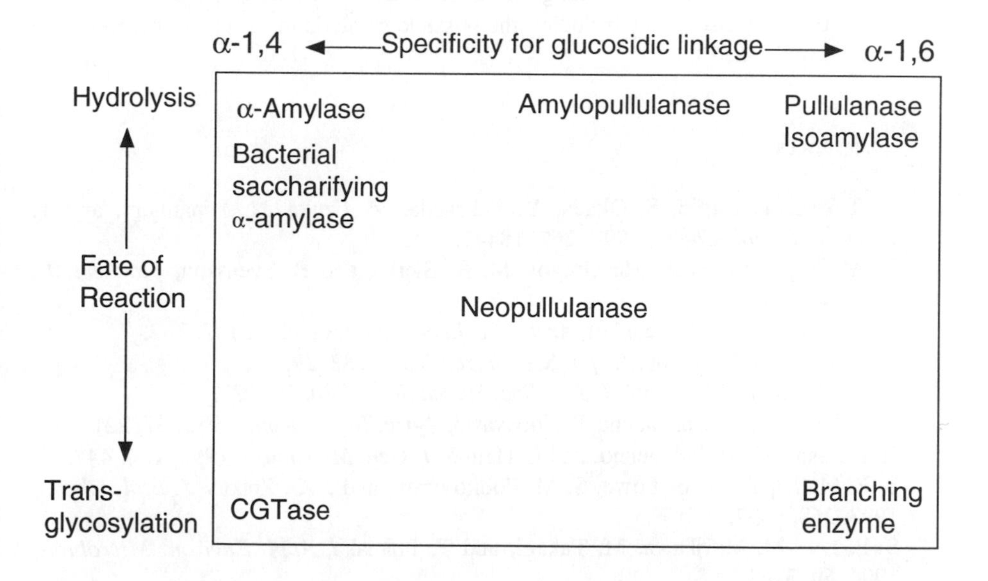

Figure 2 Schematic representation of relationship of enzymes that belong to α-amylase family according to types of reaction[45].

amylase family enzymes from an ancestor protein, we should rule out the proteins which have similar topological structures, but do not have functions as enzyme, from our concept of "enzyme family"; α-amylase family. When we speculate the function of protein only from the primary structure, this debatable point is often raised. This would be a big task ahead of us; scientists for post-genome age.

7 CONCLUDING REMARKS

The main stream of enzymology has been distinguishing and classifying enzymes into different species in detail and has been a history of finding new enzymes. However, increasing the number of information for the primary structures of enzymes and comparison of them, and continuous efforts for the crystallographic analyses may change the conventional tendency. This change may have been done by the prediction of Chothia[26]; "Not so many basic folding pattern of protein exist as the number of enzymes", the proposal of Henrissat[44] "Sequence-based classification of glycosyl hydrolases", and our proposal of the concept of α-amylase family[1] "Not only the structural similarity but also the common catalytic mechanism lie in the significant number of amylases and the related enzymes". These new ideas have been individually proposed almost at the same time, and therefore hint the curtain raiser for the coming age of enzyme engineering.

8 ACKNOWLEDGEMENTS

I am grateful to Professor T. Imanaka for helpful discussions as well as to Drs. H. Kaneko, H. Takata, and S. Okada for their thoughtful comments. Our comprehensive review of the concept of α-amylase family including the possible catalytic mechanism of the enzyme will shortly be published[45].

References

1. H. Takata, T. Kuriki, S. Okada, Y. Takesada, M. Iizuka, N. Minamiura, and T. Imanaka, *J. Biol. Chem.*, 1992, **267**, 18447.
2. H. M. Jespersen, E. A. MacGregor, M. R. Sierks, and B. Svensson, *Biochem. J.*, 1991, **280**, 51.
3. E. J. Hehre and D. G. Genghof, *Arch. Biochem. Biophys.*, 1971, **142**, 382.
4. S. Kitahata and S. Okada, *J. Jpn. Soc. Starch Sci.*, 1982, **29**, 13.
5. S. Okada and K. Mizokami, *J. Jpn. Soc. Starch Sci.*, 1980, **27**, 127.
6. Y. Sakano, J. Fukushima, and T. Kobayashi, *Agric. Biol. Chem.*, 1983, **47**, 2211.
7. H. Melasniemi, M. Paloheimo, and L. Hemiö, *J. Gen. Microbiol.*, 1990, **136**, 447.
8. S. P. Mathupala, S. E. Lowe, S. M. Podkovyrov, and J. G. Zeikus, *J. Biol. Chem.*, 1993, **268**, 6332.
9. S.-P. Lee, M. Morikawa, M. Takagi, and T. Imanaka, *Appl. Environ. Microbiol.*, 1994, **60**, 3764.
10. F. Friedberg, *FEBS Lett.*, 1983, **152**, 139.
11. R. Nakajima, T. Imanaka, and S. Aiba, *Appl. Microbiol. Biotechnol.*, 1986, **23**, 355.

12. F. Binder, O. Huber, and A. Böck, *Gene*, 1986, **47**, 269.
13. A. Amemura, R. Chakraborty, M. Fujita, T. Noumi, and M. Futai, *J. Biol. Chem.*, 1988, **263**, 9271.
14. T. Kuriki, S. Okada, and T. Imanaka, *J. Bacteriol.*, 1988, **170**, 1554.
15. T. Imanaka and T. Kuriki, *J. Bacteriol.*, 1989, **171**, 369.
16. T. Kuriki, H. Takata, S. Okada, and T. Imanaka, *J. Bacteriol.*, 1991, **173**, 6147.
17. Y. Matsuura, M. Kusunoki, W. Harada, and M. Kakudo, *J. Biochem.*, 1984, **95**, 697.
18. T. Kuriki and T. Imanaka, *J. Gen. Microbiol.*, 1989, **135**, 1521.
19. D. Borovsky, E. E. Smith, and W. J. Whelan, *Eur. J. Biochem.*, 1975, **59**, :615.
20. T. Romeo, A. Kumar, and J. Preiss, *Gene.*, 1988, **70**, 363.
21. T. Baba, K. Kimura, K. Mizuno, H. Etoh, Y. Ishida, O. Shida, and Y. Arai, *Biochem. Biophys. Res. Commun.*, 1991, **181**, 87.
22. C. Klein and G. E. Schulz, *J. Mol. Biol.*, 1991, **217**, 737.
23. M. Kubota, Y. Matsuura, S. Sakai, and Y. Katsube, *Denpun Kagaku*, 1991, **38**, 141.
24. Y. Katsuya, Y. Maezaki, M. Kubota, and Y. Matsuura, *J. Mol. Biol.*, 1998, **281**, 885.
25. D. W. Banner, A. C. Bloomer, G. A. Petsko, D. C. Philips, C. I. Pogson, I. A. Wilson, P. H. Corran, A. J. Furth, J. D. Milman, R. E. Offord, J. D. Priddle, and S. G. Waley, *Nature*, 1975, **255**, 609.
26. C. Chothia, *Nature*, 1992, **357**, 543.
27. S. Janacek, B. Svensson, and B. Henrissat, *J. Mol. Evol.*, 1997, **45**, 322.
28. T. Nagashima, S. Toda, K. Kitamoto, K. Gomi, C. Kumagai, and H. Toda, *Biosci. Biotech. Biochem.*, 1992, **56**, 207.
29. K. Takase, T. Matsumoto, H. Mizuno, and K. Yamane, *Biochim. Biophys. Acta.*, 1992, **1120**, 281.
30. M. Vihinen, P. Ollikka, J. Niskanen, and P. Mantsala, *J. Biochem.*, 1990, **107**, 267.
31. M. Søgaad, A. Kadziola, R. Haser, and B. Svensson, *J. Biol. Chem.*, 1993, **268**, 22480.
32. S. M. Podkovyrov, D. Burdette, and J. G. Zeikus, *FEBS Lett.*, 1993, **317**, 259.
33. A. Nakamura, K. Haga, S. Ogawa, K. Kuwano, K. Kimura, and K. Yamane, *FEBS Lett.*, 1992, **296**, 37.
34. H. Takata, T. Takaha, T. Kuriki, S. Okada, M. Takagi, and T. Imanaka, *Appl. Environ. Microbiol.*, 1994, **60**, 3096.
35. T. Kuriki, H. P. Guan, M. Sivak, and J. Preiss, *J. Protein Chem.*, 1996, **15**, 305.
36. H. J. Swift, L. Brady, Z. S. Derewenda, E. J. Dodsen, G. G. Dodsen, J. P. Turkenburg, and A. J. Wilinson, *Acta Crystallogr. Sec. B.*, 1991, **47**, 535.
37. M. Qian, R. Haser, and F. Payan, *J. Mol. Biol.*, 1993, **231**, 785.
38. M. Machius, G. Wiegand, and R. Huber, *J. Mol. Biol.*, 1995, **246**, 545.
39. Y. Morishita, K. Hasegawa, Y. Matsuura, Y. Katsube, M. Kubota, and S. Sakai, *J. Mol. Biol.*, 1997, **267**, 661.
40. B. Strokopytov, D. Penninga, H. J. Rozeboom, K. H. Kalk, L. Dijkhuizen, and B. W. Dijkstra, *Biochemistry*, 1995, **34**, 2234.
41. K. Watanabe, Y. Hata, H. Kizaki, Y, Katsube, and Y. Suzuki, *J. Mol. Biol.*, 1997, **269**, 142.
42. T. Kuriki, *Trends Glycosci. Glycotechnol.*, 1992, **4**, 567.
43. B. Svensson, *Plant Mol. Biol.*, 1994, **25**, 141.
44. B. Henrissat, *Biochem. J.*, 1991, **280**, 309.
45. T. Kuriki and T. Imanaka, *J. Biosci. Bioeng.*, 1999, May issue.

Structure of the Catalytic Domains of Carbohydrate Modifying Enzymes

ACETYL XYLAN ESTERASE II FROM *Penicillium purpurogenum*: STRUCTURE AND PROPERTIES

Jaime Eyzaguirre
Laboratorio de Bioquímica, Pontificia Universidad Católica de Chile, Santiago, Chile
and
Debashis Ghosh
Hauptman-Woodward Medical Research Institute, Buffalo, New York, U.S.A.

1 INTRODUCTION

Xylan is the principal constituent of hardwood hemicelluloses, and one of the most abundant biomolecules on Earth. As shown in Figure 1, it is a heteropolysaccaride composed of a linear chain of xylose residues bound by ß(1->4) linkages; depending on its source, it has a number of different substituents on carbon 2 or 3.[1] Among these substituents, found particularly in xylans from hardwoods and annual plants, are acetate groups linked to the backbone by ester bonds.

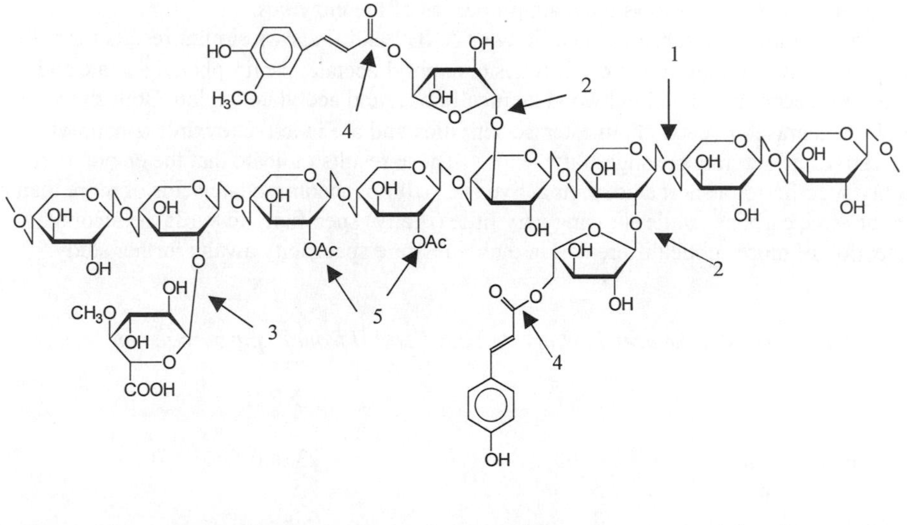

Figure 1: *The structure of xylan and site of action of the xylanases.* 1: endoxylanases; 2: α-L-arabinofuranosidases; 3: glucuronidases; 4: cinnamoyl esterases; 5: acetyl xylan esterases.

The biodegradation of xylans is a complex process with the participation of several different enzymes, collectively known as xylanases. They are produced by fungi and bacteria and are mostly extracellular.[2] Their site of action is summarized in Figure 1. Xylanases are finding increasing use in a number of biotechnological processes particularly the pulp, paper, food, feed and textile industries.[3]

One important group of enzymes within the xylanases are the acetyl xylan esterases (AXE) (E.C. 3.1.1.6.), which are responsible for the hydrolysis of the acetate substituents.[4] Several AXEs have been purified and characterized.[5]

2 PROPERTIES OF THE ACETYL XYLAN ESTERASES FROM *P. purpurogenum*

Our research group has been interested for several years in the study of the xylanolytic system, and we have used as a model organism a locally isolated strain of the filamentous fungus *Penicillium purpurogenum.*[6] This fungus produces a number of different xylanases, among them acetyl xylan esterase.[7] If the fungus is grown in different carbon sources, maximum production of AXE activity is found with chemically acetylated xylan (containing 35 % acetate).[7] No activity is detected when cellulose is used as carbon source; this contrasts with the results obtained with fungi such as *Schizophyllum commune*[8] and *Trichoderma reesei*,[9] where cellulose is a good inducer of AXE activity. Further studies of the expression and regulation of AXE are necessary to understand this discrepancy.

When a supernatant of a culture in acetylated xylan was subjected to gel filtration chromatography in Bio-Gel P 300, two peaks with acetyl esterase (AE) and AXE activities were separated. These two enzymes (called AXE I and II) have been purified to homogeneity.[7] Table 1 shows the main properties of the enzymes.

Initial work on the substrate specificity of AXE I and II shows similar results for both. They are active on the synthetic substrates α-naphtyl acetate, p-nitrophenyl acetate and xylose tetraacetate and on birchwood hemicellulose, and acetylated xylan. Both show no feruloyl esterase or acetyl cholinesterase activities and are inactive towards α-naphtyl derivatives of butyrate or longer fatty acids.[7] These results indicate that the enzymes are highly specific for a short acid; their active site will not accommodate acids of more than two or three carbons, while showing very little (if any) specificity towards the alcohol. Detection of more subtle differences in their substrate specificity awaits further study.

Table 1 *Properties of the acetyl xylan esterases I and II from P. purpurogenum*

	AXE I	AXE II
Molecular weight[a]	48 000	23 000
Isoelectric point[a]	7.5	7.8
pH optimum[a]	5.3	6.0
Temperature optimum[a]	50°	60°
Binding to cellulose	yes[b]	no[c]
Binding to xylan	no[b]	no[c]

[a] Data obtained from [7]. [b] Unpublished. [c] Data obtained from [10].

Additional differences between these two AXEs have been documented.[7] They show no antibody cross-reactivity and their amino terminal sequences show no similarity. These results suggest that they are the products of different genes.

3 SEQUENCE OF ACETYL XYLAN ESTERASE II

By means of the Edman degradation and by nucleotide sequencing of the cDNA and genomic DNA, the complete amino acid sequence of AXE II has been obtained.[10] A signal peptide of 27 residues has been determined, and the mature enzyme consists of 207 residues (Figure 2). The sequence of the mature enzyme does not include any tryptophans, while 10 cys residues are found.

Similarity searches of the databases showed a high level of identity (67%), evenly distributed in the sequence, with the catalytic domain of AXE from *Trichoderma reesei*.[11] It is interesting to note that *T. reesei* has only one AXE, which possesses a cellulose binding domain (CBD). This domain is not present in AXE II. The high similarity of AXE II and the catalytic domain of the *T. reesei* enzyme suggests a common ancestor gene, to which the CBD gene has been added by gene fusion. As indicated above, AXE I does bind to cellulose. Work is in progress to determine the sequence of AXE I; this will confirm the presence of a CBD and possible similarities with the CBD of the *T. reesei* enzyme.

4 THE THREE-DIMENSIONAL STRUCTURE OF AXE II

AXE II has been crystallized. The crystals are highly ordered, tightly packed and diffract at better than 1 Å resolution at 85 °K.[12] Isomorphous derivatives have been prepared by iodination, and the structure has been determined at 1.1 Å resolution.[13] The protein shows the alpha/beta hydrolase fold,[14] similar to that of cutinase,[15] an esterase which hydrolyzes cutin, a polyester component of the waxy layer of plant's cuticle. The secondary structure has a central parallel ß-sheet with α helices on each side. Figure 3 shows the topology of AXE II and, for comparative purposes, that of cutinase. A high similarity can be observed. All cys residues of AXE II are forming five S-S bridges. This contrasts to the cutinase which only possesses 4 cys and two such bridges. Figure 4 shows a comparison of the AXE II and cutinase sequences and the position of the S-S bridges. AXE II has, therefore, a more rigid structure than cutinase; this may account for the stricter substrate specificity of AXE II (which will only accommodate acetate), while cutinase allows for longer fatty acids (up to C-18).

5 THE CATALYTIC MECHANISM OF AXE II

Chemical modification experiments with phenyl methyl sulfonyl fluoride show that AXE II is strongly inactivated by this compound.[16] This reaction is characteristic of serine hydrolases,[17] suggesting that AXE II is a serine esterase. Serine hydrolases possess a characteristic catalytic triad: ser, his asp (or glu).[18] This triad can be located in the three-dimensional structure of AXE II: Ser 90, His 187 and Asp 175 (Figure 5),[13] and it is common to that of the cutinases and of acetyl xylan esterase from *Trichoderma*.[10] The

CGAGATGCTTCACGGAAACATGGAACAAATTTTTGTTTCGAATCATCAGTTTAAGAGAGT 60
TCTTCTATAAATCATTGACTAAGTCTCGCGGTGAAGGGAAGTATTCACGAGCATTATCTC 120
TATCATCCAAGCCTTGATTACCACCATCTATTTTGACCCTTTTCAGTTATCAACGTAACA 180
AGAAAA**ATG**CATTCCAAGTTTTTCGCAGCCTCTCTTCTTGGCCTTGGCGCAGCTGCCATC 240
M H S K F F A A S L L G L G A A A I

CCCCTCGAGGGGGTCATGGAGAAACGCAGCTGTCCTGCGATCCACGTTTTCGGTGCCCGT 300
P L E G V M E K R S C P A I H V F G A R

GAAACCACTGCCTCTCCAGGGTATGGCTCCTCCAGCACCGTCGTCAATGGCGTTCTCAGC 360
E T T A S P G Y G S S S T V V N G V L S

GCTTACCCGGGATCCACCGCCGAGGCAATCAACTACCCAGCTTGCGGTGGACAATCTTCA 420
A Y P G S T A E A I N Y P A C G G Q S S

TGCGGGGGCGCCAGCTATTCCAGCTCAGTCGCTCAGGGTATTGCCGCTGTGGCATCCGCT 480
C G G A S Y S S S V A Q G I A A V A S A

GTGAACTCATTCAATTCTCAGTGCCCGAGCACCAAGATAGTCTTGGTCGGATACTCTCAG 540
V N S F N S Q C P S T K I V L V G Y **S** Q

GGTGGTGAAATCATGGACGTGGCCCTGTGCGGCGGCGGTGATCCCAACCAGGGATACACC 600
G G E I M D V A L C G G G D P N Q G Y T

AACACCGCTGTACAGCTGTCCTCATCGGCTGTCAACATGGTCAAAGCGGCCATTTTCATG 660
N T A V Q L S S S A V N M V K A A I F M

GGCGACCCAATGTTCCGAGCGGGACTTTCGTATGAGGTTGGAACTTGCGCGGCTGGCGGT 720
G D P M F R A G L S Y E V G T C A A G G

TTCGACCAACGCCCGGCTGGCTTCTCTTGCCCCTCGGCTGCTAAGATCAAGTCTTACTGC 780
F D Q R P A G F S C P S A A K I K S Y C

GATGCTTCGGACCCGTACTGCTGCAACGGTAGTAATGCGGCGACTCATCAGGGTTATGGA 840
D A S **D** P Y C C N G S N A A T **H** Q G Y G

TCTGAGTATGGTTCTCAGGCATTGGCTTTCGTCAAGAGCAAGCTGGGT**TAA**AGCATGGGA 900
S E Y G S Q A L A F V K S K L G -

TGATACACCTCTGCTAACCCGATATTCAATACCTGAAAGCTCCAACGTAAAGACTGTTCT 960
CAAGGATTATCTAGTAAATAGCTTAGCAAAGCTTCAACAACATCAAAATTCTCATCAGAA 1020
AAAAAAAAAAAAAAAAAAAA . 1040

Figure 2 *cDNA and amino acid sequence of AXE II*. The initiation and termination codons of the open reading frame are in bold. The signal peptide residues have been underlined. The catalytic triad residues (ser90, asp175 and his187) are in bold.

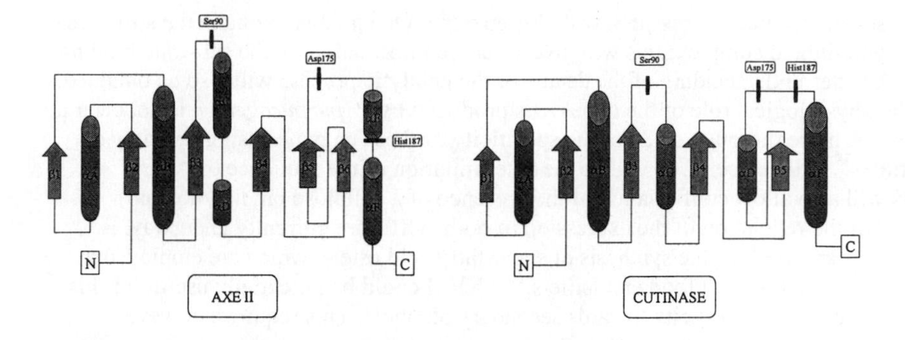

Figure 3 *Topology and connectivity of AXE II and of cutinase from Fusarium solani.* Alpha helices are represented as cylinders and beta strands as arrows. The location of the residues forming the catalytic triad is marked.

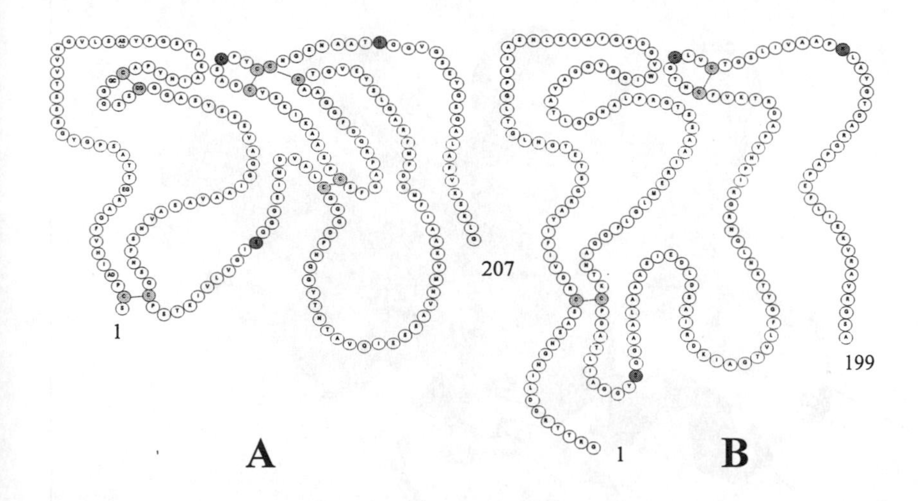

Figure 4 *The sequences of AXE II (A) and of cutinase from F. solani (B) indicating the position of the S-S bridges.* The residues belonging to the catalytic triad are in grey.

importance of Ser 90 for catalysis has recently been confirmed by site-specific mutagenesis of AXE II: the replacement of this serine by alanine leads to the synthesis of a totally inactive protein.[16]

6 PERSPECTIVES

The availability of a high-resolution structure for AXE II and a mutagenesis system opens several avenues of research with this enzyme. One goal is to obtain the structure of an enzyme-ligand complex; this will give valuable information on the substrate binding sites. A better understanding of the details of the catalytic process will also be obtained.

The physiological role of the two AXEs produced by *P. purpurogenum* is not clear at present. A more thorough study of the specificity of these enzymes using a variety of substrates[19] will be useful, as will be the determination of the sequence of AXE I (this will also allow confirmation of the presence of a cellulose binding domain). Studies of the regulation of the expression of both AXEs are currently underway.

Lipases are used for the synthesis of sugar fatty acid esters, which are employed as industrial detergents and food emulsifiers.[20] AXE II could be potentially useful for this purpose, due to its specificity towards secondary alcohols. This requires, however, relaxation of its strict specificity for acetate, so that it can accept longer chain fatty acids. This can be attempted by site-specific mutagenesis of appropriate residues related to substrate binding.

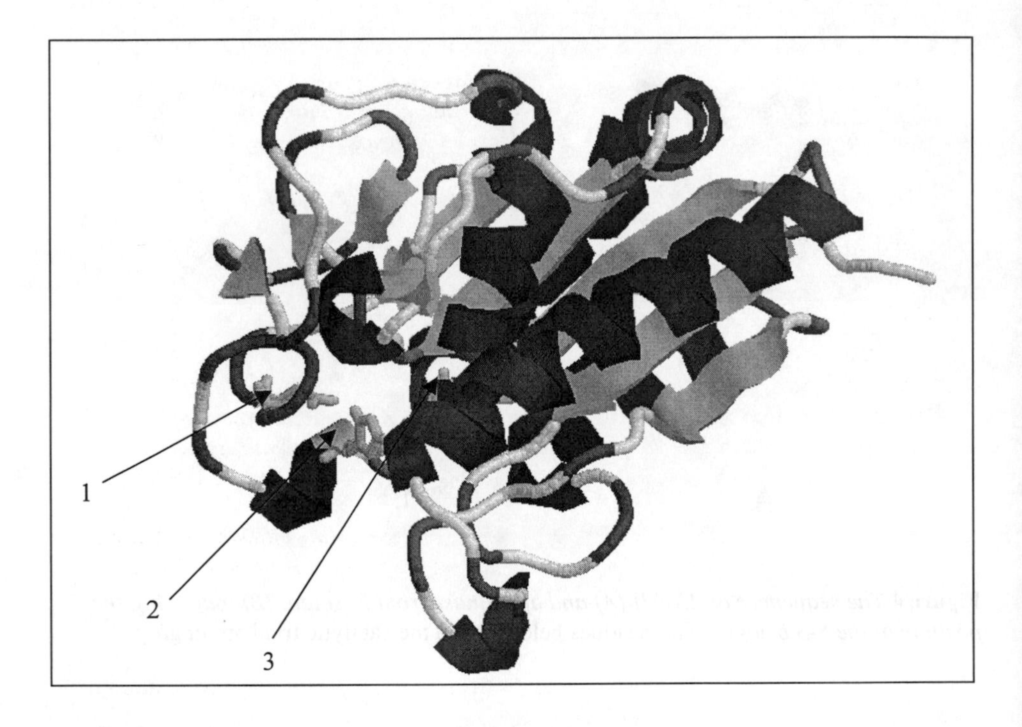

Figure 5 *Three-dimensional structure of AXE II.* The residues forming the catalytic triad are shown: 1, Asp175; 2, His 187; 3, Ser 90. The structure is represented using the computer program RASMOL.[21]

ACKNOWLEDGEMENTS. We wish to acknowledge the participation of the following co-workers in this research: From the Universidad Católica de Chile; A. Peirano, F. Herrera, V. Caputo, R. Gutiérrez, L. Egaña. From the Hauptman-Woodward Medical Research Institute; W. Duax, W. Pangborn, M. Erman, N. Li, B.M. Burkhart, V.Z. Pletnev, and D.J. Thiel. From the Karolinska Institutet, Stockholm, Sweden: E. Cederlund, L. Hjelmquist and H.. Jörnvall. From the University of Chile: J. Steiner. Financial support was obtained from DIPUC, FONDECYT (N° 1930673, 1960241 and IR 7960006), UNIDO (91/065) and the Swedish Medical Research Council (project 3532).

References

1. J.P. Joselau, J. Comptat and K. Ruel, *Progress Biotechnol.,* 1992, **7**, 1.
2 P. Biely, *Trends Biotechnol.*, 1985, **3**, 286.
3. K:K:Y: Wong and J.N. Saddler, in *Hemicellulose and Hemicellulases* (M.P. Coughlan and G.P. Hazlewood, eds), Portland Press, London, 1993, p. 127.
4. G. Williamson, P.A. Kroon and C. B. Faulds, *Microbiology*, 1998, **144**. 2011.
5. L.P. Christov and B.A. Prior, *Enzyme Microb. Technol.*, 1993, **15**, 460.
6. J. Steiner, C. Socha and J. Eyzaguirre, *World J. Microbiol. Biotechnol.*, 1994, **10**, 280.
7. L. Egaña, R. Gutiérrez, V. Caputo, A. Peirano, J. Steiner and J. Eyzaguirre, *Biotechnol. Appl. Biochem.*, 1996, **24**, 33.
8. P. Biely, J. Puls and H. Schneider, *FEBS Lett.,* 1985, **186**, 80.
9. K. Poutanen and M. Sundberg, *Appl. Microbiol. Biotechnol.,* 1988, **28**, 419.
10. R. Gutiérrez, E. Cederlund, L. Hjelmquist, A. Peirano, F. Herrera, D. Ghosh, W. Duax, H. Jörnvall and J. Eyzaguirre, *FEBS Lett.*, 1998, **423**, 35.
11. E. Margolles-Clark, M. Tenkanen, H. Söderlund and M. Pentillä, *Europ. J. Biochem.*, 1996, **237**, 553.
12. W.Pangborn, M. Erman, N. Li, B.M. Burkhart, V.Z. Pletnev, W.L. Duax, R. Gutiérrez, A. Peirano, J. Eyzaguirre, D.J. Thiel and D. Ghosh, *Proteins*, 1996, **24**, 523.
13. D. Ghosh, M. Erman, M. Sawicki, P. Lala, D.R. Weeks, N. Li, W. Pangborn, D.J. Thiel, H. Jörnvall, R. Gutiérrez and J. Eyzaguirre, *Acta Crystall. Ser. D,* 1999, **55**, 779-784.
14. D.L. Ollis, E. Cheah, M. Cygler, B. Dijkstra, F. Frolow , S.M. Franken, M. Harel, S.J. Remington, I. Silman, J. Schrag, J.L. Sussman, K. H.G. Verschueren and A. Goldman, *Protein Eng.*, 1992, **5**, 197.
15. S. Longhi, M. Czjzek, V.Lamzin, A. Nicolas and C. Cambillau, *J. Mol. Biol.,* 1997, **268**, 779.
16. F. Herrera, A. Peirano, D. Ghosh, H. Jörnvall and J. Eyzaguirre, 1999, submitted for publication.
17. J. Kraut, *Annu. Rev. Biochem.,* 1977, **46**, 331.
18. W. Köller and P.E. Kolattukudy, *Biochemistry,* 1982, **21**, 3083.
19. P. Biely, G.L. Côté, L. Kremnicky, R.V. Greene and M. Tenkanen, *FEBS Letters*, 1997, **420**, 121.
20. D.B. Sarney and E.N. Vulfson, *Trends Biotechnol.*, 1995, **13** 164.
21. R.A. Sayle and E.J. Milner-White, *Trends Biochem. Sci.*, 1995, **20**, 374.

THREE-DIMENSIONAL STRUCTURES, CATALYTIC MECHANISMS AND PROTEIN ENGINEERING OF β-GLUCAN HYDROLASES FROM BARLEY

M. Hrmova[1], R. J. Stewart[1], J. N. Varghese[2], P. B. Høj[3], and G. B. Fincher[1]

1 Department of Plant Science, University of Adelaide, Waite Campus, Glen Osmond SA 5064, Australia
2 Biomolecular Research Institute, 343 Royal Parade, Parkville Vic 3052, Australia
3 Australian Wine Research Institute, Hartley Grove, Urrbrae SA 5064, Australia

1 INTRODUCTION

(1→3,1→4)-β-Glucans are cell wall components characteristic of members of the higher plant family Poaceae. These polysaccharides consist of linear chains of up to 1000 or more β-glucosyl residues, linked through (1→3)- and (1→4)-β-glycosidic linkages. In the starchy endosperm of barley, (1→3,1→4)-β-glucans constitute about 70% by weight of cell walls and their structures have been defined in detail. Single (1→3)-β-glucosyl residues are generally separated by two or three (1→4)-β-glucosyl residues; this arrangement accounts for approximately 90% of the polysaccharide. In the remaining 10%, blocks of up to 10 adjacent (1→4)-β-glucosyl residues are detected[1]. Overall, the (1→3)-β-glucosyl residues are irregularly spaced along the polysaccharide. As a result, the extended (1→3,1→4)-β-glucan chains do not aggregate into fibrillar structures, but might participate in gelation reactions in the matrix phase of the cell wall[2].

1.1 Hydrolysis of (1→3,1→4)-β-Glucans in Germinated Grain

In germinated cereal grains, hydrolytic enzymes depolymerise storage proteins, starch and residual nucleic acids in the starchy endosperm, and hydrolysis products are translocated to the embryo, where they support the growth of the young seedling. The (1→3,1→4)-β-glucans are hydrolyzed completely in this process and the glucose finally released represents a significant proportion of the total chemical energy available for seedling growth[3].

The complete depolymerization of (1→3,1→4)-β-glucans to glucose requires a battery of hydrolytic enzymes. Endohydrolases, the most important of which are probably the (1→3,1→4)-β-glucan 4-glucanohydrolases (EC 3.2.1.73), release a series of oligosaccharides following the hydrolysis of (1→4)-β-glucosyl linkages that are immediately adjacent, on the reducing terminal side, to (1→3)-β-glucosyl residues[1]. Thus, the (1→3,1→4)-β-glucan endohydrolases exhibit a strict substrate specificity that requires juxtaposed (1→3)- and (1→4)-β-glucosyl residues (Figure 1).

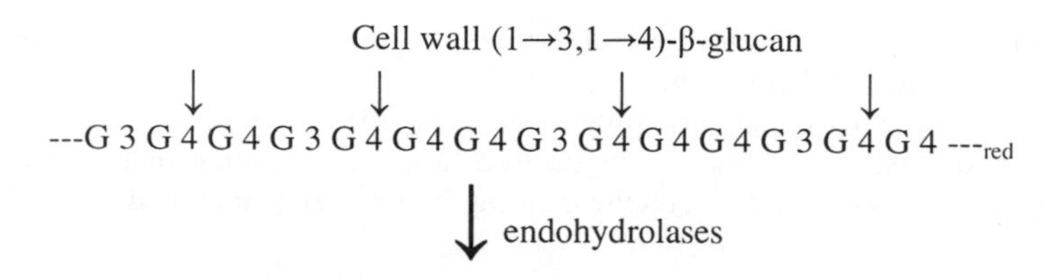

Figure 1 *Substrate specificity of (1→3,1→4)-β-glucan endohydrolases. β-Glucosyl residues are indicated by G, 3 and 4 indicate the linkage positions, and 'red' is the reducing end of the polysaccharide. Small arrows show the glycosidic linkages hydrolyzed.*

Oligosaccharides released by the (1→3,1→4)-β-glucan endohydrolases are further hydrolyzed to glucose, but the enzymes responsible have not been identified unequivocally. The two major candidates are β-glucosidases (EC 3.2.1.21)[4] and a group of broad specificity β-glucan exohydrolases which are difficult to assign to existing Enzyme Commission groups[5]. The enzymic properties and substrate specificities of these enzymes will be described in sections below.

1.2 Hydrolysis of (1→3,1→4)-β-Glucans in Elongating Coleoptiles

Cell wall (1→3,1→4)-β-glucans of barley coleoptiles decrease significantly during coleoptile growth, in a process that has been linked to wall loosening during auxin-induced cell elongation[6]. However, (1→3,1→4)-β-glucan endohydrolases of the EC 3.2.1.73 group are not detected in elongating barley coleoptiles[7], where the β-glucan exohydrolase appears to be the most abundant wall-degrading enzyme[8]. It is not yet clear how an exohydrolase could loosen the matrix of the wall and allow cell expansion, without the participation of an endohydolase.

1.3 Industrial Importance of Barley (1→3,1→4)-β-Glucans

In common with many other matrix phase polysaccharides from cell walls of higher plants, (1→3,1→4)-β-glucans are important determinants of quality in the commercial utilization of cereals or cereal residues for food and fibre, and can adversely affect a range of industrial processes. In the brewing industry, (1→3,1→4)-β-glucans extracted with hot water from malt form aqueous solutions of high viscosity. These can slow filtration rates in the brewery and can contribute to the formation of undesirable hazes in the final product. Thus, levels of both (1→3,1→4)-β-glucans and (1→3,1→4)-β-glucan endohydrolases that are capable of reducing the viscosity of malt extracts represent important malt quality parameters for the brewing industry.

Similarly, (1→3,1→4)-β-glucans can cause difficulties in stockfeeds for monogastric animals, particularly pigs and poultry. The (1→3,1→4)-β-glucans extracted from barley or other cereal components in these stockfeed formulations increase the viscosity of digesta in the alimentary tract and can greatly reduce the efficiency of digestion. This, in turn, reduces growth rates of the animals. In contrast,

(1→3,1→4)-β-glucans are considered desirable in human nutrition, where they are important components of dietary fibre.

In Section 5 of this paper, experiments are described in which a (1→3,1→4)-β-glucan endohydrolase from barley is engineered to overcome some of the difficulties imposed by (1→3,1→4)-β-glucans on the malting, brewing and stockfeed industries.

2 THREE-DIMENSIONAL STRUCTURES AND CATALYTIC MECHANISMS OF (1→3,1→4)-β-GLUCAN ENDOHYDROLASES

There are two (1→3,1→4)-β-glucan endohydrolases (EC 3.2.1.73) in barley and their properties are compared in Table 1. The three-dimensional structure of isoenzyme EII has been determined by X-ray crystallography at 2.2-2.3 Å resolution[9]. The enzyme is a family 17 glycosyl hydrolase[10] that adopts an $(\beta/\alpha)_8$ barrel fold. Catalytic amino acids are located within a deep cleft that extends across the surface of the enzyme; this cleft is likely to include amino acid residues responsible for substrate-binding. The open cleft on the surface of the enzyme is typical of many polysaccharide endohydrolases and presumably allows the essentially random binding and hydrolysis of internal glycosidic linkages that are characteristic of this class of hydrolase.

The catalytic amino acids of the barley (1→3,1→4)-β-glucanase isoenzyme EII have been determined by chemical procedures. The catalytic acid responsible for protonation of the glycosidic oxygen atom during hydrolysis was identified as E288 by carbodiimide-mediated labelling[11]. However, the carbodiimide method has certain limitations and Jenkins and colleagues[12] subsequently suggested that the catalytic acid was more likely to be E93. The catalytic nucleoptile of the barley (1→3,1→4)-β-glucan endohydrolase is probably E232[11], which is about 8 Å from E288[9] and somewhat closer to E93[12].

Table 1 *Properties of Barley (1→3,1→4)-β-Glucan Endohydrolases*

Property	*Isoenzyme I*	*Isoenzyme II*
Apparent molecular weight	30,000	32,000
Isoelectric point	8.5	10.6
Substrate specificity		
(1→3,1→4)-β-glucans	active	active
laminarin, cellulose	no activity	no activity
4-nitrophenyl β-glucoside	no activity	no activity
Anomeric configuration	retaining	retaining
Glycosyl hydrolase family	17	17
Protein fold	$(\beta/\alpha)_8$ barrel	$(\beta/\alpha)_8$ barrel
Probable catalytic acid	E93/E288	E93/E288
Probable catalytic nucleoptile	E232	E232
Substrate-binding subsites	5-8	5-8

3 THREE-DIMENSIONAL STRUCTURES AND CATALYTIC MECHANISMS OF BARLEY β-GLUCOSIDASES

Two β-glucosidases have been purified from extracts of germinated barley grain and characterized [4,13], and designated isoenzymes βI and βII. Their properties are compared in Table 1.

The barley β-glucosidases hydrolyze 4-nitrophenyl glucoside and are therefore referred to as β-glucosidases, but detailed examination of their substrate specificity revealed that their preferred substrates were cellodextrins, and that the rate of hydrolysis increases as the degree of polymerization of the cellodextrin increases from 3-6[5]. The enzymes remove single glucose units from the non-reducing termini of these substrates, and anomeric configuration of released glucose is retained[5]. Their substrate specificities and action patterns are more typical of polysaccharide exohydrolases of the (1→4)-β-glucan glucohydrolases group (EC 3.2.1.74), than of β-glucosidases of the EC 3.2.1.21 group[5]. This conclusion was supported by the observation that the isoenzyme βII has six glucosyl-binding subsites[13].

When the amino acid sequence of the barley enzyme was 'fitted' to the three-dimensional structure of a cyanogenic β-glucosidase from white clover[14], the presence of a deep shaft or pocket was revealed[13]. The probable catalytic amino acid residues E181 and E391 were located close to the bottom of the pocket, together with other amino acid residues that are highly conserved in the so-called 4/7 superfamily of $(\beta/\alpha)_8$ glycosyl hydrolases[12]. The pocket would clearly accommodate not only the relatively straight ((1→4)-β-oligoglucosides of the cellodextrin series, but are also the oligomeric products of (1→3,1→4)-β-oligoglucosides with a single (1→3)-β-linkage at their reducing ends (Figure 1). These properties and the presence of the enzymes in the starchy endosperm suggest that they are prime candidates for the participation in the conversion of (1→3,1→4)-β-glucans to glucose in germinated barley grain.

Table 2 *Properties of Barley β-Glucosidases*

Property	*Isoenzyme βI*	*Isoenzyme βII*
Apparent molecular weight	62,000	62,000
Isoelectric point	8.9	9.0
Substrate specificity		
4-nitrophenyl β-glucoside	active	active
laminarin, (1→3,1→4)-β-glucans	no activity	no activity
(1→4)-β-oligoglucosides	active	active
Anomeric configuration	retaining	retaining
Glycosyl hydrolase family	family 1	family 1
Protein fold	$(\beta/\alpha)_8$ barrel	$(\beta/\alpha)_8$ barrel
Probable catalytic acid	E181	E181
Probable catalytic nucleoptile	E391	E391
Subsite-binding subsites	6	6

4 THREE-DIMENSIONAL STRUCTURES AND CATALYTIC MECHANISMS OF BARLEY β-GLUCAN EXOHYDROLASES

The β-glucan exohydrolases from barley, which hydrolyze the non-reducing terminal glycosidic linkage of β-glucans, are difficult to classify because of their relatively broad substrate specificities. They hydrolyze (1→2)-, (1→3)-, (1→4)- and (1→6)-β-glucosyl linkages in a range of oligo- and polysaccharides and release the non-reducing terminal glucose unit[15]. The relative rate of hydrolysis is highest for the polymeric (1→3)-β-glucan, laminarin, from *Laminaria digitata*. However, many other substrates are also hydrolyzed rapidly[15]. The properties of two β-glucan exohydrolases from barley are compared in Table 3, but it is likely that up to 9 isoforms exist (A. J. Harvey and G. B. Fincher, unpublished data).

From a structural viewpoint, the barley β-glucan exohydrolases are members of the family 3 group of glycosyl hydrolases[10]. The crystal structure of isoenzyme ExoI has recently been solved by X-ray crystallography[16]. The enzyme is a two-domain globular protein of 605 amino acid residues. The first 357 residues make up a $(\beta/\alpha)_8$ barrel, which is connected to the second domain by a 16-residue linker, or hinge. The second domain consists of residues 374-559 arranged in a six-stranded β sheet, with three α-helices on either side of the sheet. A long antiparallel loop is found in the COOH-terminal region of the enzyme[16].

The active site of the enzyme is located in a shallow pocket at the interface of the two domains. The entrance to the pocket is bound by tryptophan residues, and the pocket resembles a 'coin slot' in appearance. A tightly-bound glucose molecule in the pocket is believed to be the product of the reaction that has not diffused away from the enzyme surface after catalysis. The glucose remained in the active site pocket throughout enzyme purification and crystallization, and its anomeric C1 oxygen atom is appropriate located between the catalytic acid (E491) and the catalytic nucleophile (D285)[16].

Table 3 *Properties of Barley β-Glucan Exohydrolases*

Property	*Isoenzyme ExoI*	*Isoenzyme ExoII*
Apparent molecular weight	69,000	71,000
Isoelectric point	7.8	8.0
Substrate specificity		
(1→3)-, (1→3,1→4)-, (1→3,1→6)-β-glucans	active	active
(1→2)-, (1→3)-, (1→4)-, (1→6)- β-diglucosides	active	active
aryl β-glucosides	active	active
Anomeric configuration	retaining	retaining
Glycosyl hydrolase family	family 3	family 3
Protein fold (2 domains)	$(\beta/\alpha)_8$ barrel and $(\alpha/\beta)_6$ sandwich	$(\beta/\alpha)_8$ barrel and $(\alpha/\beta)_6$ sandwich
Probable catalytic acid	E491	E491
Probable catalytic nucleophile	D285	D284
Subsite-binding subsites	2-3	2-3

The retention of the glucose in the active site pocket allowed a detailed analysis of amino acid residues involved in binding of the non-reducing residue of the substrate[16]. These amino acid residues are located on both domains of the enzyme. Furthermore, the two catalytic amino acid residues D285 and E491 also arise from different domains of the enzyme. These arrangements raise the possibility that both the structure of the active site and substrate- or product-binding could change during hydrolysis. Thus, the 16 residue loop (residues 358-373) that connects the two domains could act as a hinge which might allow the domains to separate or come together during the reaction.

The catalytic pocket is only about 13 Å deep. Subsite mapping with both laminaridextrins and cellodextrins showed that the enzyme has 2, or possibly 3, glucosyl-binding subsites (M. Hrmova and G. B. Fincher, unpublished data). These observations might explain the relatively broad specificity of the barley β-glucan exohydrolase. If the catalytic pocket is deep enough to bind only two glucosyl residues of the substrate's non-reducing terminus and if the remainder of the oligomeric or polymeric substrate protrudes away from the enzyme's surface, substrate binding could be relatively independent of substrate shape. This, in turn, would mean that enzyme substrate binding would be relatively independent of linkage type in the β-glucan substrate, and thus would account for the broad substrate specificity.

Finally, detailed kinetic analyses of the barley β-glucan exohydrolase indicated that there might be two binding sites for (1→3,1→4)-β-glucans on the enzyme. Assuming that one binding site is near the catalytic site, the most likely region for the second (1→3,1→4)-β-glucan binding site involves residues 417-440 and 484-504 of the second domain[16].

5 ENGINEERING INCREASED THERMOSTABILITY INTO THE BARLEY (1→3,1→4)-β-GLUCAN ENDOHYDROLASE

As mentioned in Section 1, barley (1→3,1→4)-β-glucan endohydrolases are key enzymes in the malting and brewing industries, where they are important for the removal of high molecular weight, wall (1→3,1→4)-β-glucans in malt extracts. Both (1→3,1→4)-β-glucanase isoenzymes are unstable at the temperatures used during kilning (up to 85°C) or during mashing (usually 65°C). Enzyme levels in the germinated grain might be high enough for the normal physiological requirements of starchy endosperm mobilization, but if the enzyme is irreversibly denatured during kilning or mashing, there may not be sufficient activity remaining to hydrolyse troublesome (1→3,1→4)-β-glucans in the brewery.

There are several approaches which might be taken in addressing the loss of (1→3,1→4)-β-glucanase activity at elevated temperatures. Firstly, heat stable enzymes from microbial sources could be added to the mash or expressed directly from transgenic barley into which the corresponding microbial gene had been inserted[17]. Secondly, random mutagenesis of the isolated barley gene, followed by selection for heat stable mutants, could be attempted. This approach has been successfully applied for the generation of more thermostable forms of barley β-amylase[18]. Thirdly, thermostable mutants might be generated through the design, based on three-dimensional structural data, of a more stable enzyme and the implementation of that design by specific, site-directed mutagenesis. Here, the third approach has been adopted for the design and production of a more heat stable barley (1→3,1→4)-β-glucanase enzyme.

Amino acid substitutions with the potential to enhance heat stability of the barley (1→3,1→4)-β-glucanase isoenzyme EII were identified by reference to the three-dimensional structure of the enzyme[9], to thermal mobility data generated during X-ray crystallography, to previously published work on thermostability of hydrolytic enzymes[19], and by comparison with a pathogenesis-related (1→3)-β-glucanase from barley which has an almost identical three-dimensional fold as the (1→3,1→4)-β-glucanase[9] but which is significantly more heat-stable[1]. Selected substitutions were distributed along the length of the polypeptide chain, although care was taken to avoid changes in the catalytic regions. Ion pairs were created, together with new hydrogen bonds and the incorporation of amino acids with enlarged side chains to 'fill up' hydrophobic spaces within the enzyme. Surface lysine residues were removed and residues with the potential to reduce the entropy of unfolding were introduced. Site-directed mutagenesis was effected using a near full-length cDNA and mutations were checked by nucleotide sequence analysis. The mutant cDNAs were inserted in the pET vector, oriented so that the polyhistidine tag was added to the COOH-terminal of the expressed enzyme. Following expression in *E. coli*, mutant enzymes were purified in one step by affinity chromatography on NT-NTA columns (Qiagen). Heat stabilities of mutant enzymes were compared with expressed 'wildtype' enzyme either by measuring residual activity of purified enzymes after heating for 15 minutes at temperatures from 35°C to 60°C, or by monitoring over time activity added to simulated malt mashes at 55°C or 65°C.

Enhanced thermostability was only observed when amino acid substitutions were effected in the COOH-terminal region the enzyme, which is characterized by an unstructured loop of approximately 10 amino acid residues[9]. When multiple mutations were introduced into this COOH-terminal loop, no additive effects were observed (R. J. Stewart and G. B. Fincher, unpublished data). Mutants exhibited a maximum increase of thermostability of approximately 4°C compared with unmodified, expressed enzyme, and retained activity up to 6x longer than the wildtype enzyme in activity decay assays at 55°C or 65°C. Amino acid substitutions in other regions of the enzyme decreased heat stability slightly or had no effect.

6 SUMMARY AND CONCLUSIONS

The three-dimensional structures of the endo-and exohydrylases from barley that combine to completely depolymerize cell wall (1→3,1→4)-β-glucans to glucose have been defined. The endohydrolases of the EC 3.2.1.73 group bind polymeric substrates in an open cleft that extends across one surface of the enzyme[9]. The exohydrolases bind their substrates in pockets which extend into the core of the enzyme. In the case of barley β-glucosidases this pocket is relatively deep and straight, and can be reconciled with the enzyme's apparent preference for (1→4)-β-oligoglucosides of the cellodextrin series. The substrate appears to be bound with its non-reducing terminus wedged in the base of the pocket; catalytic amino acids are positioned near the bottom of the pocket, such that single glucose units are hydrolysed from the non-reducing terminus of the substrate[13].

The barley β-glucan exohydrolases also bind their substrates in a pocket, but in this case the pocket is not deep and probably accommodates only 2 glucosyl residues[16]. The shallow nature of the pocket might account for the relatively broad specificity of this class of enzyme. Again, catalytic residues are positioned such that single glucose molecules are removed from the non-reducing terminus of substrates. It appears that the glucose unit that is cleaved from the substrate remains tightly bound to the enzyme. How this glucose residue is displaced by the incoming substrate is now known, but spatial movement of the two domains of the enzyme might be required[16]. Furthermore, the possible role of the second domain of the enzyme in binding wall (1→3,1→4)-β-glucan has not yet been defined.

The detailed three-dimensional structural data of the barley (1→3,1→4)-β-glucanase have been applied to enhance the performance of the enzyme in the malting and brewing processes. Thus, rational protein design has enabled the synthesis of mutant forms of the enzyme, in which single amino acid substitutions have significantly increased heat stability, both *in vitro* and in simulated industrial processes.

ACKNOWLEDGEMENTS

This work has been supported by grants from the Australian Research Council (to G. B. Fincher and P. B. Høj) and the Grains Research and Development Corporation of Australia (to G. B. Fincher). We gratefully acknowledge the ongoing support of Professor P. M. Colman and the earlier contributions of Drs T. P. J. Garrett and L. Chen.

REFERENCES

1. P. B. Høj and G. B. Fincher, *Plant J.*, 1995, **7**, 367.
2. N. C. Carpita and D. M. Gibeaut, *Plant J.*, 1993, **3**, 1.
3. P. Morrall and D. E. Briggs, *Phytochemistry*, 1978, **17**, 1495.
4. R. Leah, J. Kigel, I. Svendsen and J. Mundy, *J. Biol. Chem.*, 1995, **270**, 15789.
5. M. Hrmova, A. J. Harvey, J. Wang, N. J. Shirley, G. P. Jones, B. A. Stone, P. B. Høj and G. B. Fincher, *J. Biol. Chem.*, 1996, **271**, 5277.
6. N. Sakurai and Y. Masuda, *Plant Cell Physiol.*, 1978, **19**, 1225.
7. N. Slakeski and G. B. Fincher, *Plant Physiol.*, 1972, **99**, 1226.
8. T. Kotake, N. Nakagawa, K. Takeda and N. Sakurai, *Plant Cell Physiol.*, 1997, **38**, 194.
9. J. N. Varghese, T. P. J. Garrett, P. M. Colman, L. Chen, P. B. Høj and G. B. Fincher, *Proc. Natl. Acad. Sci. USA*, 1994, **91**, 2785.
10. B. Henrissat, *Biochem. Soc. Trans.*, 1998, **26**, 153.
11. L. Chen, G.B. Fincher and P.B. Høj, *J. Biol. Chem.*, 1993, **263**, 13318.
12. J. Jenkins, L. Leggio, G. Harris and R. Pickersgill, *FEBS Lett.*, 1995, **362**, 281.
13. M. Hrmova, E. A. MacGregor, P. Biely, R. J. Stewart and G. B. Fincher, *J. Biol. Chem.*, 1998, **273**, 11134.
14. T. Barrett, C. G. Suresh, S.P. Tolley, E. T. Dodson and M. A. Hughes, *Structure*, 1995, **3**, 951.
15. M. Hrmova and G. B. Fincher, *Carbohydr. Res.*,1998, **305**, 209.
16. J. N. Varghese, M. Hrmova and G. B. Fincher, *Structure*, 1999, **7**, 179.
17. L. G. Jensen, O. Olsen, O. Kops, N. Wolf, K. K. Thomsen and D. Vonwettstein, *Proc. Natl. Acad. Sci. USA*, 1994, **93**, 3487.
18. Y. Okada, N. Yoshigi, H. Sahara and S. Koshino, *Biosci. Biotech. Biochem.*, 1995, **59**, 1152.
19. B.W. Matthews, *FASEB*, 1996, **10**, 35.

STRUCTURE OF A NUCLEOTIDE-DIPHOSPHO-SUGAR TRANSFERASE: IMPLICATIONS FOR THE SYNTHESIS OF POLYSACCHARIDES

Gideon J. Davies and Simon J. Charnock

Structural Biology Laboratory
Department of Chemistry
University of York
York, YO10 5DD
U.K.

1 INTRODUCTION

1.1 The enzymatic synthesis of glycosidic bonds

The synthesis of the glycosidic-bond in carbohydrates is, in terms of quantity, the most significant enzyme-catalysed reaction on Earth. Glycosyltransferases come in a variety of guises. Many utilise nucleotide-diphospho (NDP) sugars as the glycosyl donor whilst others use lipid-pyrophosphate sugars or sugar pyrophosphates. All of these enzymes catalyse the transfer of a sugar group from the (activated) sugar donor to the acceptor. Typically the acceptor is a second sugar which leads to the formation of di-, oligo-, or poly-saccharides. Both inverting and retaining glycosyltransferases are known. It is widely believed that the transfer reaction involves a general base, to activate the acceptor species, and that divalent metal ions (such as Mg^{2+} or Mn^{2+}) assist departure of the nucleotide-diphosphate leaving-group. A putative mechanism for an inverting transferase is given in **Scheme 1**.

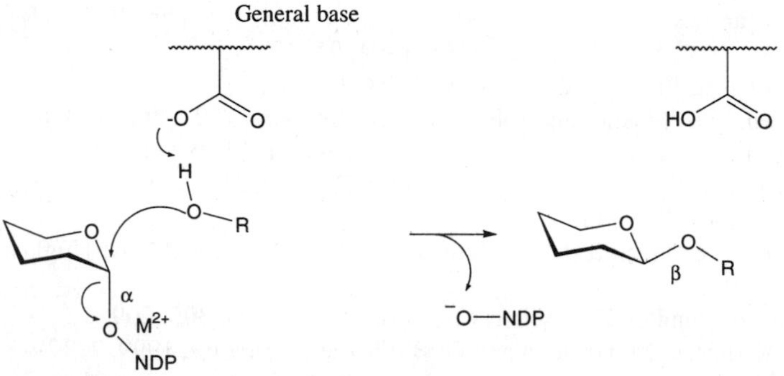

Scheme 1 *Enzymatic glycosyl transfer utilising an activated sugar donor. The reaction shown uses an α-linked NDP sugar to form a β-linked product, hence, in this case the reaction is with inversion of stereochemistry.*

Glycosidic-bond synthesis in aqueous solution needs to overcome the thermodynamic problems associated with 55M water. This limits the use of glycoside hydrolases acting "in reverse", although some enzymes found in glycoside hydrolase families, such as the cyclodextrin glycosyltransferase in "α-amylase" family 13, can still prove effective in the synthesis of glycosidic bonds, **Scheme 2**. "True" glycosyltransferases utilise the activated sugar donor in order to overcome these problems.

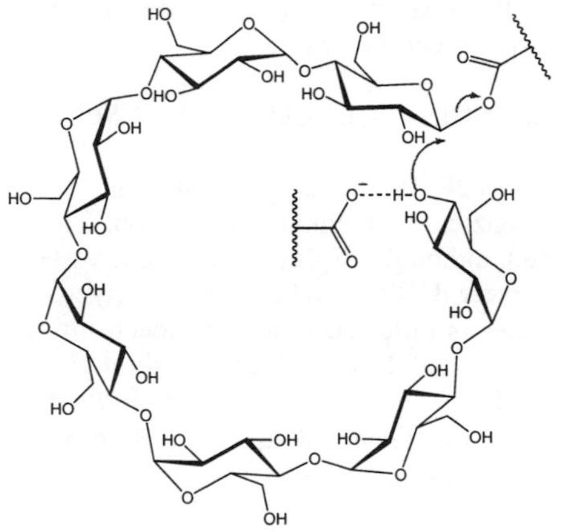

Scheme 2 *The enzymatic formation of cyclodextrins by CGTases shown above, is an example of enzymatic glycosidic-bond synthesis by a member of a glycoside "hydrolase" family.*

The enzymatic synthesis of glycosidic-bonds is central to many key processes. The production of "biomass" such as starch, glycogen, cellulose and chitin, is catalysed by NDP-dependent glycosyltransferases, as is the formation of lipopolysaccharides, cell-surface antigens, the glycosylation of proteins and numerous other processes. The enzymes catalysing such reactions have attracted investigation both as drug targets and for their potential applications in the chemoenzymatic synthesis of oligosaccharides[1, 2].

1.2 Sequences and families for NDP-dependent glycosyltransferases

Recent advances in molecular biology, notably the wealth of sequences generated by genome-sequencing projects, have resulted in an enormous number of reading-frames encoding glycosyltransferases. Unfortunately, this treasure-trove of sequences is not paralleled by knowledge of the structures, mechanisms or even the substrates for these proteins. Sugar donor and acceptor species have been defined for very few enzymes indeed. At the structural level, three-dimensional structures have been reported for just two NDP-dependent sugar transferases: the DNA-modifying β-glucosyltransferase from bacteriophage T4[3] and the sporulation-specific sugar transferase SpsA from *Bacillus subtilis*[4] described here.

This situation is in marked contrast to work on the glycoside hydrolases. The majority of glycoside hydrolase sequences encode enzymes of known, or predictable, function. Of the 71 sequence families, three-dimensional structural representatives are known for approximately 30 of these. Since the discovery of "clans" of related enzyme families[5-7], the structures for many of the remaining sequence families may be predicted with a high degree of confidence.

Despite the paucity of knowledge for glycosyltransferases, however, the derived amino-acid sequences for the NDP-dependent enzymes have been successfully classified on the basis of amino-acid sequence similarities. Currently, 31 families have been identified[8] and are available on the www at URL: http://afmb.cnrs-mrs.fr/~pedro/CAZY/db.html. Early indications suggest, as was observed for the glycoside hydrolase families, that enzymes within a given family will share a common catalytic mechanism (although see section 3.3).

2 FAMILY 2 NDP-DEPENDENT GLYCOSYLTRANSFERASES

One of the largest families of NDP-dependent glycosyltransferases is family 2, but the bulk of its members encode enzymes of unknown function. The majority of enzymes which have been characterised, although displaying a wide diversity of sugar substrates, appear to use almost exclusively UDP or *d*TDP as the glycosyl-donor species. The diverse array of sugar substrates is reflected in a wide spectrum of biological functions ranging from the synthesis of cellulose and chitin through to the intricacies of lipopolysaccharides and bacterial cell-surface *O*-antigens, **Figure 1**. Family 2 members are inverting enzymes which are expected to use nucleotide-diphospho-α-D-sugars to generate β-linked products, as shown in **Scheme 1**.

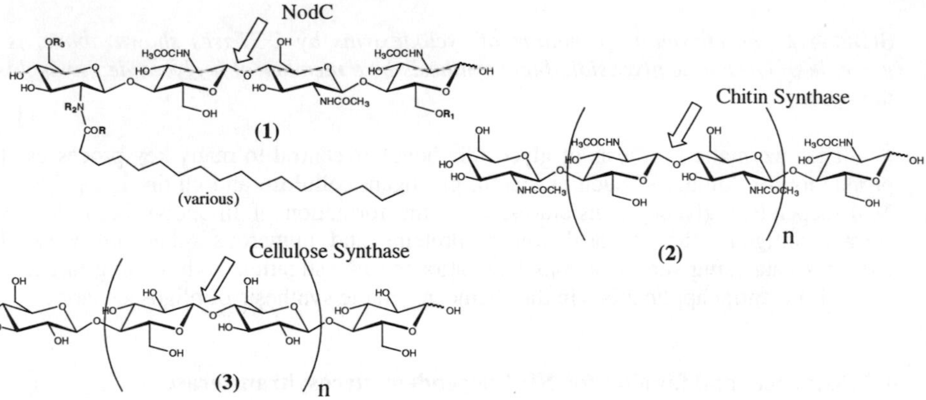

Figure 1 *Family 2 contains a variety of different glycosyltransferases, such as those involved in the formation of lipooligosaccharide root-nodulation factors (1), chitin (2), cellulose (3) and bacterial O-antigens (various, not shown).*

2.1 SpsA from *Bacillus subtilis*

Whilst many of family 2 enzymes are known to be membrane-bound, SpsA from *B. subtilis*, may be expressed as a soluble protein in *E. coli*[9]. It therefore makes an ideal "precursor" system for the analysis of complex structure-function relationships in glycosyltransferase family 2.

The *spsA* gene encodes a glycosyltransferase implicated in the production of the mature spore-coat during the spore response of *B. subtilis*[10, 11]. It encodes a protein of 256 amino-acids whose substrate specificity is undefined but whose sequence has been classified into family 2. Expression of *spsA* in sporulating Bacilli is under control of a σ^K-dependent promoter activated by the GerE protein[10]. SpsA may be involved in any aspect of the glycosylation of the spore-coat and sequence comparison gives little indication of its specificity with an extremely diverse array of activities associated with the closest homologues. Sequence similarity is found with enzymes involved in teichoic and colanic acid synthesis, galactose and rhamnose transfer, lipopolysaccharide production and the synthesis of cellulose and chitin.

2.2 3-D Structure of SpsA

The structure of SpsA was solved by X-ray crystallography at a resolution of 1.5Å[4]. It has an overall size of approximately 45 x 40 x 36Å. The 256 amino-acids of SpsA fold into two domains. The first (N-terminal) domain, residues 2-100, forms a typical nucleotide-binding domain of 4 central parallel β-strands flanked on both sides by two α-helices, **Figure 2**. The remainder of the structure, residues 101-256, is composed of a mixed β-sheet flanked by three helices on one side and one on the other. The C-terminal domain presents an open groove which may be the binding-site for the acceptor. A disordered loop, residues 218-231 spans the open groove and could provide the means to prevent abortive hydrolysis of the nucleotide-diphospho donor, as has been observed on mechanistically-related enzymes such as the phosphoribosyltransferases[12]. The native enzyme structure, in the presence of 200mM $MgCl_2$ also revealed a hexacoordinate magnesium ion, in the active-site, coordinated *via* five solvent water molecules and a cryoprotectant glycerol molecule, **Figure 3**.

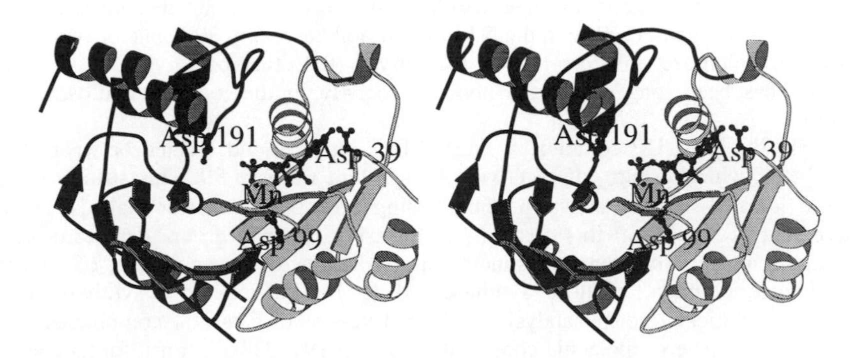

Figure 2 *Structure of the family 2 glycosyltransferase, SpsA, from B. subtilis. The structure is in divergent stereo with the nucleotide-binding domain (residues 2-100) shaded in light grey and the C-terminal domain in dark grey. This figure was prepared with the MOLSCRIPT program*[13].

To date, the only other available structure for an NDP-sugar transferase is the DNA-modifying β-glucosyltransferase from bacteriophage T4[3]. The nucleotide-binding domain of SpsA bares topological similarity with the T4 enzyme but this similarity does not extend to the mode of nucleotide-binding, the conformation of the nucleotide or the presentation of potential catalytic residues in the vicinity of the distal phosphate.

2.3 Nucleotide-binding and sequence conservation

In addition to the native enzyme, SpsA was also studied in the presence of both Mg-UDP and Mn-UDP. In both the Mg-UDP and Mn-UDP complexes the UDP binds to the nucleotide-binding domain in a deep cleft. The ribose adopts a C2'-*endo* conformation with the base equatorial to the ring, markedly different from the pseudo-axial orientation for the base in the bacteriophage T4 β-glucosyltransferase. The marked difference between the Mn-UDP and Mg-UDP complexes is at the distal phosphate where Mn^{2+} appears to be the correct metal for phosphate coordination, **Figure 3.**

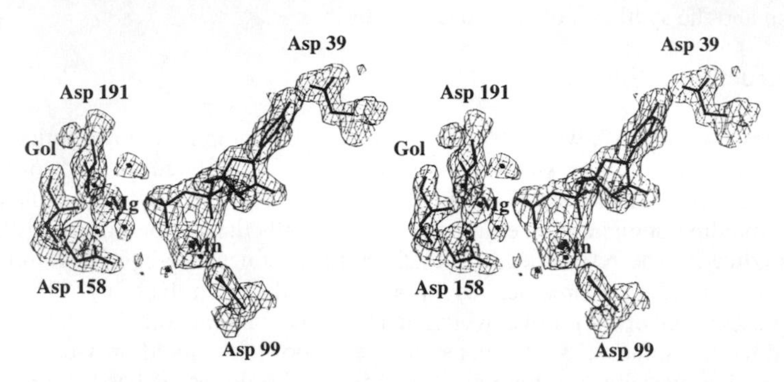

Figure 3. *Observed electron density for Mn-UDP binding to SpsA. Residues mentioned in the text are indicated as is a cryoprotectant glycerol molecule (labelled 'Gol'), and a second metal which is a magnesium ion coordinated to the terminal phosphate.*

The uracil moiety binds both through aromatic stacking with Tyr 11 and through a number of hydrogen-bonding interactions. Asp 39, an invariant residue at the end of strand β-2, hydrogen-bonds with the N3 of the uracil base. Asp 39 is one of two invariant aspartates which are signature motifs in family 2 (the other being Asp 99) and whose function has been long speculated upon, especially in the related cellulose synthase system[14].

In the Mn-UDP complex, a single Mn^{2+} ion is found located between the two phosphates, which in turn interacts with the protein *via* Asp 99. This strictly invariant residue lies on β-strand 4, as was predicted using hydrophobic cluster analysis[15]. It seems likely that the role of this residue is to assist leaving-group departure through coordination of the divalent manganese ion. Mutation of the equivalent of Asp 99 in *Acetobacter xylinium* cellulose synthase led to loss of cellulose synthase activity, implying a critical role in catalysis[14]. The active-site of SpsA is complicated by the presence of a glycerol molecule coordinated to Asp 191. This, in turn, forms one of the axial ligands of a second metal, a hexacoordinate Mg^{2+} ion (equivalent to that seen in the native structure), which interacts with the distal phosphate of UDP. Recent work on a mechanistically similar enzyme, the hypoxanthine phosphoribosyltransferase (HPRT) from *Trypanosoma cruzi*, surprisingly implicated a two metal-ion mechanism for glycosyl transfer with both Mn^{2+} and Mg^{2+} coordinated to the distal phosphate of the leaving-group[12]. In the HPRT case the two metals are believed to be catalytic, but it is unclear whether SpsA, or other family 2 glycosyltransferases of known specificity, utilise a similar two metal mechanism.

The Mn-UDP complex of SpsA may be interpreted in light of the known sequence conservation within family 2. The vast majority of sequence conservation in family 2 is in the N-terminal nucleotide-binding domain, residues 2-100. Three clusters of invariance are of interest, **Figure 4**. These involve residues in direct contact with the NDP donor. Tyr 11 is involved in stacking with the nucleotide base, Asp 39 in UDP binding and Asp 99 sits adjacent to the distal phosphate where it coordinates the leaving-group Mn^{2+} ion.

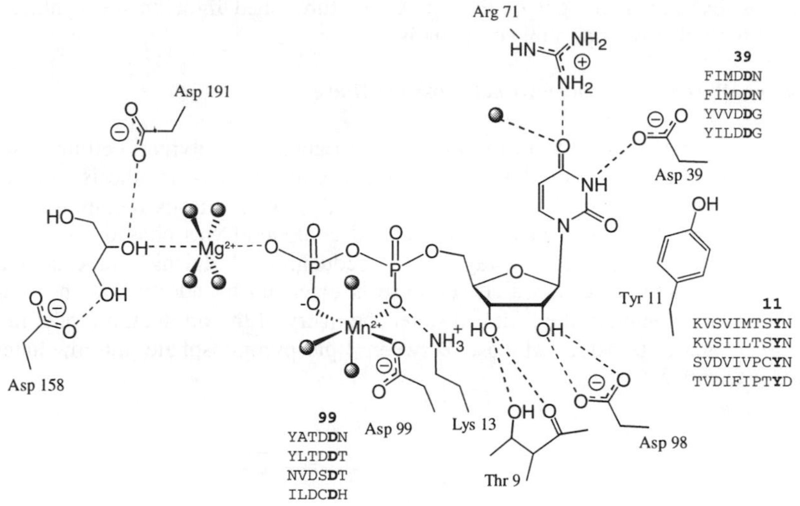

Figure 4 *Schematic representation of the interactions of SpsA with UDP-Mn. Sequence conservation around Tyr 11, Asp 39 and Asp 99 is shown. The four sequences given are SpsA from B. subtilis, CgeD from B. subtilis, NodC from Rhizobium sp. and cellulose synthase from Acetobacter xylinus.*

A candidate for a general base is more difficult to identify. The structure reveals 4 potential bases: the "triad" of Asp 158, His 159 and Cys 160 and Asp 191. None of the residues of the Asp, His, Cys triad appear to be conserved, although sequence alignments in this region are difficult to interpret with any great certainty. Geometrically Asp 191 is in an ideal location and may interact with the glycerol molecule in a way analogous to its coordination of the natural sugar acceptor, as is frequently the case with glycoside hydrolases (for example see ref. 16). Additionally, iterative sequence searches with ψ-BLAST[17] suggest that this residue may be conserved throughout family 2, but again sequence alignments may not be interpreted with great confidence. Further structures of distantly related members of family 2 of the glycosyltransferases are required to permit further identification of residues invariant in the 3-D structure.

NDP-sugar transferases are active in aqueous solution yet they avoid abortive transfer of water to the NDP sugar. The phosphoribosyltransferases which utilise a pyrophosphate-activated donor face the same problem of transfer to water and utilise a mobile loop which completes the active-site only upon formation of the ternary complex[12]. It is tempting to speculate that the disordered loop of SpsA plays a similar role.

3 IMPLICATIONS FOR THE SYNTHESIS OF POLYSACCHARIDES

SpsA shows sequence similarity with other family 2 members, as well as low, but significant, similarities with enzymes from other glycosyltransferase families. These similarities, examined in light of the SpsA structure, shed light on the synthesis of a variety of biologically important compounds.

3.1 The synthesis of cellulose by cellulose synthase

The enzymatic synthesis of the world's most abundant polymer, cellulose, remains a controversial area. Two schools of thought exist for its synthesis. Pulse-chase experiments, using (α-linked) UDP-[^{14}C]Glc together with various membrane fractions from *A. xylinum*, demonstrated that the labelled compound was chased into the water-soluble fraction of the polymer, away from its reducing-end, during synthesis. This was interpreted with a scheme in which cellulose is extended by addition at the *reducing-end*[18]. In order to maintain the β-linked stereochemistry of the product, a model in which cellulose synthase transfers glucose between lipid-pyrophosphate intermediates was invoked, **Scheme 3**.

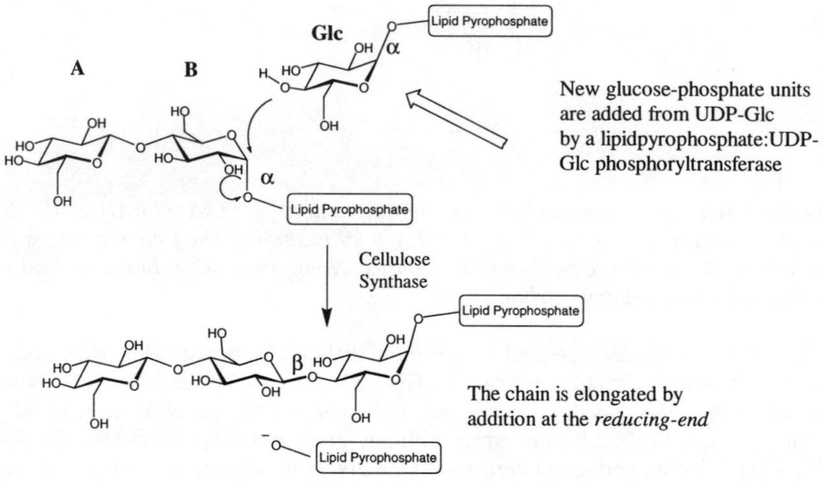

Scheme 3 *Cellulose synthesis according to Han and Robyt*[18] *– cellulose synthase traffics glucose between lipid-phosphate intermediates with elongation at the reducing-end of the growing polymer.*

Conversely, electron crystallographic analysis of cellulose crystals which were both eroded at their non-reducing end by CBH II and Ag-stained at their reducing-end, demonstrated the orientation of the chains ("parallel-up") with respect to the long axis of the crystal[19]. Following this, the direction of cellulose synthesis could be demonstrated by microdiffraction and tilt experiments alone. These experiments indicate a model in which cellulose is extended at the *non-reducing end*, **Scheme 4**. Similar experiments also suggest a non-reducing end elongation for chitin synthesis[20].

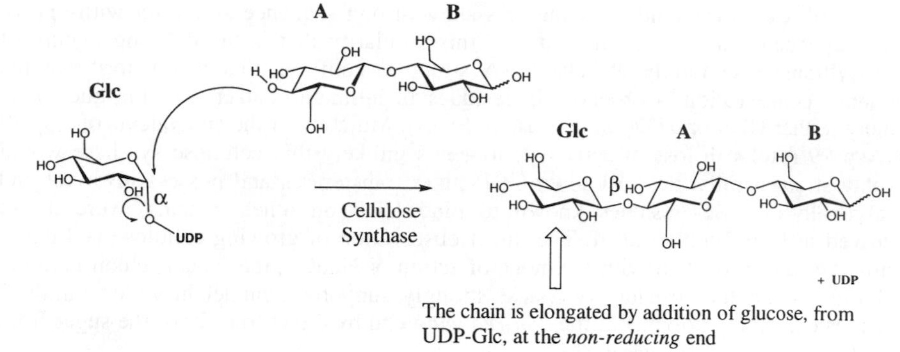

Scheme 4 *Cellulose synthesis according to Koyama and colleagues*[19] *- cellulose is formed by the addition of glucose at the non-reducing end of the polymer by direct transfer from UDP-Glc. A similar model has been proposed for the synthesis of β-chitin*[20].

Recently, elegant work using *p*NPGlcNAc as an artificial glycosyl acceptor species has conclusively demonstrated that the family 2 nodulation factor, NodC, elongates the polymer at the non-reducing end (equivalent to **Scheme 4**) by direct transfer to O-4 of the acceptor[21], **Scheme 5**. Lipid intermediates are not used.

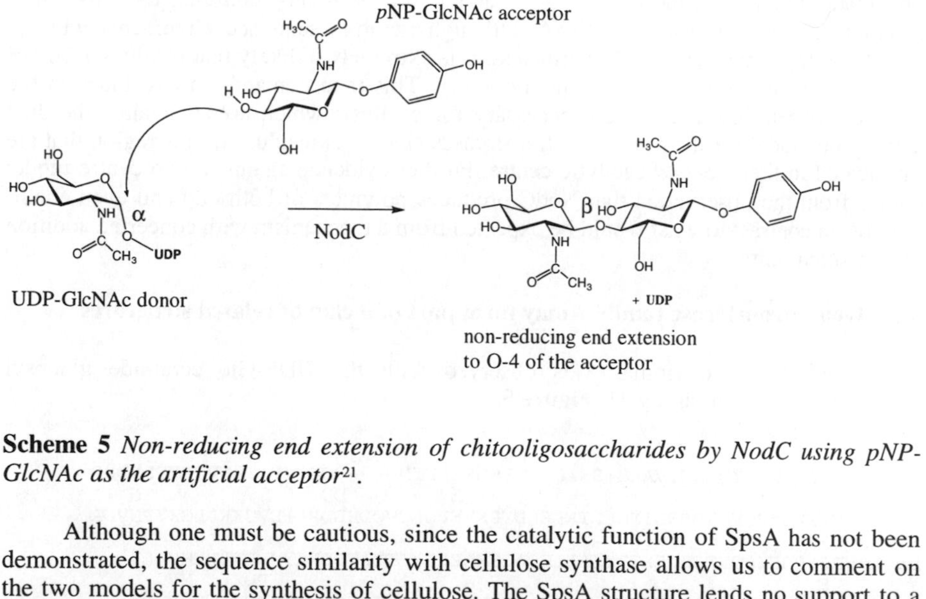

Scheme 5 *Non-reducing end extension of chitooligosaccharides by NodC using pNP-GlcNAc as the artificial acceptor*[21].

Although one must be cautious, since the catalytic function of SpsA has not been demonstrated, the sequence similarity with cellulose synthase allows us to comment on the two models for the synthesis of cellulose. The SpsA structure lends no support to a model involving lipid intermediates or elongation at the reducing-end.

In order to maintain the correct stereochemistry, the Han and Robyt model demands that cellulose synthase is not a UDP-Glc transferase, but instead is involved in the shuttling of glucose between lipid pyrophosphate intermediates. Whilst similar mechanisms exist for other polymers, their role in cellulose synthesis must be questioned. Cellulose synthase sequences show strong sequence similarity with known NDP-dependent transferases, indeed it is this similarity that is the defining feature of glycosyltransferase family 2[8]. The SpsA structure confirms that the majority of the sequence conservation is observed in residues in intimate contact with the nucleotide moiety (either UDP or *d*TDP in the case of SpsA). Mutation of the equivalents of Asp 39 or Asp 99 results in loss of activity[14]. It seems unlikely that cellulose synthase would display strong similarities to known UDP-sugar transferases and possess invariant and catalytically-essential residues known to bind UDP on other systems, were it not involved in UDP binding itself. The direct observation of growing cellulose and chitin chains[19, 20], taken together with the mode of action of NodC[21], the sequence conservation in family 2 and the structure of SpsA[4] strongly supports a model in which family 2 enzymes extend the polymer at the *non-reducing* end by direct transfer of the sugar from the UDP-donor to the acceptor.

The non-reducing end elongation model has also been developed to incorporate a two-catalytic centre model for the "processive" transferases such as the cellulose and chitin synthases and NodC[14,15,19,22]. Adjacent sugar monomers in the products display a two-fold relationship which may result from the presence of two catalytic sites within the same polypeptide, each with the appropriate orientation. Such a two catalytic centre model could also account for the synthesis of hyaluronan, a polymer of alternating *N*-acetylglucosamine and glucuronic acid. Key sequence motifs, notably three conserved aspartates, were believed to represent the "two" catalytic centres. Whilst the SpsA structure and the sequence similarity with cellulose synthase supports a model featuring elongation at the non-reducing end, it gives little support to this two catalytic-centre model. The first two aspartates in the "processive transferase motifs" correspond unambiguously to Asp 39 and Asp 99 of SpsA and are involved in uracil and Mn-phosphate binding, respectively. One cannot say with any confidence if the third aspartate has an equivalent in SpsA, although iterative sequence alignments with ψ-BLAST[17] implicate Asp 191. Nevertheless, it is extremely unlikely that residues 1-100 of SpsA alone can form an intact catalytic entity. This then demands that residues in the region between 100 and 200 are necessary for catalysis which argues against the dual active-site model for the processive transferases since it is residues in this region that are implicated in their *second* catalytic centre. Further evidence against a two-centre model comes from the observation that NodC produces polymers of both odd and even chain-length[21] in contrast to what would be expected from a mechanism with concerted addition of two sugar units.

3.2 Glycosyltransferase family 2 may form part of a clan of related structures

Low level sequence similarity is observed with the UDP-Glc ceramide glucosyl transferases found in family 21, **Figure 5**.

```
     2
SpsA PKVSVIMTSYNKSDYVAKSISSILSQTFSDFELFIM----DDNSNEETLNVIRPFLNDNR
     P VS                           E        DD              N
Cera PGVSLLKPLKGVDPNLINNLETFFELDYPKYEVLLCVQDHDDPAIDVCKKLLGKYPNVDA

SpsA VRFYQSDISGVKERTEKTRYAALINQAIEMAEGEYITYATDDNIYMPDRLLKMVREL
       F      G                  E A    I          PD L  MV
Cera RLFIGGKKVGINPKINNLMP------GYEVAKYDLIWICDSGIRVIPDTLTDMVNQM
```

Figure 5 *(a) Sequence similarity (calculated with ψ-BLAST[17]) between SpsA and the family 21 enzyme human UDP-Glc ceramide glucosyltransferase.*

These enzymes catalyse the first step in the synthesis of glycosphingolipids from ceramide, **Figure 6**. The product, glucosylceramide (**4**), forms the core of over 300 different glycosphingolipids.

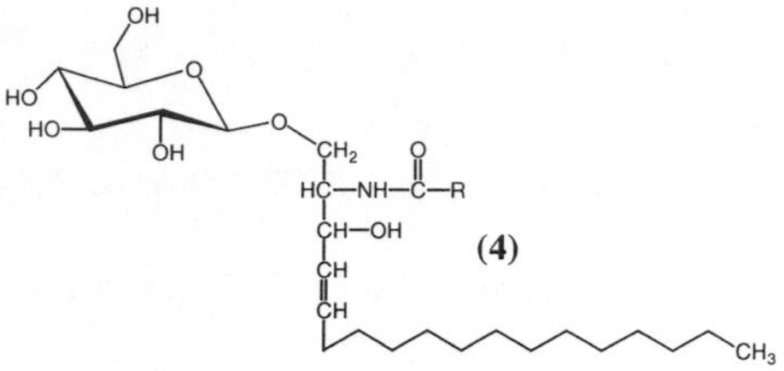

(4)

UDP-Glc + Ceramide ⟶ UDP + glucosylceramide (4)

Figure 6 *The reaction catalysed by UDP-glucose ceramide glucosyltransferase.*

Although sequence similarities as low as this can often be misleading, these results suggest that family 21 may show structural and catalytic similarity with the family 2 enzymes[8]. Although the sequence similarity is at an extremely low-level, it may be significant that it involves residues, such as Asp 39 (SpsA numbering), implicated in binding of the UDP moiety of the donor substrate. Such speculation will need to be verified by three-dimensional analyses.

3.3 The SpsA nucleotide-binding domain may be shared with other families which display the opposite stereochemistry

One of the most provocative and intriguing sequence similarities displayed by SpsA is that with the family 27 enzymes[23], **Figure 7**. These enzymes are involved in the synthesis of "mucin-like" *O*-glycosylation of polypeptides. The reaction involves the transfer of the *N*-acetylgalactosamine moiety of UDP-GalNAc to a peptide hydroxyl, **Figure 8.** This reaction is performed with net *retention* of anomeric configuration. An α-linked UDP-GalNac gives rise to α-linked *O*-glycosylation.

```
     2
SpsA PKVSVIMTSYNKS-DYVAKSISSILSQTFSDF--ELFIMDDNSNEETLNV-IRPFLNDNR
     P  SV     N           S           E    DD S    L
Gnt  PTTSVVIVFHNEAWSTLLRTVHSVINRSPRHMIEEIVLVDDASERDFLKRPLESYVKKLK

SpsA VRFYQSDISGVKERTEKTRYAALINQAIEMAEGEYITYATDDNIYMPDRLLKMVRELDTH
     V         V                     G  IT            L         H
Gnt  VPVH------VIRMEQRSGLIRARLKGAAVSRGQVITFLDAHCECTAGWLEPLLARI-KH
```

Figure 7 *Sequence alignment of SpsA with the family 27 enzyme polypeptide GalNAc transferase-T1*

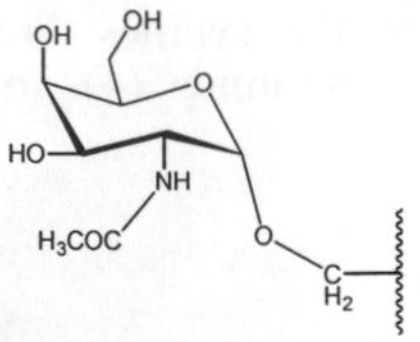

(5)

UDP-GalNAc+ polypeptide ⟶ UDP + *O*-linked GalNAc-polypeptide (5)

Figure 8 *The reaction catalysed by family 27 enzymes involves net retention of anomeric configuration.*

Again, it is elements of the nucleotide-binding domain, particularly those residues involved in binding of the UDP in SpsA, that are invariant. Despite the lack of a 3-D structure for a family 27 enzyme, elegant "threading" analyses correctly identified the topology of the "GT1" domain equivalent to the SpsA nucleotide-binding domain[24]. This, in harness with the known sequence conservation, allowed Hagen and co-workers to identify and mutate likely catalytic residues. Asp 156, equivalent to Asp 39 of SpsA, was mutated to glutamine resulting in a protein with essentially no catalytic activity. This lends support to the proposals that these domains play a related function in the two families. The similarity between the inverting transferases of family 2 and the retaining transferases from family 27 opens up the intriguing prospect that common nucleotide-binding "modules" could have migrated during evolution and be present on enzymes displaying different catalytic stereochemistry.

4 SUMMARY

The importance of glycosyltransferases, particularly the nucleotide-diphospho-sugar dependent enzymes, serves only to emphasise our lack of catalytic and structural knowledge on these systems. In this paper we have described the first structure for a nucleotide-diphospho-sugar transferase from glycosyltransferase family 2. Other family 2 members, including those of well-defined substrate specificity, display a much higher complexity than SpsA, often including multiple domains and membrane-spanning regions. Although the sugar donor and acceptor specificities of SpsA remain unknown, the structure provides an initial glimpse into the active centre of family 2 glycosyltransferases and will act as the precursor to the study of these other more challenging systems.

The SpsA structure provokes a reevaluation of the current models for the synthesis of polysaccharides, such as cellulose and chitin. The reducing-end elongation *via* lipid pyrophosphates seems unlikely given the sequence similarity of cellulose and chitin synthases with SpsA and the role of the invariant residues in UDP binding. The SpsA structure argues in favour of the non-reducing end elongation but does not lend support to a two catalytic centre model for "processive transferases". Sequence searches using the SpsA template, do identify proteins containing two repeats of an SpsA-related module suggesting that the "two catalytic centre within one peptide" model will be valid in some other cases. The SpsA structure reveals provocative similarities to enzymes involved in ceramide synthesis and the *O*-glycosylation of peptides. These will have to be verified with three-dimensional analyses and this work is currently in progress.

5 ACKNOWLEDGEMENTS

The authors would like to thank Bernard Henrissat for useful discussions and the BBSRC for supporting the Structural Biology Laboratory in York. G.J.D is a Royal Society University Research Fellow.

References

1. G.-J. Boons, *Tetrahedron,* 1996, **52,** 1095.
2. F. Reichel and G.-J. Boons, *Chemistry in Britain,* 1998, **34**, 43.
3. A. Vrielink, W. Rüger, H.P.C. Driessen and P.S. Freemont, *EMBO. J.*, 1994, **13**, 3413.
4. S.J. Charnock and G.J. Davies, *Biochemistry,* 1999, **38, in press**
5. B. Henrissat, I. Callebaut, S. Fabrega, P. Lehn, J.-P. Mornon and G. Davies, *P.N.A.S. (USA),* 1995, **92**, 7090 .
6. B. Henrissat and A. Bairoch, *Biochem. J.,* 1996, **316**, 695 .
7. B. Henrissat and G.J. Davies, *Curr. Op. Struct. Biol.*, 1997, **7**, 637.
8. J.A. Campbell, G.J. Davies, V. Bulone and B. Henrissat, *Biochem. J.*, 1997, **326**, 929.
9. S.J. Charnock and G.J. Davies, *Acta. Cryst.*, 1999, **D 55**, 677.
10. P. Stragier and R. Losick, *Annu. Rev. Genet.*, 1996, **30**, 297 .
11. S. Roels and R. Losick, *J. Bacteriol.*, 1995, **177**, 6263 .
12. P.J. Focia, S.P. Craig III and A.E. Eakin, *Biochemistry,* 1998, **37**, 17120.
13. P.J. Kraulis, *J. Appl. Cryst.*, 1991, **24**, 946 .
14. I.M. Saxena and R.M. Brown, *Cellulose,* 1997, **4**, 33.
15. I.M. Saxena, R.M. Brown, M. Fevre, R.A. Geremia and B. Henrissat, *J. Bacteriol.*, 1995, **177**, 1419.
16. W.P. Burmeister, S. Cottaz, H. Driguez, S. Palmieri and B. Henrissat, *Structure,* 1997, **5**, 663.
17. S.F. Altschul, T.L. Madden, A.A. Schäffer, J. Zhang, Z. Zhang, W. Miller and D.J. Lipman, *Nucl. Acid. Res.*, 1997, **25**, 3389.
18. N.S. Han and J.F. Robyt, *Carbohydr. Res.*, 1998, **313**, 125.
19. M. Koyama, W. Helbert, T. Imaj, J. Sugiyama and B. Henrissat, *P.N.A.S. (USA),* 1988, **94**, 9091.
20. J. Sugiyama, C. Boisset, M. Hashimoto and T. Watanabe, *J. Mol. Biol.*, 1999, **286**, 247.
21. E. Kamst, J. Bakkers, N.E.M. Quaedvleig, J. Pilling, J.W. Kijne, B.J.J. Lugtenberg and H.P. Spaink, *Biochemistry,* 1999, **38**, 4045.
22. N. Carpita and C. Vergara, *Science,* 1998, **279**, 672.
23. J.A. Campbell, G.J. Davies, V. Bulone and B. Henrissat, *Biochem. J.*, 1997, **329**, 917.
24. F.K. Hagen, B. Hazes, R. Raffo, D. deSa and L.A. Tabak, *J. Biol. Chem.*, 1999, **274**, 6797.

CRYSTAL STRUCTURES OF POLYGALACTURONASE AND PECTIN METHYLESTERASE

R. W. Pickersgill and J. A. Jenkins

Institute of Food Research
Reading Laboratory
Earley Gate
Whiteknights Road
Reading RG6 6BZ
UK

1 INTRODUCTION

We have recently solved the structures of two pectin modifying enzymes, polygalacturonase and pectin methylesterase. Pectin methylesterase has a profound influence on the functional properties of pectin and the demethylated pectin then becomes the substrate for polygalacturonase and pectate lyase. The bacterial enzymes are plant virulence factors while the plant equivalents are involved in cell development, growth, ripening and scenescence.

2 POLYGALACTURONASE

Polygalacturonases are classified as family 28 glycosyl hydrolases[1]; they catalyse the cleavage of the α-1,4 glycosidic bond in polygalacturonate and in the homogalactan regions of pectin[2-4] with inversion of the anomeric carbon[5]. Plant polygalacturonases are important in fruit ripening, whereas the microbial enzymes are involved in pathogen attack. We have solved the structure of the *Erwinia carotavora* polygalacturonase. The structure comprises a right-handed parallel β-helix protein[6]. Unlike pectate lyase[7,8] and pectin lyase[9,10] there are ten, not seven complete turns and four not three β-strands per turn. Previous results had suggested that there was a histidine directly involved in activity. Comparison of the structure of polygalacturonase with that of rhamnogalacturonase A[11] and examination of sequence conservation within the polygalacturonase family reveals that the active site amino acids are aspartates 202 and 223[6] (Fig. 1). There is a histidine close by and it is clear that a blocking group[12] or a lysine substitution[13] at this position would affect activity. These carboxylates are closer than the accepted distance of about 9.5Å for an inverting glycosyl hydrolyase[14,15]. We suggest that the separation is less, 4.5Å, because the requirements for cleavage of an α-linked rather than β-linked polysaccharide are different in that protonation of the glycosidic oxygen and nucleophilic attack at the anomeric carbon can be from the same side of the glycosidic bond[6] (Fig. 2).

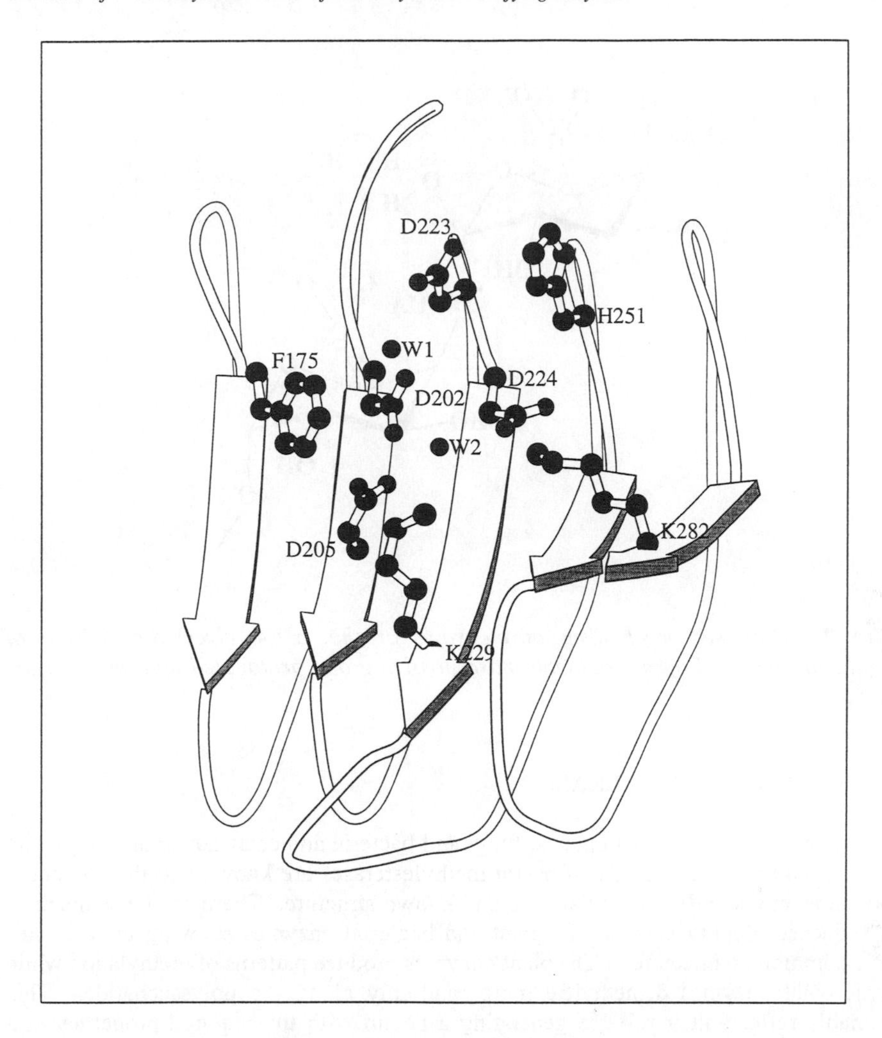

Figure 1 *The active site of Erwinia carotovora polygalacturonase is formed by the surface of parallel β-sheet one and the adjacent loops. β-strands four through eight of the sheet are shown. Aspartates 202 and 223 are conserved in all polygalacturonases and in the related enzyme rhamnogalacturonase A (RGaseA). A third aspartate is also present in polygalacturonases (D224), the equivalent residue is a glutamate in RGaseA.*

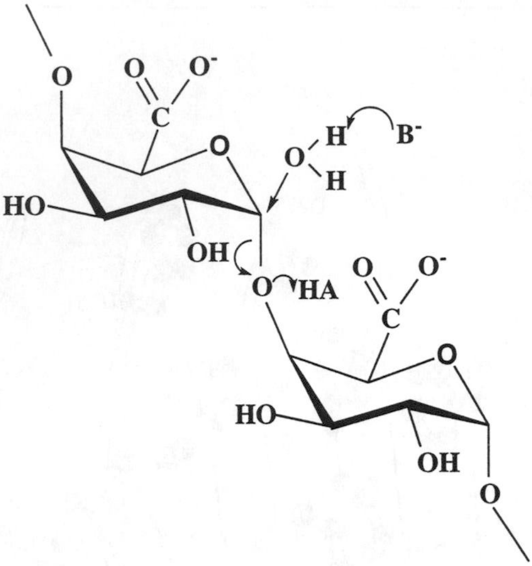

Figure 2 *Proposed mechanism of cleavage of the α-1,4 glycosidic linkage of polygalacturonic acid by polygalacturonase involving one protonated and one charged aspartate.*

3 PECTIN METHYLESTERASE

Pectin methylesterases occur in plants, fungi and bacteria and catalyse the demethylation of pectin. Only a single family of pectin methylesterases are known and the sequences do not show any significant similarity to any known structure. There can be as much as 30% sequence identity between the plant and bacterial enzymes showing the enzymes share a common architecture[16]. The plant enzymes produce patterns of methylation while the microbial enzymes demethylate more randomly along the polysaccharide. This presumably reflects their roles in generating a pectin with the required properties in a growing pollen tube for example or as an agent of bacterial attack in plant disease. The product of the action of pectin methylesterase is the substrate for polygalacturonase and pectate lyase. We have solved the structure of *Erwinia chrysanthemi* pectin methylesterase using standard heavy atom methods and averaging the two molecules in the asymmetric unit. The structure comprises a right-handed parallel β-helix domain that closely resembles that of pectate lyase[17]. There is an aromatic stack inside the β-barrel of pectin methylesterase but it is at a different location from that seen in pectate and pectin lyase[7-10]. There is no internal asparagine ladder in pectin methylesterase.

The active site is in a similar position to that of the other pectin modifying enzymes and comprises two aspartates in an arrangement reminiscent of but not identical to that of the aspartyl proteases (Fig. 3). One aspartate makes a hydrogen bond to an adjacent arginine and is the therefore likely to be unprotonated. The other aspartate is likely to be protonated. A water molecule adjacent to the unprotonated aspartate may be activated by transferring its proton to the aspartate, the generated hydroxyl attacking the carbonyl-carbon. Simultaneous protonation of one of the oxygens results in the formation of a tetrahedral intermediate which collapses with the release of methanol and demethylation of one pectin residue (Fig. 4).

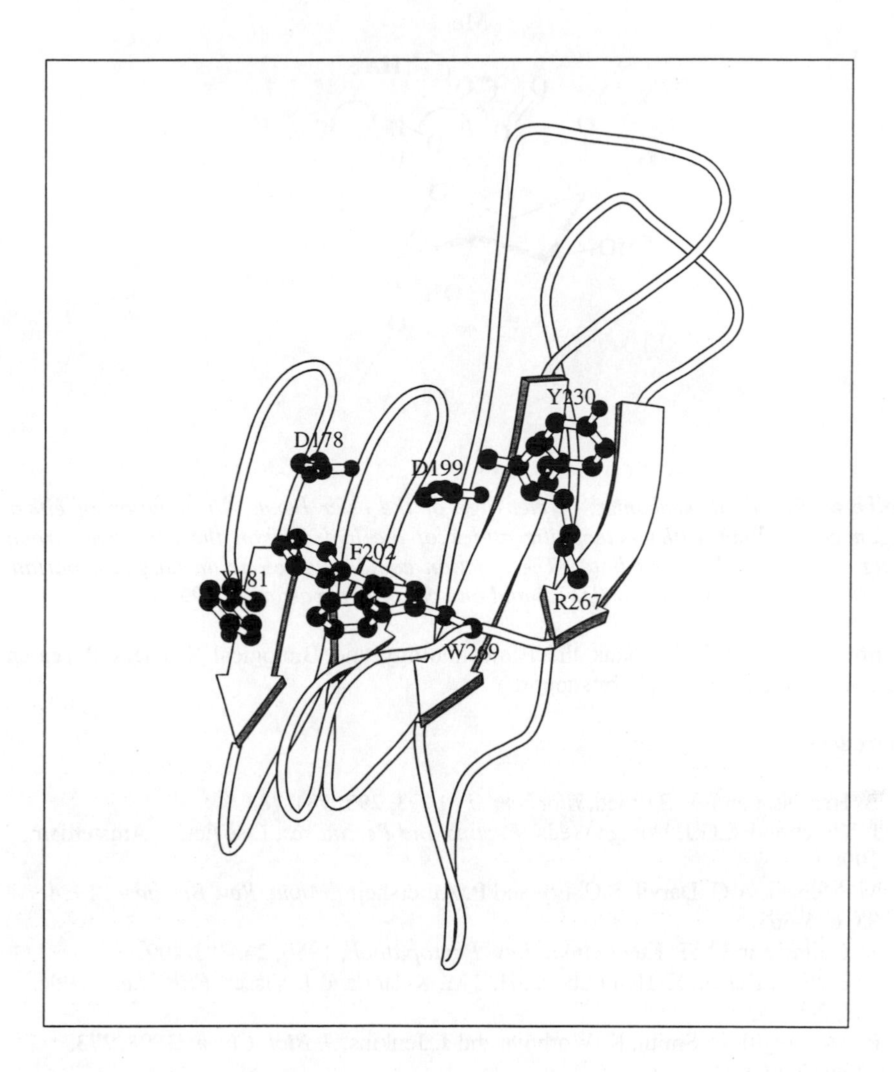

Figure 3 *The active site of Erwinia chrysanthemi pectin methylesterase is on the surface of parallel β-sheet one and the adjacent loops. β-strands four through seven of the sheet are shown. Aspartates 178 and 199 and arginine 267 are conserved in all pectin methylesterases. A number of close aromatic residues that could be involved in binding methylated pectin are also shown.*

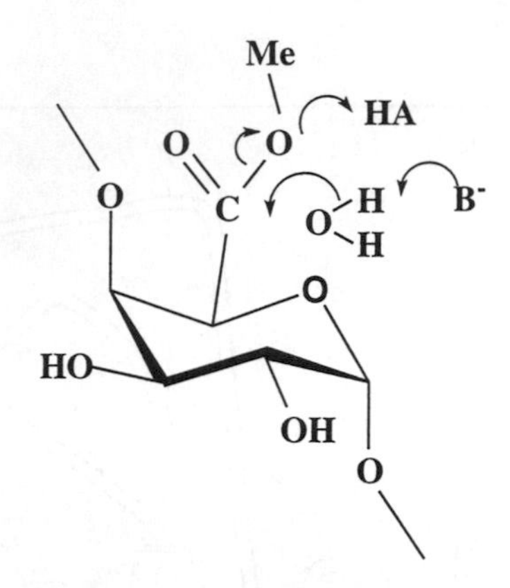

Figure 4 *Proposed mechanism of cleavage of the ester bond. Protonation of either oxygen could occur with nucleophilic attack of the hydroxyl on the carbonyl carbon giving a tetrahedral intermediate. The reaction could involve one uncharged aspartate (D178) only or one uncharged (D178) and one charged aspartate (D199).*

Acknowledgement. We thank the Biotechnology and Biological Sciences Research Council of the UK (BBSRC) for support.

References

1. B. Henrissat and A. Bairoch, *Biochem. J.*, 1993, **293**, 781-788.
2. J. Visser and A.G.J. Voragen (eds) *Pectins and Pectinases*, Elsevier, Amsterdam, 1996.
3. M. McNeil, A.G. Darvil, S.C. Fry and P. Albersheim, *Annu. Rev. Biochem.*, 1984, **53**, 625-663.
4. A. Collmer and N.T. Keen, *Annu. Rev. Phytopathol.*, 1986, **24**, 383-409.
5. P. Biely, J. Benen, K. Heinrichova, H.C.M. Kester and J. Visser, *FEBS Lett.*, 1996, **382**, 249-255.
6. R. Pickersgill, D. Smith, K. Worboys and J. Jenkins, *J. Biol. Chem.*, 1998, **273**, 24660-24664.
7. M.D. Yoder, N.T. Keen and F. Jurnak, *Science*, 1993, **260**, 1503-1507.
8. R. Pickersgill, J. Jenkins, G. Harris, W. Nasser and J. Robert-Baudrouy, *Nat. Struct. Biol.*, 1994, **1**, 717-723.
9. O. Mayans, M. Scott, I. Connerton, T. Gravesen, J. Benen, J.Visser, R. Pickersgill and J. Jenkins, *Structure*, 1997, **5**, 677-689.
10. J. Vitali, B. Schick, H.C.M. Kester, J. Visser and F. Jurnak, *Plant Physiol.*, 1998, **116**, 69-80.
11. T.N. Petersen, S. Kauppinen and S. Larsen, *Structure*, 1997, **5**, 533-544.

12. M.N. Rao, A.A. Kembhavi and A. Pant, *Biochim. Biophys. Acta*, 1996, **1296**, 167-173.
13. C. Caprari, B. Mattei, M.L. Basile, G. Salvi, V. Crescenzi, G. Delorenzo and F. Cervone, *Mol. Plant-Microbe Interact.*, 1996, **9**, 617-624.
14. J.D. McCarter and S.G. Withers, *Curr. Opin. Struct. Biol.*, 1994, **4**, 885-892.
15. G. Davies and B. Henrissat, *Structure*, 1995, **3**, 853-859.
16. F. Laurent, A. Kotoujansky, G. Labesse and Y. Bertheau, *Gene*, 1993, **131**, 17-25. Jenkins, J., Worboys, K., Smith, D., Mayans, O. & Pickersgill, R. (1999) in preparation.

CRYSTAL STRUCTURE OF THE PULLULAN-HYDROLYZING α-AMYLASE FROM *THERMOACTINOMYCES VULGARIS* R-47, AND MANIPULATION OF THE SUBSTRATE SPECIFICITIES

Takashi Tonozuka and Yoshiyuki Sakano

Department of Applied Biological Science
Tokyo University of Agriculture and Technology
Fuchu, Tokyo 183-8509, Japan

1 INTRODUCTION

Numerous amylases and related enzymes have been reported. Although they hydrolyze the same substrates such as starch, their products are different. Therefore, in addition to having commercial importance, amylases and related enzymes can also provide useful information for the study of protein engineering. Pullulan-hydrolyzing enzymes (Fig. 1) would be good materials for such studies.

1.1 Pullulan-Hydrolyzing Enzymes

Although most α-amylases scarcely hydrolyze pullulan, various pullulan-hydrolyzing enzymes have been reported recently as follows.

(i) Enzymes which hydrolyze the (α1→6)-glucosidic linkages of pullulan (denoted as linkage 1 in Fig. 1) to produce maltotriose: i-i) pullulanase, which also hydrolyzes the (α1→6)-glucosidic linkages of starch[1]; i-ii) amylopullulanase (also called α-amylase-pullulanase), which also hydrolyzes the (α1→4)-glucosidic linkages of starch to produce mainly maltose[2]. Most amylopullulanases reportedly have a single catalytic center. However, Hatada et al. reported an amylopullulanase which has two catalytic centers[3].

(ii) Enzymes which hydrolyze the (α1→4)-glucosidic linkages of pullulan (denoted as linkage 2 in Fig. 1) to produce panose: ii-i) α-amylases from *Thermoactinomyces vulgaris* R-47, TVA I[4,5] and TVA II[6], and from several *Bacilli*[7], which hydrolyze starch more effectively than pullulan, to produce mainly maltose; ii-ii) neopullulanase, which hydrolyzes starch less efficiently[8,9].

(iii) An enzyme which hydrolyzes the (α1→4)-glucosidic linkages of pullulan (denoted as linkage 3 in Fig. 1) to produce isopanose: isopullulanase, which does not hydrolyze starch[10].

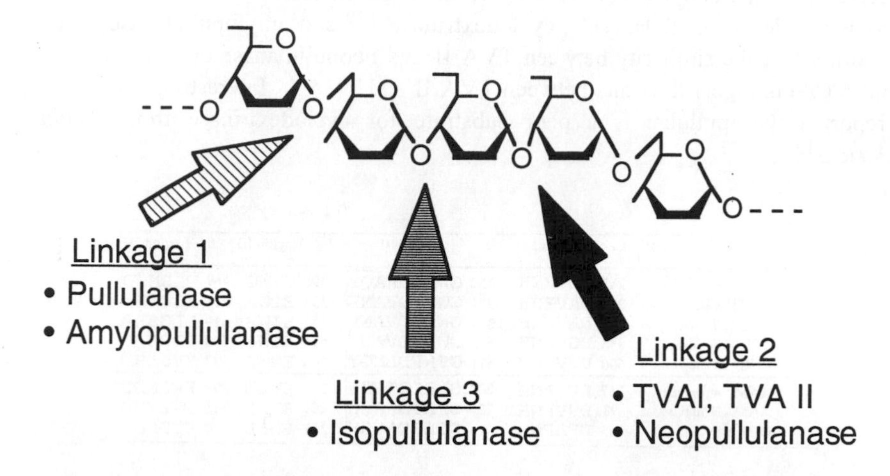

Figure 1 *Schematic Action Pattern of Pullulan-hydrolyzing Enzymes for Pullulan*
Bold arrows are the enzymatic cleavage points.

1.2 α-Amylases from *Thermoactinomyces vulgaris* R-47

Thermoactinomyces vulgaris R-47 produces two pullulan-hydrolyzing α-amylases, TVA I and TVA II. TVAs efficiently hydrolyze specific linkages of pullulan to produce panose. In addition, TVAs hydrolyze cyclodextrins[11,12] and the (α1→6)-glucosidic linkage of isopanose[13]. TVAs also strongly catalyze transglycosylation at both (α1→4)– and (α1→6)-glucosidic linkages[14].

Although the properties of TVA I and TVA II are similar, they show some differences: i) TVA I strongly hydrolyzes starch while TVA II shows outstanding kinetic values for small oligosaccharides[15]; ii) TVA II efficiently hydrolyzes α-, β- and γ-cyclodextrins while TVA I does not prefer α- and β-cyclodextrins[12,15]; iii) TVA II accumulates a large quantity of transglycosylation products while TVA I accumulates less[16].

2 STRUCTURES OF TVA I AND TVA II

2.1 Primary Structures of TVA I and TVA II

The genes of TVA I[6] and TVA II[15] were cloned, and the primary structure was deduced. The length of their primary structures are similar. The primary structure of TVA I is composed of 666 amino acid residues including an N-terminal signal sequence consisting of 29 amino acid residues. The primary structure of TVA II is composed of 585 amino acid residues with no signal sequence. The overall primary structure of TVA I and TVA II are similar, although the identity is only 30 %. Both TVAs resembled neopullulanase[17], cyclodextrinase[18,19] and amylopullulanase[20]. It is noteworthy that the similarity between TVA II and neopullulanase or cyclodextrinase (about 50 %) is higher than that between TVA II and TVA I. Interestingly, Oguma et al. reported that pullulan is a poor substrate for cyclodextrinase from *Bacillus sphaericus*[19].

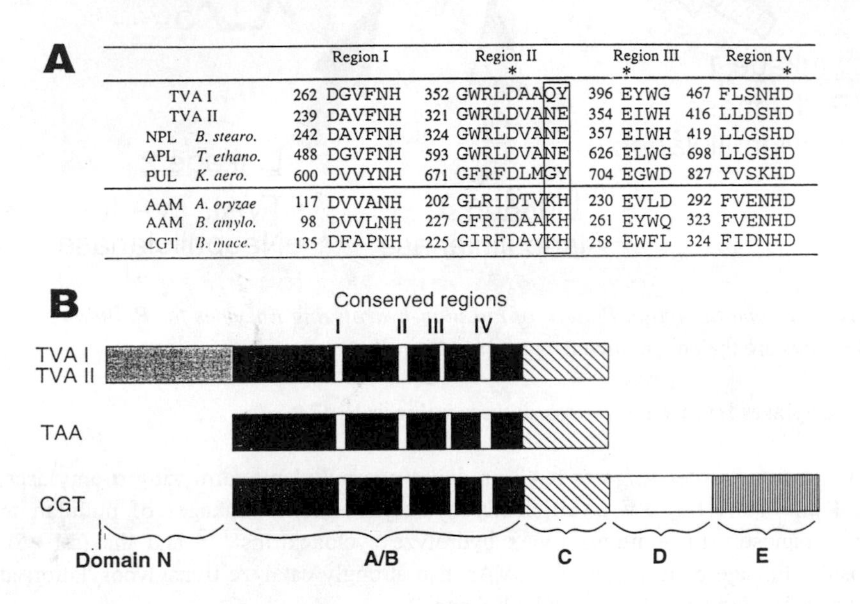

A

		Region I	Region II *	Region III *	Region IV *
TVA I		262 DGVFNH	352 GWRLDAAQY	396 EYWG	467 FLSNHD
TVA II		239 DAVFNH	321 GWRLDVANE	354 EIWH	416 LLDSHD
NPL	*B. stearo.*	242 DAVFNH	324 GWRLDVANE	357 EIWH	419 LLGSHD
APL	*T. ethano.*	488 DGVFNH	593 GWRLDVANE	626 ELWG	698 LLGSHD
PUL	*K. aero.*	600 DVVYNH	671 GFRFDLMGY	704 EGWD	827 YVSKHD
AAM	*A. oryzae*	117 DVVANH	202 GLRIDTVKH	230 EVLD	292 FVENHD
AAM	*B. amylo.*	98 DVVLNH	227 GFRIDAAKH	261 EYWQ	323 FVENHD
CGT	*B. mace.*	135 DFAPNH	225 GIRFDAVKH	258 EWFL	324 FIDNHD

Figure 2 *Comparison of Primary Structures of TVAs and Related Enzymes*
(A) Comparison of the four conserved regions (I–IV) of TVA I and related enzymes. The abbreviations are as follows: NPL, neopullulanase; APL, amylopullulanase; PUL, pullulanase; AAM, α-amylase; CGT, cyclodextrin glucanotransferase. The enzyme sources are abbreviated as follows: *B. stearo., Bacillus stearothermophilus; T. ethano., Thermoanaerobacter ethanolicus; K. aero., Klebsiella aerogenes; A. oryzae, Aspergillus oryzae; B. amylo., Bacillus amyloliquefaciens; B. mace., Bacillus macerans.* The three catalytic amino acid residues of the α-amylase family, two Asp and one Glu, are indicated by asterisks. The numbering of the sequences of the enzymes starts at the *N*-terminal amino acid of each mature enzyme. The latter part of region II is also indicated. TVAs, NPL, APL, and PUL are pullulan-hydrolyzing enzymes, whereas AAM and CGT are enzymes that do not hydrolyze pullulan. (B) Schematic alignment of TVA II and related enzymes. Enzyme sources are as follows: TAA, α-amylase from *Aspergillus oryzae* (Taka-amylase A); CGT, cyclodextrin glucanotransferase from *Bacillus circulans.* The four vertical bars (I–IV) indicate the conserved regions of amylases. The primary structures are aligned according to the position of the conserved regions.

The primary structures were aligned with those of α-amylase family enzymes (Fig 2). α-Amylase family enzymes have four conserved regions (I–IV) in their primary structures[21], and three acidic residues, two Asp and one Glu, located in the conserved regions function as catalytic residues (Fig. 2). Except for in isopullulanase[22], the four conserved regions have been observed in the primary structures of all the pullulan-hydrolyzing enzymes as well (Fig. 2A). The most notable feature is that TVAs have an extra sequence in the N-terminal region (Fig. 2B), thus the N-terminal region was predicted to form a domain structure[6,15].

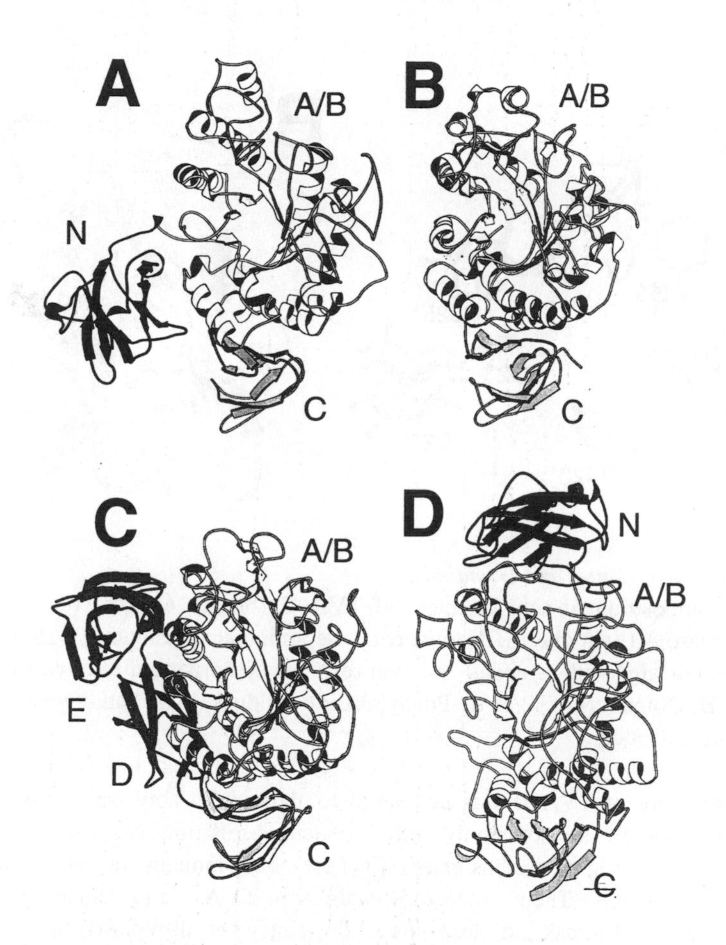

Figure 3 *Three-dimensional Structures of α-Amylase Family Enzymes*
Three-dimensional structures of TVA II (A), Taka-amylase A (B), CGTase (C) and isoamylase (D). The PDB IDs are 1BVZ, 6TAA, 1CDG and 1BF2, respectively. Domains N, A/B, C, D and E are shown in black, white, light gray, dark gray and dark gray, respectively.

2.2 Three-dimensional Structure of TVA II

The X-ray structure of TVA II was determined[23,24], and the overall structure of TVA II was shown (Fig. 3A). TVA II is composed of three or four domains. Domain A/B has a (β/α barrel structure with a small component which is called a domain B[25]. Domain C contains the C-terminus, and it consists of six β-strands. The notable structural feature of TVA II is domain N. Domain N contains the N-terminus, and consists of seven β-strands. Domain N is isolated from other domains, and interacts only with the sixth α-helix of domain A.

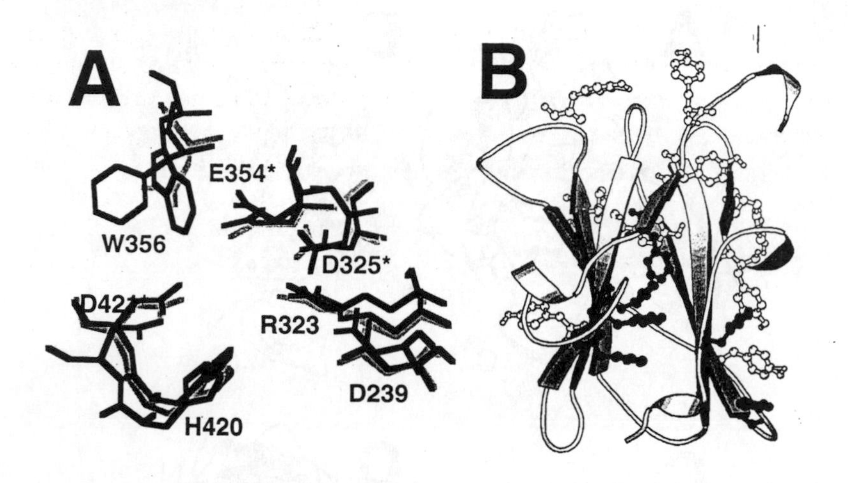

Figure 4 *Active Center and Domain N*
(A) Superimposition of active centers of TVA II (black), CGTase (dark gray) and Taka-amylase A (light gray). The three catalytic amino acid residues, D325, D421, and E354 are indicated with asterisks. Several residues comprising the active site are also shown. (B) Domain N of TVA II. Phenylalanine residues (black) and tyrosine residues (white) are indicated.

The structure of TVA II was compared to those of α-amylase family enzymes. Domains A/B and C are commonly found in α-amylase family enzymes[26-28] (Fig 3A–D). Cyclodextrin glucanotransferase (CGTase) has domains D and E in the C-terminus[27] (Fig. 3C). The position of domain N in TVA II and domains D and E in CGTase are not identical, but close. Recently, Katsya et al. reported that isoamylase also has a domain in the N-terminus, in addition to domains A/B and C[28]. Domain N of both enzymes mainly consists of -strands. However, the number of -strands, the positions of the domains, and interaction with neighbor domains were completely different (Fig. 3D).

2.3 Active Center and Domain N

α-Amylase family enzymes have three catalytic amino acid residues, two aspartate residues and one glutamate acid residue. In TVA II, they are D325, E354 and D421. The catalytic residues and some adjacent residues of TVA II, Taka-amylase A, and CGTase were superimposed (Fig. 4A). The distance between the catalytic residues of TVA II is greater than in other enzymes. For instance, the distance between two C of D325 and D421 is 9.6 ≈, whereas that of the corresponding residues of Taka-amylase A and CGTase is 7.8 ≈ and 8.6 ≈, respectively. This wide cleft may be suitable for the binding of pullulan and cyclodextrins.

Domain N consists of seven β-strands, and the first and seventh β-strands are divided by bulges (Fig. 4B). Domain N forms a distorted incomplete β-barrel structure, and if one more strand is located in front of the open barrel, the barrel structure is a complete and closed form. Thus many hydrophobic residues in the barrel are located on the exterior to form the solvent accessible surface. In particular, many phenylalanine residues are located in the barrel. There are many reports that aromatic residues such as phenylalanine, tyrosine, and trytophan residues of carbohydrate-hydrolyzing enzymes function in the substrate recognition and binding[29,30]. Other than the phenylalanine residues, domain N is also rich in tyrosine residues. Domain E of CGTase, which also consists of several β-strands and bulges with many aromatic residues, was reported to bind to two cyclodextrins[31]. Although the role of domain N for the enzyme activity has been unclear, the hydrophobicity and these residues may play an important role in the substrate recognition and binding.

3 MANIPULATION OF THE SUBSTRATE SPECIFICITIES

3.1 Construction of Mutated TVA I

One of the goals of protein engineering is the conversion of the substrate specificity of an enzyme into that of other enzymes. In this study, we focused on region II among the four regions conserved in the α-amylase family (Fig. 2A, box). In the enzymes which do not hydrolyze pullulan, a Lys-His (KH) sequence is highly conserved in the latter part of region II, whereas this KH sequence is not conserved in the pullulan-hydrolyzing enzymes. Thus we hypothesized that the latter part of region II plays an important role in the hydrolysis of pullulan.

The sequences of region II of TVA I, TVA II and amylopullulanase were compared further (Fig. 5A). The latter part of regions II of TVA II and amylopullulanase is VANE, whereas that of TVA I is AAQY. Also TVA I has an extra 11 amino acids, ANGNNGSDVTN, located after the C-terminus of region II. Thus mutated TVA I were constructed as follows: (i) Del11, in which the 11 amino acid sequence, ANGNNGSDVTN, was deleted; (ii) AQY/VNE, in which A357, Q359 and Y360 in the latter part of region II, AAQY, were replaced by Val, Asn and Glu respectively, giving the VANE sequence; (iii) Del11+AQY/VNE, in which the sites of mutation introduced were a combination of (i) and (ii)[32].

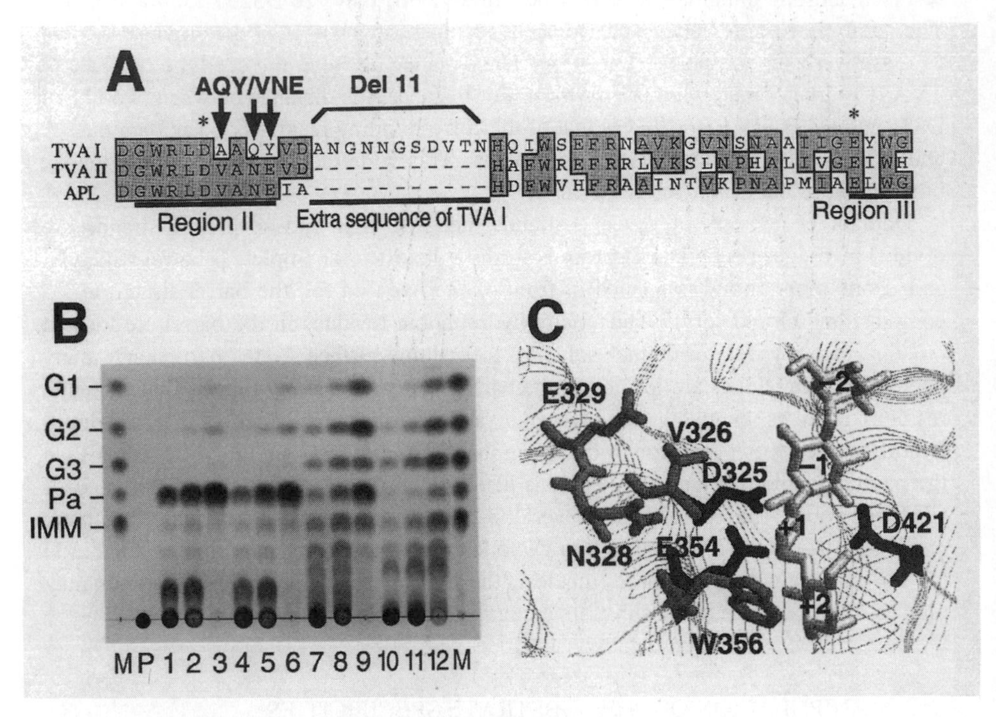

Figure 5 *Construction and Properties of Mutated TVA I*
(A) Comparison of amino acid sequences of regions II and III of TVA I, TVA II, and *Thermoanaerobacter ethanolicus* amylopullulanase. Identical residues are boxed. Bars represent gaps introduced during the alignment process. The catalytic amino acid residues are indicated by asterisks. The conserved regions, II and III, and the extra sequence of TVA I are underlined. Ala357, Gln359, and Tyr360 of TVA I are indicated. (B) TLC showing the actions of wild and mutated TVA I on pullulan. The reaction mixtures contained 500 μ l of 1% pullulan and 0.65 units of wild-type or mutated TVA I. Lane M, oligosaccharide markers: G1, glucose; G2, maltose; G3, maltotriose; Pa, panose; IMM, 6^3-α-glucosylmaltotriose. Lane P, pullulan. Lanes 1—3, 4—6, 7—9, and 10—12 are samples incubated with the wild-type, Del11, AQY/VNE, and Del11+AQY/VNE enzymes, respectively. Lanes 1, 4, 7, and 10 are samples at 1 h, lanes 2, 5, 8, and 11 are samples at 3 h, and lanes 3, 6, 9, and 12 are samples at 20 h. (C) Putative maltotetraose binding model of TVA II. Numbers -2 to +2 indicate glucose units. The enzymatic cleavage point is located between glucose -1 and +1. The catalytic amino acid residues (black) and several residues comprising the active site (dark gray) are also indicated.

3.2 Substrate Specificities of Mutated Enzymes

Pullulan was incubated with the wild-type and mutated TVA I. The wild-type enzyme liberated panose from pullulan. In contrast, the action pattern of the mutated enzyme was markedly different from that of the wild-type enzyme. The enzyme Del11+AQY/VNE mainly produced maltotriose from pullulan and the spot of panose was decreased (Fig. 5B). Namely, the mutated enzymes mainly hydrolyzed (α1→6)-glucosidic linkages of pullulan (linkage 1 in Fig. 1). The action patterns for starch of the wild-type and mutated enzymes were almost identical. These findings indicated that the Del11+AQY/VNE enzyme could be regarded as an amylopullulanase. The activities of the mutated enzyme for (α1→6)-glucosidic linkages of for example isopanose were virtually unaltered, while those for (α1→4)-glucosidic linkages were about 100 times smaller than those of the wild-type enzyme[32].

To investigate why the substrate specificities of mutated enzymes were altered, a putative substrate-binding model was constructed. The three catalytic residues D325, E354 and D421 and several adjacent residues are shown in Fig. 5C. The W356 directly interacts with the reducing-end side of the substrate. Because the catalytic center of TVA II is wider than in other α-amylase family enzymes, both of the substrates for which the cleaving points are (α1→4)–glucosidic linkage and (α1→6)–glucosidic linkage could be entered into the catalytic center. One possible explanation is that the direction of the side chain of W356 is different depending on whether the cleaving point is (α1→4)–glucosidic linkage or (α1→6)–glucosidic linkage. The adjacent residues of W356, the corresponding mutated residues of TVA II, V326, N328 and E329, may effect the direction of the side chain of W356. In fact, when W356 of TVA II was substituted with other amino acid residues, the transglycosylation activities of the mutated TVA II were altered (details to be published elsewhere).

This work was supported in part by the Ministry of Agriculture, Forestry and Fisheries of Japan in the framework of the Pioneering Research Project in Biotechnology.

References

1. H. Bender and K. Wallenfels, *Biochem. Z.*, 1961, **334**, 79.
2. M. V. Ramesh, S. M. Podkovyrov, S. E. Lowe and J. G. Zeikus, *Appl. Environ. Microbiol.*, 1994, **60**, 94.
3. Y. Hatada, K. Igarashi, K. Ozaki, K. Ara, J. Hitomi, T. Kobayashi, S. Kawai, T. Watanabe and S. Ito, *J. Biol. Chem.*, 1996, **271**, 24075.
4. M. Shimizu, M. Kanno, M. Tamura, and M. Suekane, *Agric. Biol. Chem.*, 1978, **42**, 1681.
5. Y. Sakano, S. Hiraiwa, J. Fukushima and T. Kobayashi, *Agric. Biol. Chem.*, 1982, **46**, 1121.

6. T. Tonozuka, M. Ohtsuka, S. Mogi, H. Sakai, T. Ohta and Y. Sakano, *Biosci. Biotechnol. Biochem.*, 1993, **57**, 395.
7. H.-J. Cha, H.-G. Yoon, Y.-W. Kim, H.-S. Lee, J.-W. Kim, K.-S. Kweon, B.-H. Oh and K.-H. Park, *Eur. J. Biochem.*, 1998, **253**, 251.
8. T. Kuriki, S. Okada and T. Imanaka, *J. Bacteriol.*, 1988, **170**, 1554.
9. K. A. Smith and A. A. Salyers, *J. Bacteriol.*, 1991, **173**, 2962.
10. Y. Sakano, M. Higuchi and T. Kobayashi, *Arch. Biochem. Biophys.*, 1972, **153**, 180.
11. Y. Sakano, Y. Sano and T. Kobayashi, *Agric. Biol. Chem.*, 1985, **49**, 3391.
12. Y. Shimura, T. Tonozuka and Y. Sakano, *J. Appl. Glycosci.*, 1998, **45**, 393.
13. Y. Sakano, J. Fukushima and T. Kobayashi, *Agric. Biol. Chem.*, 1983, **47**, 2211.
14. T. Tonozuka, H. Sakai, T. Ohta and Y. Sakano, *Carbohydr. Res.*, 1994, **261**, 157.
15. T. Tonozuka, S. Mogi, Y. Shimura, A. Ibuka, H. Sakai, H. Matsuzawa Y. Sakano and T. Ohta, *Biochim. Biophys. Acta*, 1995, **1252**, 35.
16. T. Tonozuka, A. Ibuka, H. Sakai, H. Matsuzawa and Y. Sakano, *J. Appl. Glycosci.*, 1996, **43**, 95.
17. T. Kuriki and T. Imanaka, *J. Gen. Microbiol.*, **135**, 1989, 1521.
18. S. M. Podkovyrov and J. G. Zeikus, *J. Bacteriol.*, **1992**, 174, 5400.
19. T. Oguma, A. Matsuyama, M. Kikuchi and E. Nakano, *Appl. Microbiol. Biotechnol.*, 1993, **39**, 197.
20. S. P. Mathupala, S. E. Lowe, S. M. Podkovyrov and J. G. Zeikus, *J. Biol. Chem.*, 1993, **268**, 16332.
21. R. Nakajima, T. Imanaka and S. Aiba, *Appl. Microbiol. Biotechnol.*, 1986, **23**, 355.
22. H. Aoki, Yopi and Y. Sakano, *Biochem. J.*, 1997, **323**, 757.
23. S. Kamitori, T. Satou, T. Tonozuka, Y. Sakano, H. Matsuzawa and K. Okuyama, *J. Struct. Biol.*, 1995, **114**, 229.
24. S. Kamitori, S. Kondo, K. Okuyama, T. Yokota, Y. Shimura, T. Tonozuka and Y. Sakano, *J. Mol. Biol.*, 1999, *in press.*
25. H. M. Jespersen, E. A. MacGregor, M. R. Sierks and B. Svensson, *Biochem. J.*, 1991, **280**, 51.
26. Y. Matsuura, M. Kusunoki, W. Harada and M. Kakudo, *J. Biochem.*, 1984, **95**, 697.
27. C. Klein and G. E. Schulz, *J. Mol. Biol.*, 1991, **217**, 737.
28. Y. Katsuya, Y. Mezaki, M. Kubota and Y. Matsuura, *J. Mol. Biol.*, 1998, **281**, 885.
29. I. Matsui, S. Yoneda, K. Ishikawa, S. Miyairi, S. Fukui, H. Umeyama and K. Honda, *Biochemistry*, 1994, **33**, 451.
30. D. Penninga, B. Strokopytov, H. J. Rozeboom, C. L. Lawson, B. W. Dijkstra, J. Bergsma and L. Dijkhuizen, *Biochemistry*, 1995, **34**, 3368.
31. R. M. A. Knegtel, B. Strokopytov, D. Penninga, O. G. Faber, H. J. Rozeboom, K. H. Kalk, L. Dijkhuizen and B. W. Dijkstra, *J. Biol. Chem.*, 1995, **270**, 29256.
32. A. Ibuka, T. Tonozuka, H. Matsuzawa and H. Sakai, *J. Biochem.,* 1998, **123**, 275.

ROLES OF CATALYTIC RESIDUES IN α-AMYLASES AS EVIDENCED BY THE STRUCTURE ANALYSES OF THE PRODUCT-COMPLEXED MUTANT MALTOTETRAOSE-FORMING AMYLASES

Yoshiki Matsuura[a], Kazuya Hasegawa[a¶], Michio Kubota[b]

[a]Institute for Protein Research, Osaka University, Suita, Osaka 565-0871, Japan
[b]Hayashibara Biochemical Laboratories Inc., Amaseminami, Okayama 700-0834, Japan
[¶]Present address: Protonic NanoMachine Project, ERATO, 3-4 Hikaridai, Seika, 619-0237, Japan

1 INTRODUCTION

Maltotetraose-forming α-amylase (G4-amylase) from *Pseudomonas stutzeri* is an exo-amylase whose crystal structures have been investigated for uncomplexed wild-type G4-2 (the catalytic part of G4-1 enzyme)[1] and product-complexed forms of the catalytic glutamic acid mutated to glutamine (E219Q)[2]. Previous studies revealed that the enzyme contains a $(\beta/\alpha)_8$ barrel structure that is frequently found in other endo-type α-amylases[3,4,5,6,7,8], isoamylase[9], cyclodextrin glucanotransferases[10,11] and also in β-amylase[12]. Further, these studies revealed a highly conserved three-dimensional arrangement of the three catalytic residues[10] that locate at the C-terminal end of a β-strand in the barrel. Subsequently, it is possible to discuss the general catalytic mechanism in the α-amylase family enzymes[13,14].

The maltopentaose-cocrystallized structure of mutated G4-amylase revealed the binding mode of each of the non-reducing end four glucose residues (product) of the cleaved substrate; and the mechanism of recognition of the non-reducing end of amylose, which determines the exo-wise cleavage of this enzyme; and a deformation of a sugar ring at the cleaved reducing end[2].

We have further crystallized the product complexes of the four different single mutants with respect to the three catalytic residues Glu219, Asp193 and Asp294; *i. e.*, E219G, D193N, D193G and D294N. The complexed structures of these mutants were studied by single crystal X-ray diffraction at 2 Å resolutions for the purpose of comparatively elucidating the detailed sugar-bound structures at the active site. We describe these structures with discussing the possible roles of these residues in the catalytic and substrate binding mechanism, which may be generalized to apply to other α-amylase family enzymes.

2 RESULTS AND DISCUSSION

2.1 Structures of product-complexed mutated enzyme active sites

In the present structural analyses, single mutations have been introduced in the catalytic residues, which surround the reducing end glucose unit (Glc4) of the bound maltotetraose. Thus, major structural changes have occurred in the region of the enzyme active site and Glc4. The average r.m.s. deviations of the atoms from E219Q in the reducing end glucose unit Glc4 (0.71 Å) were significantly larger than that in the non-reducing end glucose units Glc1-3 (0.27 Å). A superimposed picture of the five complexed mutant enzymes at the active sites is shown in Fig. 1, depicting only the reducing end glucose unit of the bound maltotetraose and the three catalytic amino acid residues. As shown in this figure, the Glc4 rings are deformed, adopting a half-chair conformation in all mutants except the D294N complex. The average positions of Glc4 with respect to the enzyme active site are similar to those of the E219Q complex[2] in the present E219G and D294N complexes. However, in D193G the position moved down toward the bottom of the cleft, and in D193N it moved away from the bottom.

The side chains at the position 219 in D193N, D193G, and D294N rotated about the Cα-Cβ bond compared to their position in E219Q, probably because of differences among the side chain groups. In E219G, water molecule Wat518 was incorporated in the space of the side chain of Glu219, having been hydrogen-bonded to Arg233NH1, Ile241O and Asp244OD1. Similarly, in D193G water molecule Wat730 was incorporated in the space of the side chain of Asp193, being hydrogen bonded to O6 of Glc4. Hydrogen-bonds and close contacts surrounding Glc4 residue in the complexed mutants are summarized in Table 1. As shown in this Table, the following hydrogen-bonds always exist: between O1 (Glc4) and side chain atoms of the residues at the position 219 except E219G; between O2, O3 (Glc4), and the side chain atoms at the position 294; and between O6 (Glc4) and the side chain atoms of the position 193 except D193G. In addition, the O6 atoms make hydrogen-bonds between respective His117NE2's in E219Q, E219G, and D193G, but the distances in D193N and D294N are a little too long (3.5 and 3.6 Å, respectively) for them to make hydrogen-bond. The O5-atom in the ring of Glc4 seems to have interactions with the side chain atoms at the position 193, except in D193G, where Wat730 locates close (3.0 Å), though it is uncertain that these atoms form hydrogen-bonds[15].

In analyzing the D294N complex structure, we first used the crystal mounting technique that is used with other crystals for collecting intensity data. However, by using this data we observed high B-values in the atoms composing maltotetraose, probably because of their low occupancies. We then collected the intensity data for this complex by soaking a crystal in a capillary in a solution containing 10 mg/ml maltopentaose during data collection. After refinement of the structure of D294N, we obtained lower B-values for the maltotetraose atoms in the case of soaking, of average value 27.6 $Å^2$, compared to 38.8 $Å^2$ for the unsoaked crystal. For the purpose of comparison with other complexes, we used the structure of D294N from the soaked crystal.

2.2 On the roles of catalytic residues

The three catalytic residues Glu219, Asp193, and Asp294 in the present enzyme, and corresponding residues in α-amylase family enzymes[13], are always conserved in amino acid sequences and in all known three-dimensional structures, as shown in Fig. 2[9]. This fact suggests that these catalytic residues share their individual roles and/or cooperate with each other in the catalytic process. Of these residues, Glu219 in this enzyme is currently considered to work as the acid catalyst in the first step of the general acid-base mechanism. In the present analyses, one of the side chain atoms of this residue always (except in E219G) hydrogen-bonded to O1 of Glc4 (Table 1), which supports this supposition. Further, this residue is supposed to be non-ionized in the

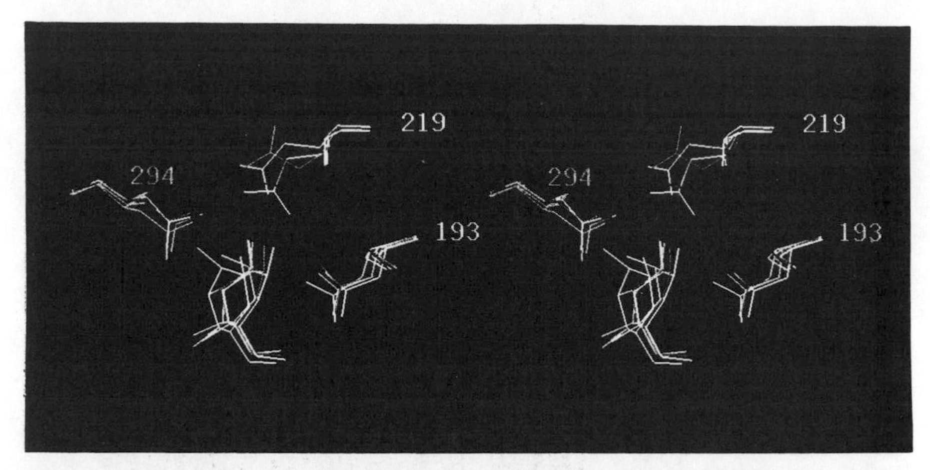

Figure 1 *Stereo plot of the superimposition of the reducing end glucose units (Glc4) of the bound maltotetraoses and the three catalytic residues in the five complexed mutants; E219Q, E219G, D193G, D193N and D294N.*

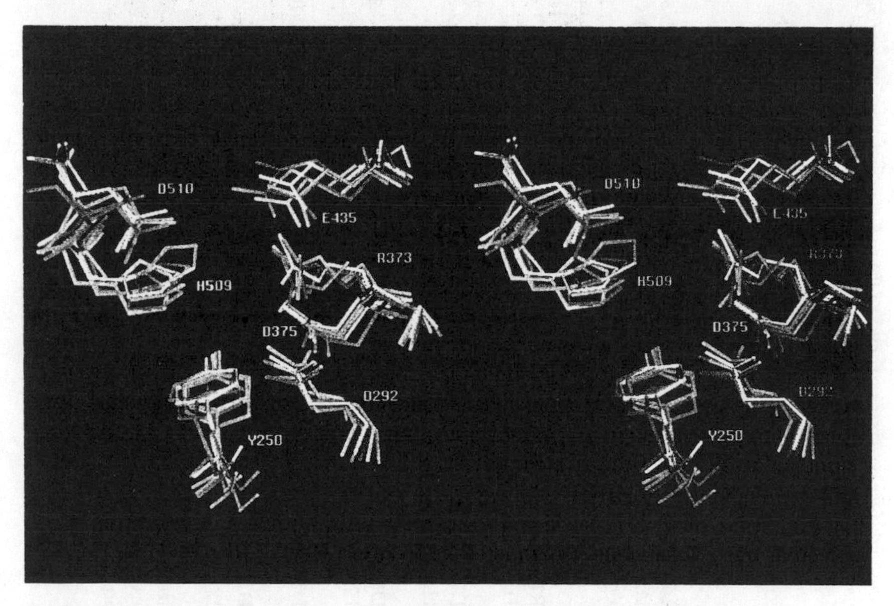

Figure 2 *Stereo plots of three-dimensionally conserved residues in the active sites. The models for isoamylase[9], TAA[8], PPA[3], BA2[5], G4A[2] and CGT[10,11], respectively.*

Table 1 *Hydrogen-bonds and close contacts surrounding the reducing end glucose unit (Glc4) in the product complexed mutants. The atom names and distances (in Å) are indicated in parentheses.*

	E219Q[2]	E219G	D193N	D193G	D294N
O1	Q219(NE2: 3.4) D193(OD1: 2.9) Wat591(2.5)	D193(OD1: 2.8) Wat733(2.6)	E219(OE1,2: 3.0, 3.1) Wat731(2.7)	E219(OE1: 2.5) R191(NH1: 3.2)	E219(OE2: 3.3) Wat755(3.2)
O2	D294(OD1: 2.6) R191(NH2: 3.2) H293(NE2: 3.2)	D294(OD1: 2.7) R191(NH2: 3.3) H293(NE2: 3.0)	E219(OE2: 3.1) D294(OD1: 3.3)	D294(OD1: 2.6) R191(NH2: 3.0) H293(NE2: 3.1)	N294(OD1: 2.7) R191(NH2: 3.2) H293(NE2: 3.1)
O3	D294(OD2: 2.8) H293(NE2: 3.1)	D294(OD2: 2.8) H293(NE2: 2.9)	D294(OD2: 2.7) Wat725(2.3)	D294(OD2: 2.6) H293(NE2: 3.2)	N294(ND2: 2.9) H293(NE2: 3.1)
O6[(a)]	D193(OD2: 3.0) H117(NE2: 3.3)	D193(OD2: 3.0) H117(NE2: 3.1)	N193(OD1: 3.1) H117(NE2: 3.5)	H117(NE2: 3.2) Wat730(3.1)	D193(OD2: 2.7) H117(NE2: 3.6)
O5[(b)]	D193(OD1,2: 3.4, 3.5)	D193(OD1,2: 3.4, 3.7)	N193(OD1,ND2: 3.8, 3.0)	Wat730(3.0)	D193(OD1,2:3.3,3.5)
C1[(c)]	D193(OD1: 2.6)	D193(OD1: 2.8)	N193(OD1: 2.9)	—	D193(OD1: 3.1)

[(a)] Long distance pairs between H117NE2 are also listed for comparison.
[(b)] This atom posesses potential hydrogen bond ability, and may form weak hydrogen bonds between the atoms listed.
[(c)] This atom is not possible to form hydrogen bond, but listed here because of unusually short interatomic distances between the side chain atoms of D193.

catalytically optimum pH, working for protonation of the glucosidic O1-atom to cleave the C1-O4 bond at the position between subsites -1 and +1[15]. The second step of the catalysis is the formation of the intermediate of the reaction. The side chain of Asp193 is pointing toward the O1-C1-O5 triad in the cleaving reducing end (position -1) glucose unit (Glc4), and is hydrogen-bonded to O1 and also possibly to O5. In the present complex structure analyses, as shown in Table 1, it should be noted that the carboxyl oxygen atom OD1 of Asp193 lies unusually close to the C1 atom, the distances between them ranging from 2.6 to 3.1 Å, considerably shorter than those of van der Waals contact. It is difficult to imagine that the C1-atom has the potential to form a typical hydrogen bond, despite the fact that it has an attached hydrogen atom. However, the Glc4 ring is distorted in all complexed structures herein except D294N. This suggests that the C1 attached hydrogen atom might be displaced somehow in a stereochemically distorted position, making it possible for C1 and Asp193OD1 atoms to form such close contact. One possible interpretation is that C1 is to some extent positively charged in the product complexed state, and the contact between C1 and OD1 atoms demonstrate ionic bond character, on the other hand, it can also be said that this contact bears a covalent character. In any case, considering the close contact of these atoms, Asp193 is the most likely candidate for the role of being involved in the reaction intermediate complex formation in the catalytic process. With respect to the nature of the chemical bond[17], in which any bond has an intermediate character between the two extremes, ionic or covalent, it is still difficult to declare whether the intermediate formation proceeds by a single displacement via the oxocarbonium ion[18] or by a double displacement mechanism via covalent bond[19] between the C1 and OD1 of Asp193. The isolation of a covalently bonded compound[20] may not be direct identifying evidence for the formation of covalent intermediate in the reaction, because once the reaction intermediate is isolated, it must form a covalent bond whatever the true intermediate is. The oxocarbonium ion intermediate could exist in only a strongly stereochemically restricted reaction field. Furthermore, there is also a possibility that the intermediate may in itself cause a transition from ionic to covalent in bond character.

The fact that even in the D193G complex crystal, where side chain atoms do not exist, the maltopentaose added during the crystallization process was also hydrolysed, producing maltotetraose, implies that the intermediate formation between amino acid residues is not necessarily or absolutely indispensable. A probable explanation for this finding may be that some other chemical unit such as an ion or a water molecule from the solvent, might be working as a replacement for the intermediate formation counterpart.

As described in the previous section, in the D294N complex structure, the maltotetraose showed a low occupancy in binding onto the cleft. This finding suggests that the mutation of Asp294 to Asn lowered the binding constant of maltotetraose, even though the side chain atoms of Asn294 maintained hydrogen-bonds between both O2 and O3 of Glc4. This suggestion is supported by the experiment in which the occupancy was raised (B-values lowered) by soaking the crystal in a substrate solution during data collection. Further, in this mutant complex, the deformation of Glc4 was not observed. These findings, taken together, make highly probable that Asp294 is working predominantly to firmly bind the substrate, inducing the ring distortion of the reducing end glucose unit (Glc4) in order to lower the reaction potential of the hydrolysis.

References

1. Y. Morishita *et al., J. Mol. Biol.*, 1997, **267**, 661.
2. Y. Yoshioka, K. Hasegawa, Y. Matsuura, Y. Katsube and M. Kubota, *J. Mol. Biol.*, 1997, **271**, 619.
3. M. Qian, R. Haser and F. Payan, *Protein Sci.*, 1995, **4**, 747.
4. G. Machius, G. Wiegand and R. Huber, *J. Mol. Biol.*, 1995, **246**, 545.
5. A. Kadziola, J. Abe, B. Svensson and R. Haser, *J. Mol. Biol.*, 1994, **239**, 104.
6. Z. Fujimoto *et al., J. Mol. Biol.*, 1998, **277**, 393.
7. S. Strobl *et al., J. Mol. Biol.*, 1998, **278**, 617.
8. Y. Matsuura, M. Kusunoki, W. Harada and M. Kakudo, *J. Biochem.*, 1994, **95**, 697.
9. Y. Katsuya, Y. Mezaki, M. Kubota and Y. Matsuura, *J. Mol. Biol.*, 1998, **281**, 885.
10. M. Kubota, Y. Matsuura, S. Sakai and Y. Katsube, *Denpun Kagaku*, 1991, **38**, 141.
11. B. Strokopytov *et al., Biochemistry*, 1996, **35**, 4241.
12. B. Mikami, M. Degano, E. J. Hehre and C. Sacchettini, *Biochemistry*, 1994, **33**, 7779.
13. H. Takata *et al. J. Biol. Chem.*, 1992, **267**, 18447.
14. B. Svensson, *Plant Molecular Biol.*, 1994, **25**, 141.
15. G. A. Jeffrey, *Acta Crystallog.*, 1990, **B46**, 89.
16. G. J. Davies, K. S. Wilson and Henrissat, *Biochem. J.*, 1997, **321**, 557.
17. L. Pauling, 'The Nature of the Chemical Bond' 3rd ed., chap.3, Cornell Univ. Press, New York, 1960.
18. D. C. Phillips, *Proc. Roy. Acad. sci.*, 1967, **57**, 484.
19. D. E. Koshland, *Biol. Rev.*, 1953, **28**, 416.
20. R. Kuroki, L. H. Weaver and W. Matthews, *Science*, 1993, **262**, 2030.

MOLECULAR MODELLING OF COMPLEXES BETWEEN α-AMYLASES AND AMYLOSE FRAGMENTS OF HIGH DP

G. André and V. Tran

Laboratoire de Physico-Chimie des Macromolécules, INRA
BP 71627 - 44316 NANTES Cedex 03 - FRANCE

1 INTRODUCTION

α-Amylases (EC 3.2.1.1) are amylolytic enzymes catalysing the α(1-4) glycosidic linkages of starch components (amylose or amylopectin) and derived products. These enzymes are widely distributed among living organisms and the architectural knowledge of the region where its natural substrate is bound, is essential for an efficient modelling of the corresponding complexes. Native enzymes or complexes with inhibitors provide main structural features about the global organisation of these enzymes and the probable arrangement of the substrate just before the catalytic action. Thus, the Protein Data Bank (PDB) is the most important source of information. For several years, α-amylases of different origins have been solved with enough accuracy to undertake a global molecular modelling approach of their interactions with substrate fragments of high degree of polymerisation (DP > 10). For mammalian species, it is possible to access to α-amylases from porcine pancreas[1-10] in the native form or complexes with protein inhibitors and more interestingly, with pseudo-saccharide inhibitors. Other α-amylases from human pancreas[11,12] and from human saliva[13,14] are accessible. The α-amylase of yellow meal worm is also available[15,16]. Concerning bacteria, three of them are accessible, from *Alteromonas haloplanctis* [17-19] with a psychrophilic activity, from *Bacillus subtilis*[20-23] and from *Bacillus licheniformis*[24]. For plant species, only the barley α-amylase has been solved [25-28]. Finally, the fungal species are represented by *Aspergillus niger*[29] and *Aspergillus oryzae* (TAKA amylase)[30-31].

For docking calculations, a series of papers have been published on the barley α-amylase of high pI (AMY2). The glucose ring deformation has first been analysed [32] in terms of glycosidic torsion angle shifts and amylose chain propagation. For this purpose, new maps of maltose have been drawn, taking into account the stable "chair" form or one of the six "skew" forms either for the non-reducing or the reducing ring. Then, the systematic docking of a maltopentaose fragment [33] has been carried out, yielding to a unique docking solution basically characterised by two strong stacking bindings for rings A and C (with residues Y51 and W206 respectively). The glucopyranose ring labelling used in the present text is mentioned on figure 1. Finally, this docking approach has been extended to a substrate fragment of DP10[34] (from ring A^{-5} to ring E) and one docking solution has been refined with two other external stacking residues (ring A^{-5}

with Y104 and ring E with Y211). Another docking study [35] has been published for porcine pancreatic α-amylase with a pentasaccharide fragment spanning from subsite (-3) to (+2).

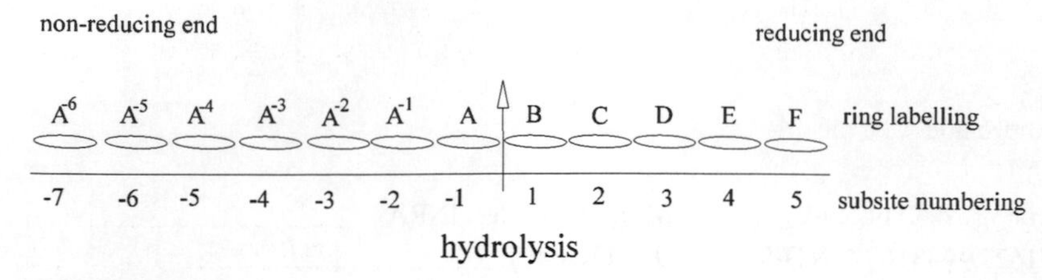

Figure 1 *Ring labelling and subsite numbering*

The purpose of the present work is to generalise the docking approach for long substrate fragments, already performed in the barley case, to other α-amylases whose coordinates are accessible from the PDB. In so doing, a better knowledge of the interaction between the enzyme and the substrate is expected because several features are still obscure such as for example, the uniqueness of the docking solution for a given α-amylase. For the enzyme, the cleft region could be functionally defined as the zone where the substrate is first trapped, then deformed and finally hydrolysed. Therefore, a better characterisation of the cleft region is needed for an accurate understanding of the enzymatic mechanism. This region should be defined from a topological point of view (i.e. the relative arrangement of enzyme segments delimiting to this region) and from a topographical point of view as well (i.e. identification and exact location of residues responsible for the substrate binding by means of stacking interactions, hydrogen bond network or Van der Waals contacts). Obviously, for the substrate, an accurate description of each ring conformation is expected. More interestingly, the chain propagation could have purely helical shape, less pronounced helical tendency or even a random behaviour. It must be underlined that this chain propagation is of special interest for the enzymatic efficiency of α-amylases on solid and crystalline substrates that could be modelled as helical chains. Depending on the results, the chain conformation could be compared to structural models of starch of type A or B with regard to the starch origin.

2 PROTOCOLS

The molecular modelling study has been carried out on Silicon Graphics computers with the MSI/Biosym packages (San Diego, CA, USA). Molecular displays and energy minimizations have been performed with InsightII, Biopolymer and Discover modules. For all calculations, the CFF91 forcefield has been selected.

2.1 Starting Enzyme Geometries

The enzymes selected for docking calculations have been listed on table I. The corresponding PDB code is mentioned on this table. For mammalian species, the α-amylase from porcine pancreas has been selected [PIG]. The α-amylases of yellow meal worm [JAE] and of the bacterium *Alteromonas haloplanctis* [AQH] have been kept as representatives of less developed organisms. The barley α-amylase [AMY] is the unique representative of the vegetal kingdom, while for fungal species, only the Taka-amylase coordinates [TAA] have been selected.

Table I *Modelled α-amylase complexes*

kingdom	PDB code	origin	reference	resolution
animal	1PIG	porcine pancreas	[1]	2.2 Å
	1JAE	yellow meal worm	[15,16]	1.6 Å
bacterium	1AQH	*alteromonas haloplanctis*	[17,18,19]	2.0 Å
plant	1AMY	barley	[25,26]	2.8 Å
fungus	7TAA	*aspergillus oryzae*	[31]	2.0 Å

2.1.1 Superposition of Catalytic Residues. For further comparisons, it was first necessary to superpose all geometries. The superposition has been carried out only on the atomic coordinates of the backbone of the three catalytic residues. This very basic assumption is nevertheless powerful because the common $(\beta/\alpha)_8$ barrel folding pattern of all α-amylases is well superimposed. More interestingly, this superposition provides a first visual definition of the cleft region when considering the (pseudo)saccharide rings of crystallised inhibitors as foot prints. On figure 2, all inhibitors of interest have been represented on the basis of the PIG backbone.

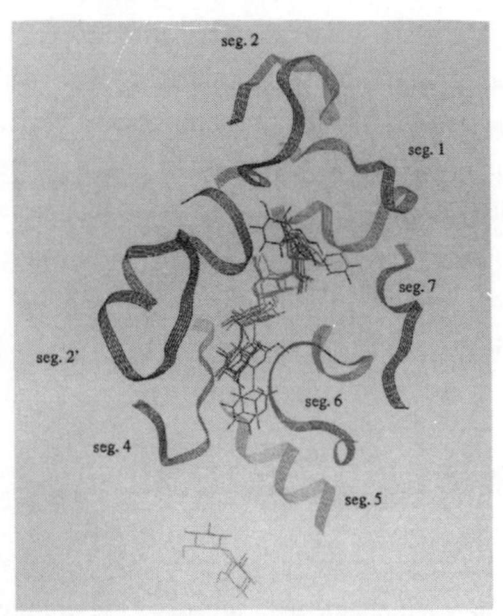

Figure 2 *Cutting out of the PIG model cleft. Segments are shown with ribbons. For sake of clarity, segment 3 is not drawn. As foot prints of the cleft region, all rings found in complexes have been superimposed on the basis of the catalytic triad backbone.*

As a net result of this basic superimposition, rings A, B and C for the substrate are very well located deep inside the catalytic cleft. Therefore, it could be assumed that there is no ambiguity concerning the location of these rings for further docking calculations. From this picture, the direction of the substrate propagation is also unambiguous from the non-reducing end to the reducing one.

2.1.2 Cutting-out of the Enzymes. The segregation between active and inactive parts of the enzyme regarding for the substrate binding, is necessary to spare CPU time for the minimisation procedures. However, in order to explore several docking solutions, it has been decided that the catalytic cleft should be much broader than the one which could be obtained from a purely visual characterisation. The main reason is that, once the positions of rings A, B and C are fixed, the other rings could significantly diverge from the main trace of the cleft. It has been decided that most of the area on the top of the architectural $(\beta/\alpha)_8$ barrel motif should be selected. In practice, this cleft region for all α-amylases is composed of 8 segments of residues. For sake of continuity, the second and the third segment have been gathered and the intermediate residues have been included in the cleft region. As another manifestation of the superimposition strength, all these segments could be satisfactorily identified even between enzymes of different origins. More interestingly, from a topological point of view, the catalytic cleft is composed of loops between architectural motifs β_i and α_i (β_2-α_2, β_3-α_3, β_4-α_4, β_5-α_5, β_6-α_6, β_7-α_7 for segments 1, 2, 3, 4, 5 and 6 respectively). The only exception concerns segment 5 which is part of helix 8.

2.2 Starting Substrate Geometries

Consistently with experimental data, the initial substrate geometries adopt a skew form for ring A. Then, it has been decided to dock a long fragment of substrate to ensure the complete occupancy of the cleft. Depending on the structural variability between selected enzymes and also on the docking paths, the degree of polymerisation to take into account could differ. Nevertheless, for sake of comparisons, all the calculations have been carried out with substrates of DP12. Several starting geometries have been preliminary tested at random to avoid false minima problems during the minimisation steps. This means that, except the identical position of rings A, B and C deep inside the cleft, other docking solutions than the one at the bottom of the cleft), have been tried for all other glucopyranose rings. The purpose of such an approach was to explore all possible docking solutions. This approach could be assimilated to a simulated annealing technique but, as any stochastic procedure, there is no guarantee that all probable docking solutions have been found.

2.3 Minimisation Procedures

Due to the cutting out of the enzymes, it was necessary to fix all atomic coordinates of the two extreme residues defining each segment to preserve the architecture of the whole protein. Inside each segment, all other coordinates were allowed to vary during the minimisation. However, in order to keep the crystalline structure as far as possible during the calculations, no side chain orientations have been initially modified that could arbitrarily favour any peculiar docking solution. All acceptable docking solutions obtained after preliminary calculations were then extensively refined with several cycles of 30000 iterations.

3 RESULTS AND DISCUSSIONS

3.1 Interaction Energies

All main structural features of docking solutions are listed in Table II. Due to the cutting out of the enzyme and consequently to the variability of these proteins from different origins, the number of selected residues is different in each case although the same cutting out criteria have been used to delimit the cleft region. Therefore, the choice of a potent energy criterion is crucial to compare the different docking solutions. In practice, even if all the minimisation procedures have been carried out using the total energy of the enzyme/substrate complex, it has been found that the interaction energy between the truncated enzyme and its substrate was the most relevant criterion capable to minimise the presence of a different number of

Table II *Results and main docking features. The interaction energies (kcal/mol) are reported on row 4. The energies in parentheses correspond to relative ones for the same α-amylase. Stacking interactions are mentioned in the second part of the table. For the glycosidic forms of the substrate (third part of the table), a rough characterisation of the corresponding zone is reported.*

PDB	**1PIG**		**1JAE**		**1AQH**		**1AMY**		**7TAA**
Type	**S1**	**S2**	**S1**	**S2**	**S1**	**S2**	**S1**	**S2**	**S1**
E	**-298**	**-272**	**-254**	**-231**	**-293**	**-282**	**-224**	**-263**	**-248**
(ΔE)	(0)	(26)	(0)	(23)	(0)	(11)	(39)	(0)	
Stacking									
A^{-6}									F149
A^{-5}								Y104	
A^{-4}									
A^{-3}									
A^{-2}									
A^{-1}	W59	W59	W57	W57	W47	W47			H80
A	Y62	Y62	Y60	Y60	Y50	Y50	Y51	Y51	Y82
B									
C	Y151	Y151	Y139	Y139	Y127	Y127	W206	W206	Y155
D							W297	W297	
E									
F									
linkage									
A^{-6} - A^{-5}	**B**	**B**	**B**	**B**	**B**	**B**	**B**	**B**	**B**
A^{-5} - A^{-4}	**B**	**B**	**A**	**C**	**B**	**B**	**B**	**B**	**B**
A^{-4} - A^{-3}	**B**	**B**	**B**	**B**	**B**	**B**	**A-B**	**A**	**B**
A^{-3} - A^{-2}	**B**	**A**	**B**	**A'**	**B**	**B**	**C**	**B**	**B**
A^{-2} - A^{-1}	**B**	**B**	**B**	**B**	**B**	**A**	**A-B**	**A**	**B**
A^{-1} - A	**A-B**	**A-B**	**A**	**A**	**A**	**B-C**	**B**	**B**	**B**
A - B	**A'**	**A'**	**A'**	**A'**	**A'**	**A'**	**A'**	**A'**	**A'**
B - C	**C**	**C**	**C**	**B**	**B**	**C**	**B**	**B**	**C**
C - D	**B**	**B**	**B**	**B**	**B**	**B**	**C**	**C**	**B**
D - E	**C**	**B**	**C**	**C**	**B**	**B**	**B**	**A**	**B**
E - F	**B**	**B**	**B**	**B**	**B**	**B**	**B**	**B**	**B**

residues for the enzymes. This energy is composed of VDW and electrostatic terms and it has been used to roughly compare docking solutions from different enzymes. Nevertheless, one must keep in mind that only the relative energies between solutions for a given enzyme are really meaningful for comparisons.

3.2 Two Types of Docking Solutions

Despite all the attempts performed to increase the number of docking solutions, only two families were found differing by the chain orientation at the non-reducing part of the substrate.

3.2.1 Enzyme Architecture. We shall first discuss the enzyme topography of the cleft region and the substrate can be used as foot prints of the cleft region. As seen on figure 2, the cleft can be characterised as a deep and rather straight valley determined by the superimposition of all inhibitors fragments, especially in the core of the cleft. From this picture, the probable propagation of the substrate at the reducing extremity is strengthened by the presence of a secondary site that can fix the glucopyranose rings in the direct continuation of the cleft and despite the enlargement of the valley. Similarly, at the non reducing end, a main trace can be easily spotted as the continuation of the deep valley at the bottom of the cleft. Consistently with our calculations, this deep valley (from non-reducing to reducing extremities) is occupied in about half of the docking cases. This trivial type -from a purely visual point of view- of docking solutions has been named S2 according to an energy criterion discussed further. The docking solution previously found in the barley α-amylase/DP10 complex corresponds to this category. However, another type of docking solutions (named S1) has been found from our calculations. In fact, this other docking type could be also predicted from figure 2. At the non-reducing extremity, and according to the last docked residues of the superimposed inhibitors, a possible path exists on the left side of the deep valley.

These two families of docking solutions could also be identified according to a topological criterion. The deep valley for type S1 goes through segment 1 and segment 2. Conversely, the narrower passage defining type S2 is comprised between segment 2 and segment 2'. Obviously, these virtual docking solutions have been defined regardless of the orientation of particular side chain residues that could definitely obstruct one of these paths.

3.2.2 Substrate Conformations. This segregation between the docking solutions when analysing the topography of the catalytic region of the enzymes is obviously the consequence of the glycosidic conformations of the substrate. The (Φ,Ψ) values between substrate rings have been converted (Table II) to a rough characterisation of the corresponding accessible zone on the maltose map [33] (**A'**, **A**, **B**, **C**, **D** or **E**)*. Most of glycosidic linkages (except between rings A and B) adopt forms **A**, **B** and **C** or an intermediate one which corresponds to a rather compact area in the maltose map. It means that no external zone such as **D** or even **E** is accessible. This clearly indicates that these substrate chains have a rather regular shape. Concerning the linkage between rings A and B, the only glycosidic form found (**A'**) is the clear consequence of the deformation of ring A (skew form) already present in the initial geometry. In fact, this ring deformation and, consequently the **A'** form for the glycosidic linkage A-B, is the response of the

* according to our previous labelling, **A'**, **A**, **B**, **C**, **D** and **E** zones grossly correspond to the following couples of (Φ,Ψ) values: (20,-160), (60,-150), (110,-130), (130,-110), (70,-80) and (80,80) respectively. Due to the continuity of some of these zones (especially between **A**, **B** and **C**), some intermediate forms could be defined by two letters.

substrate to the bent deep inside the catalytic cleft. When analysing more precisely the accessible zones for the linkages of all docking solutions, it can be concluded that the **A** zone is the most external one. From table II, the differentiation between the two kinds of docking solutions is due to the presence of this **A** form for the second type of solutions between rings A^{-3}-A^{-2} (i.e. for the PIG model) or between rings A^{-2}-A^{-1} (i.e. for the AMY model). Concerning the helical shape of the substrate, it can be concluded that the docking solutions of the first type are significantly more regular -in general with a succession of **B** forms for the glycosidic linkages- than those of the second type. In the former case (type S1), except the necessary distortion of the helical shape due to the A ring deformation, the chain mainly presents linkages of **B** form. In the latter case (type S2), another net break in this regularity appears with one **A** form at the non-reducing part of the chain. Figure 3 illustrates these two types of docking solutions.

Figure 3 *Superimposition of docking solutions of type S1 (left part) and S2 (right part). All the substrates have been superimposed on the basis of the catalytic residues backbones. For sake of clarity, the substrates are mainly represented by glucopyranose rings.*

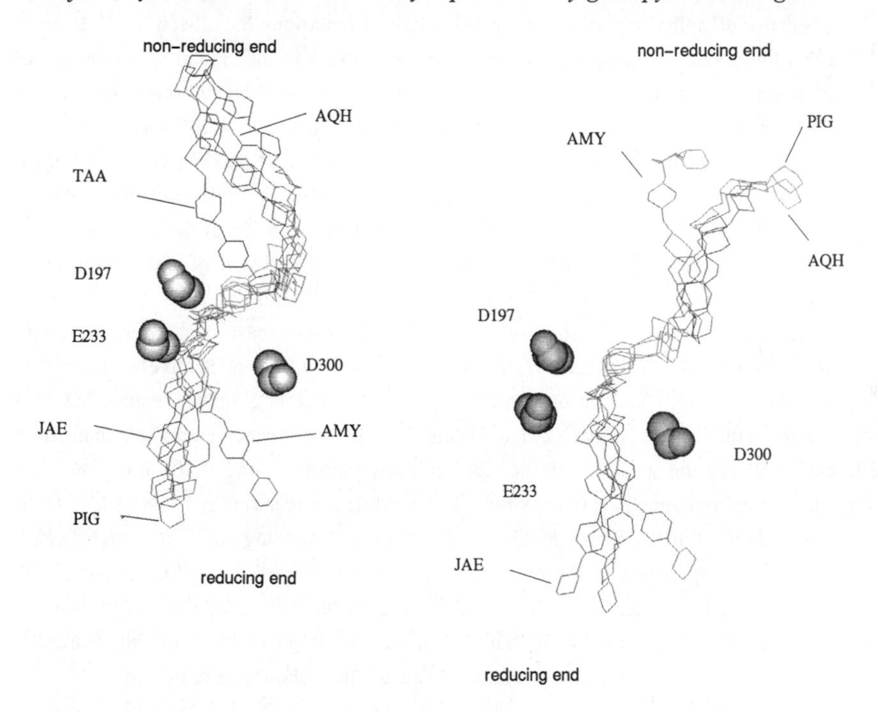

3.3 Comparisons Between Docking Solutions

3.3.1 Energy Criterion. According to the interaction energies, it can be concluded that the PIG and JAE models have two docking solutions, although those of type S2 seem less favoured (ΔE=25.9kcal.mol^{-1} and 22.7kcal.mol^{-1} respectively). This could be explained by high similarities of secondary structures, but it could also be more simply explained by topographical (3D) similarities of the catalytic clefts. For the α-amylase from *Alteromonas haloplanctis* (AQH), the two docking solutions are also present but here with much less pronounced energy difference (ΔE=11.0kcal.mol^{-1}). For the barley case (AMY), the reverse situation occurs with a favoured energy for the docking solution of type S2. In this case, the energy difference is so important (ΔE=39.2kcal.mol^{-1}) that one could have reasonable doubts about the existence of the docking solution of type S1, despite the necessary cautiousness of the energy interpretation. For the TAKA-amylase, only one docking solution is presented which corresponds to type S1. In fact, one residue (Y75) has a side chain orientation strictly preventing the propagation of the substrate in the valley characterising the second type of docking solutions. For this enzyme, the unique docking solution has a good interaction energy which means that the substrate is strongly bound.

3.3.2 Binding Modes. The stacking interactions will be discussed as the strongest binding mode in this kind of enzyme/substrate docking. Stacking interactions are reported in table II. Firstly, it must be mentioned that no side chain orientations have been significantly modified. Therefore, because all initial crystallographic side chain orientations have been kept, it can be concluded that these stacking interactions were formally present in the crystal structure. For all docking solutions, a basic pattern exists, composed of these stacking interactions between rings A, C and tyrosine residues, except for the barley case (stacking between W206 and ring C). This essential feature is consistent with the precise location of rings A, B and C (when existing) in crystal complexes and fully justifies our assumption of these initial ring positions in our starting geometries. Another subsidiary feature has been found with the stacking of the A^{-1} ring in most cases, mainly with tryptophan residues. For the only selected solution of Taka-amylase, an additional stacking has been found at the non-reducing end (F149 with ring A^{-6}).

For the most probable docking solution (type S2) in the barley case, two external stacking residues have been found (Y104 and W297 with rings A^{-5} and D, respectively at the non-reducing and the reducing ends). In our previous paper [35], another stacking residue has been mentioned at the reducing extremity (Y211 with ring E). This discrepancy could be explained by the different lengths of the substrates. In our former study (substrate of DP 10), only four rings were modelled at the reducing part (from ring B to E), while another ring has been added (F) in the present calculation (DP12). This supplementary glucopyranose ring could have significantly modified the chain propagation at the reducing end which therefore, is no longer compatible with the previous stacking interaction. Nevertheless, the stacking binding previously observed should be still valid for substrate with four glucopyranose rings at the reducing extremity. However, with longer substrates, the new scheme presented here should be preferred.

Hydrogen bonds, in addition to VdW contacts, should undoubtedly increase the enzyme/substrate binding. However, due to the initial assumptions of these calculations and because these docking solutions should certainly be still refined, it is difficult to convert all these binding modes to corresponding subsites as can be derived from experimental biochemical data. This detailed analysis is in progress and will be reported further. Therefore, it must be underlined that the presence of twelve glucopyranose rings in our calculations does not mean that each of them could generate a subsite in a biochemical sense. This long fragment of substrate was initially constructed to mimic the static interaction of these enzymes with polysaccharide chains.

4 CONCLUSIONS

On the basis of the backbones of the three catalytic residues, all enzymes have been superimposed. Then, the cleft regions have been defined with eight segments of residues essentially made of loops. From these calculations involving α-amylases from several species interacting with amylose fragments of DP 12, the docking solutions have been segregated on two types corresponding to two paths for the substrate in the non-reducing direction. But such a dual feature does not seem to occur at the reducing extremity. The docking type (S2) corresponds to a deep valley at the bottom of the cleft while the other type (S1) is less trivial to define and looks like a pass, more or less pronounced depending on the α-amylases. However, this rough topographical characterisation is strengthened by a topological criterion since both types depend on well defined paths between two adjacent segments of residues.

Obviously, the two types of docking solutions are revealed on the substrate conformations. In all cases, the A ring of the substrate presents a skew deformation (essential for the hydrolysis) that could be initiated by the strong bent of the cleft at this position. More interestingly, all glycosidic linkages adopt forms characterised by points located in a rather compact area (comprising zones **A'**, **A**, **B** and **C**) on the corresponding (Φ,Ψ) projection. Due to the more external location of zones **A'** and **A**, it can be concluded that the S1 docking solutions have a more helical shape than those of type S2. In the former case, these solutions generally present successive **B** forms for the glycosidic linkages at the non-reducing part. This regularity (yielding to an helical behaviour) is only perturbed by the **A'** form between rings A and B. Conversely, the docking solutions of type S2, present another break point in the regular organisation with the **A** form appearing between rings A^{-2}-A^{-1} or A^{-3}-A^{-2}.

Finally, among the different α-amylases, some of them have this dual opportunity to efficiently bind this long substrate while others seem to have only one docking solution. This segregation could be originated from the variability between α-amylases from different species but it could more probably be due to local side chain arrangements of residues specific to a given enzyme. Lots of improvements have to be done for more efficient comparisons. Extensions of the present work are in progress at least to corroborate these preliminary modelling results to experimental data available as for example, the number and the location of subsites on each part of the hydrolytic zone. Nevertheless, the modelling results presented here could already be relevant to engage rational modifications of these enzymes.

Bibliographic References

1. M. Machius, L. Vertesy, R. Huber and G. Wiegand, *J. Mol. Biol.*, 1996, **260**, 409.
2. G. Wiegand, O. Epp and R. Huber, *J. Mol. Biol.*, 1995, **247**, 99.
3. M. Qian, R. Haser and F. Payan, *Protein Sci.*, 1995, **4**, 747.
4. S.B. Larson, A. Greenwood, D. Cascio, J. Day and A. McPherson, *J. Mol. Biol*, 1994, **235**, 1560.

5. M. Qian, R. Haser, G. Buisson, E. Duée and F. Payan, *Biochem.*, 1994, **33**, 6284.
6. M. Qian, R. Haser and F. Payan, *J. Mol. Biol.*, 1993, **231**, 785.
7. C. Bompard-Gilles, P. Rousseau, P. Rouge and F. Payan, *Structure*, 1996, **4**, 1441.
8. C. Gilles, P. Rousseau, P. Rouge and F. Payan, *Acta. Crystallogr. Sect. D*, 1996, **52**, 581.
9. M. Qian, S. Spinelli, H. Driguez and F. Payan, *Protein Sci.*, 1997, **6**, 2285.
10. C. Gilles, J.P. Astier, G. Marchis-Mouren, C. Cambillau and F. Payan, *Eur. J. Biochem*, 1996, **238**, 561.
11. G.D. Brayer, Y. Luo and S.G. Withers, *Protein Sci.*, 1995, **4**, 1730.
12. D. Burk, Y. Wang, D. Dombroski, A.M. Berghuis, S. V. Evans, Y. Luo, S.G. Withers and G.D. Brayer, *J. Mol. Biol.*, 1993, **230**, 1084.
13. N. Ramasubbu, K.K. Bhandary, F.A. Scannapieco and M.J. Levine, *Proteins: Struct. Funct.*, 1991, **11**, 230.
14. F.A. Scannapieco, K. Bhandary, N. Ramasubbu and M.J. Levine, *Biochem. Biophys. Res. Comm.*, 1990, **173**, 1109.
15. S. Strobl, K Maskos, M. Betz, G. Wiegand, R. Huber, F.X. Gomis-Ruth and R. Glockshuber, *J. Mol. Biol.*, 1998, **278**, 617.
16. S. Strobl, F.X. Gomis-Ruth, K. Maskos, G. Frank, R. Huber and R. Glockshuber, *FEBS Lett.*, 1997, **409**, 109.
17. N. Aghajari, G. Feller, C. Gerday and R. Haser, *Protein Sci.*, 1998, **7**, 564.
18. N. Aghajari, G. Feller, C. Gerday and R. Haser, *Protein Sci.*, 1998, **7**,1481.
19. N. Aghajari, G. Feller, C. Gerday and R. Haser, *Structure*, 1998, **6**, 1503.
20. L. Fujimoto, K. Takase, N. Doui, M. Momma, T. Matsumoto and H. Mizuno, *J. Mol. Biol.*, 1998, **277**, 393.
21. H. Mizuno, Y. Murimoto, T. Tsukihara, T. Matsumoto and K. Takase, *J. Mol. Biol*, 1993, **234**, 1282.
22. K. Takase, T Matsumoto, H. Mizuno and K. Yamane, *Biochim. Biophys. Acta*, 1992, **1120**, 281.
23. K. Yamane, H. Hirata, T. Furusato, H. Yamazaki and A. Nakayama, *J. Biochem.*, 1984, **96**, 1849.
24. M. Machius, G. Wiegand and R. Huber, *J. Mol. Biol*, 1995, **246**, 545.
25. A. Kadziola, J-I. Abe, B. Svensson and R. Haser, *J. Mol. Biol*, 1994, **239**, 104.
26. A. Kadziola, M. S gaard, B. Svensson and R. Haser, 1998, *J. Mol. Biol.*, **278**, 205.
27. F. Vallee, A. Kadziola, Y. Bourne, J-I. Abe, B. Svensson and R. Haser, *J. Mol. Biol.*, 1994, **236**, 368.
28. M. S gaard, A. Kadziola, R. Haser and B. Svensson, *J. Biol. Chem*, 1993, **268**, 22480.
29. E. Boel, L. Brady, A.M. Brzozowski, Z.S. Derewenda, G.G. Dodson, V.J.Jensen, S.B. Petersen, H. Swift, L. Thim and H.F. Woldike, *Biochem.*, 1990, **29**, 6244.
30. H.J. Swift, L. Brady, Z.S. Derewenda, E.J. Dodson, G.G. Dodson, J.P. Turkenburg and A.J. Wilkinson, *Acta. Crystallogr. Section B*, 1991, **47**, 535.
31 A.M. Brzozowski and G.J. Davies, *Biochem.*, 1997, **36**, 10837.
32. G. André, A. Buléon, F. Vallée, M. Juy, R. Haser and V. Tran, *Biopolymers*, 1996, **39**, 737.
33. G. André, A. Buléon, M. Juy, N. Aghajari, R. Haser and V. Tran, *Biopolymers*, 1999, **49**, 107.
34. G. André, A. Buléon, R. Haser and V. Tran, *Biopolymers*, submitted
35. F. Casset, A. Imberty, R. Haser, F. Payan and S. Perez, *Eur. J. Biochem.*, 1995, **232**, 85.

REACTIVITY OF PSYCHROPHILIC α-AMYLASE IN THE CRYSTALLINE STATE

N. Aghajari and R. Haser

Institut de Biologie et Chimie des Protéines, UPR-412, CNRS
Laboratoire de Bio-Cristallographie
7, Passage du Vercors 69367 Lyon Cedex 07, France

1 INTRODUCTION

All over the world living organisms have to cope with extreme conditions. These extreme conditions can be high or low temperatures, high pressure, high salt concentrations, extreme pH's, and not necessarily one at a time. For example, organisms from the deep-sea waters which are exposed to cold temperatures often have to cope with high pressures as well. The α-amylase from *Alteromonas haloplanctis* studied here is adapted to cold temperatures. In fact cold environments such as polar regions and oceans which maintain an average temperature of 1 to 3°C, present over half of the earth's surface. However, little is known about the relationship between structure/function/adaptation of enzymes originating from organisms in these regions, which permanently function at cold temperatures.

This work has given rise to the crystallization and three-dimensional structure determination of two different crystal forms of native α-amylase from the psychrophilic (cold-loving) bacterium *Alteromonas haloplanctis* (2.0 Å and 2.4 Å resolution). These structures have allowed detailed analysis with the highly similar mesophilic mammalian α-amylases and resulted in the proposal of a number of plausible factors governing adaptation to cold environments on a structural level. For the two crystal forms of this psychrophilic α-amylase, several complexes with substrate analogue inhibitors have been prepared and their crystal structures established to high resolution in order to study the relationship between structure and function of this cold adapted enzyme. On the basis of these studies a mechanism for the catalytic reaction has been proposed and the influence on catalysis of a chloride activator ion situated in the vicinity of the active site of this cold adapted enzyme has been studied.

1.1 Adaptation to Low Temperatures

What is an extreme environment? This is the way in which Kushner[1] introduces "Microbial life in extreme environments". He continues: "a definition made by humans is likely to be anthropocentric; an environment that we, and microorganisms closely associated with us, find extreme.... Every environment will be extreme to some organisms".

Thermophiles, psychrophiles, halophiles, alkalophiles, acidophiles, barophiles and so on, are all terms that can be assigned to organisms which survive and grow under extreme conditions. As concerns adaptation to cold temperatures, Schmidt-Nielsen in 1902[2] proposed the term psychrophile (from the Greek psychros = cold), to describe microorganisms with the ability to grow around 0°C. This was however not the first report of a microorganism which meets the above conditions; as early as 1887 Forster[3] isolated bioluminiscent bacteria from preserved fish. After many reports on psychrophilic and psychrotrophic organisms and definitions hereof (nicely described in Morita[4]), Ingraham & Stokes[5] and Stokes[6] extended Schmidt-Nielsens definition by stating that psychrophiles should not only be able to grow at 0°C, but also grow at a rate to become macroscopically visible in about 1 - 2 weeks. The additives "strict or obligate and facultative" were furthermore assigned to the psychrophiles depending on their ability to grow most rapidly below or above 20°C. As Morita[4] and Baross & Morita[7] later found that the facultative psychrophiles not necessarily prefer low temperatures, they redefined the term psychrophile as organisms having an optimal growth temperature about or below 15°C, and a maximum growth temperature around 20°C, besides the former definition of being able to grow at or below 0°C. The difference between psychrophiles and psychrotrophs was also introduced, the latter having an optimal and maximum growth temperature around 20°C (or slightly higher for the maximum temperature), and a minimum growth temperature at or slightly above 0°C. To these definitions it can be added that some cave bacteria were found which grew at 10°C and 20°C, but not at 4°C nor at 28°C[8] so they could neither be classified as mesophilic bacteria, nor as psychrophilic or psychrotrophic. It was therefore proposed that these bacteria should be described as "stenothermic" (Greek stenos = close, narrow, little)[9]. During many years several reports on the isolation of psychrophilic microorganisms from various environments such as fish, meat, milk, vegetables, soils, continental and marine waters followed, but a great deal of them were probably not real psychrophiles according to Gounot[9]. The discussion on definitions of cold adapted organisms has however not stopped. Feller & Gerday[10] point out that there are no formal reasons to restrict the term psychrophile to bacteria or to prokaryotes and that species from yeast, algae, insects and plants can be referred to as psychrophilic if they continuously "practice" at low temperatures, e.g. 5°C. They further mention that the term psychrotolerant should be added to Morita's definition[4] in order to describe mesophilic species which are able to acclimatize at low temperatures. Feller and co-workers[11] also attract attention to the fact that there has been some confusion in the definition of the optimum temperature of psychrophilic and psychrotrophic organisms. They argue that the growth yield at this temperature is probably a more important factor than the commonly used reference growth rate, since some psychrophiles produce a higher quantity of biomass at low temperatures, although the growth rate may differ a little from that achieved at a higher temperature.

Psychrophilic or psychrotrophic species, such as microorganisms and other ectothermic species, live in polar and alpine regions or in deep-sea waters. The deep-sea is also a high pressure environment, and therefore many psychrophilic and psychrotrophic organisms are barophilic or barotolerant as well[12]. In order for a psychrophile to grow well at low temperatures, it must contain enzymes which display a high specific activity under cold temperatures[13]. Cold enzymes produced by psychrophilic microorganisms or ectothermic organisms which live in cold habitats are generally characterized by[11]:

1) a higher specific activity or turnover number, k_{cat} and a higher physiological efficiency (or catalytic efficiency), k_{cat}/K_m than their mesophilic counterparts over a temperature range from 0-30°C

2) a limited thermal stability which can be illustrated by their fast denaturation mode at moderate temperatures

3) an activity curve which is shifted towards low temperatures as compared to their mesophilic counterparts

A current hypothesis is that the first two factors should stem from an increased flexibility or plasticity of the structure of these enzymes, which should enable them to undergo conformational changes during catalysis at low temperatures. The interest in finding relations between the three-dimensional structures of psychrophilic enzymes and their function and adaptation to low temperatures is quite recent. Because of the high interest of thermophiles in industrial processes due to high stability at high temperatures, the main efforts in studying the relationships between enzyme adaptation and three-dimensional structure, were put on thermophilic enzymes. Nowadays psychrophiles are of interest to biotechnology and to a higher degree (as compared to some years ago) to fundamental research as well. Until recently, crystal structures of true psychrophilic enzymes were lacking. It should however be noticed that some three-dimensional structures of enzymes adapted to cold environments, but which can not be characterized as psychrophilic or true cold-active according to Feller and co-workers[11] have been described. These include Anionic Salmon trypsin,[14-16] pancreatic elastase from North Atlantic salmon[17] and in a comparative study in which determinants of thermophily were discussed[18] the structure of D-glyceraldehyde-3-phosphate dehydrogenase from the lobster *Homarus americanus*[19] was taken as the psychrophilic reference. However, only the article by Smalås and co-workers[15] debated the aspect of adaptation to cold environments. In the case of Salmon trypsin the number of inter-domain interactions between the N- and C-terminal domains as compared to the counterpart from bovine was decreased and the authors found that the interdomain hydrophobic interactions probably were weaker as well[15].

1.2 Crystal Structures of Psychrophilic Enzymes

In addition to that of *A. haloplanctis* α-amylase,[20,21] two structures of psychrophilic enzymes have been published. Alvarez and co-workers[22] have described the three-dimensional structures of triose-phosphate isomerase (TIM) from the psychrophilic bacterium *Vibrio marinus* in complex with sulfate and the substrate analogue 2-phosphoglycolate respectively, both structures to 2.7 Å resolution and Russell and co-workers have studied citrate synthase from the Antarctic bacterium DS2-3R[23]. Factors responsible for adaptation to low temperatures were identified in these structures.

Psychrophilic or psychrotrophic organisms have often been associated to the damage they can cause ([24] and references herein). Only recently the usefulness of cold adapted enzymes, psychrophilic/psychrotrophic microorganisms and their products in biotechnology, has been recognized; see e.g.[13,24-26]. Potential uses for psychrophilic/psychrotrophic glycosylhydrolases as α-amylases and cellulases are found in cold water washing and food-processing. Some of the advantages of psychrophiles and/or their enzymes in biotechnological applications have been summarized elsewhere[13]. These include rapid and economic process termination by mild heat treatment, higher yields from reactions involving thermosensitive components, modulation of the (stereo) specificity of enzyme catalyzed reactions and cost savings by elimination of expensive heating/cooling process-steps.

2 CRYSTAL STRUCTURE AND FUNCTION OF *A. haloplanctis* α-AMYLASE

α-Amylase from *Alteromonas haloplanctis* (henceforward referred to as AHA) belongs to family 13 of glycosyl hydrolases. Only three of the α-amylases for which the sequence is known have been classified outside family 13 (http://www.expasy.ch/cgi-bin/lists?glycosid.txt). They belong to family 57 and include the anaerobic thermophile *Dictyoglomus thermophilum* α-amylase, the thermophile *Pyrococcus furiosus* α-amylase and a putative α-amylase from *Methanococcus jannachii*.

2.1 Physico-Chemical Parameters Characterizing AHA

The Antarctic bacterium *A. haloplanctis* secretes an α-amylase composed of 453 amino acid residues with an M_r of 49340 and a pI of 5.5[27]. It was furthermore discovered that AHA is activated by chloride, and by sequence alignment it was found that AHA and porcine pancreas α-amylase (hereafter PPA) displayed a striking sequence identity of 53% which was increased to 66% when considering amino acids of the same type[28]. It should be mentioned that this sequence similarity, which also is conspicuous when comparing AHA to other mammalian α-amylases (hereafter MAA) does not exist between this psychrophilic enzyme and α-amylases of fungal, plant or other bacterial origin. On the basis of this high similarity with PPA the analyses of substrate and ion affinities were compared with the corresponding values for the mesophilic PPA.

Amylolytic activities were measured towards the synthetic substrate 4-nitrophenyl-α-D-maltoheptaoside-4,6-*O*-ethylidene ($Np(Glcp)_7OCH_3CH^{2+}$) as well as starch[28], and it was found that the turnover number, k_{cat}, at 4°C is approximately seven times higher for AHA as compared to PPA, when considering the natural substrate, and approximately 6 times higher for the synthetic substrate. The physiological efficiency, k_{cat}/K_m, follows this tendency as a result of K_m values which are rather similar for the two enzymes. A drastic shift in the apparent optimal temperature (app. 25°C for AHA compared to 55°C for PPA) which characterizes psychrophiles, was found as well. When considering the thermal stability the authors[28] found that the half time of inactivation for AHA is approximately 20 times lower than the corresponding value for PPA.

AHA was shown to be calcium as well as chloride activated[27] and removal of either of these two ions result in the reversible inactivation of the enzyme[28]. Dissociation constants, K_d, measured by Feller and co-workers[28] showed that the binding affinities for these ions were significantly lower than the same values obtained for PPA.

2.2 3D-Structure of AHA

Two different crystal forms of AHA have been grown[21,29] and their respective structures solved to 2.0 and 2.4 Å resolution[20,21]. These structures show that AHA displays an overall topology common with all known structures of α-amylases from other species, and is characterized by three domains: domain A is composed of a $(\beta/\alpha)_8$ barrel super-secondary structure; domain B protrudes out of this barrel and is inserted between its β3 and α3 secondary structure elements; domain C is a separate globular unit composed of β-strands (eight in AHA) arranged in a Greek key motif. In AHA, like in the other 3D α-amylase structures, a calcium ion is present, forming a binding site which bridges the A and B domains, and appears essential for enzyme folding, stability and optimal activity.

Interestingly, this psychrophilic α-amylase like its mammalian and at least one of its insect counterparts[30], is also chloride activated[27,31]. At the structural level, an explanation for the allosteric activation of AHA by chloride has been given and may be valid for all chloride-dependent α-amylases[20]. Detailed comparisons of AHA with MAA's point out a number of structural factors which are most probably involved in the control of molecular adaptation and activity of this enzyme at low temperatures[21]. In brief, the observed structural differences account for the cold-active nature of AHA: there are reduced inter-domain interactions giving an increase of overall flexibility; there is a higher resilience of the molecular surface resulting from a reduced number of charged residues consistent with less intra-molecular ion-pairs; obvious differences arise in loop regions, which are generally shorter and characterized by a drastic reduction in proline content.

2.3 Tris in the Active Site of AHA: an Ideal Inhibitor?

A Tris molecule was bound in the active site of AHA[20] where it acts as an inhibitor *via* strong interactions to the essential catalytic residues. Furthermore, it was shown that removal of this molecule only was possible when purification was performed in the absence of Tris buffer. All three catalytic residues are involved in this interaction (Figure 1) and binding distances are given in Table 1. From Table 1 it is noted that Glu 200 (the proton donor) and Asp 174 (the catalytic nucleophile) make very strong hydrogen bonds with atoms from the Tris molecule, two of them being between 2.3 and 2.4 Å, and therefore comparable in strength to the so-called low barrier hydrogen bonds found recently in protein-ligand systems, as cited later. The striking strength of these bonds was immediately questioned, but whatever was done in order to place the Tris molecule in different orientations, to give some more normal hydrogen bond distances, the outcome of the refinements always corresponded to the above situation. It should be added here that the Tris molecule only interacts with the three catalytic residues, the conserved arginine residue binding chloride (Arg 172) and water molecules 1196 and 1203, and makes a maximum number of hydrogen bonds *via* its nitrogen atom. According to Wang and co-workers[32] such short hydrogen bonds have been found in the structures of wild type and mutant PBP (phosphate binding protein) which were solved to 0.98 Å and 1.05 Å resolution respectively. In these structures which were very accurately determined, an Asp forms a short hydrogen bond to one of the phosphate oxygens, the values being 2.432 ± 0.007 Å and 2.435 ± 0.007 Å respectively for wild type and mutant PBP. Another study in which very short hydrogen bonds have been reported, is the 1.70 Å and 1.65 Å resolution structures of a citrate synthase in complex with carboxyl and amide analogues of acetyl coenzyme A,[33] where an Asp makes hydrogen bonds of 2.38 Å with the carboxyl analogue, and of 2.49 Å with the amide analogue.

The discovery of a Tris molecule firmly bound in the active site of AHA has allowed us to highlight, on a molecular level, why α-amylases are rather "lazy" in Tris buffers. This inefficiency was noticed earlier by industrial users, but no explanation was given to this phenomenon. It is clear now that Tris can act as a potent inhibitor of α-amylases since it fits ideally to the catalytic trio (Asp, Glu, Asp) which is strictly conserved in all amylolytic enzymes of family 13 (see e.g. Svensson[34]). Nevertheless, even though Tris forms very strong interactions with the active site residues, it can be expelled from the active site by incoming polysaccharide substrate analogues[35].

Table 1 *Interactions between the Tris molecule and its ligands*
* This residue was not reported in ref. 20 due to a 3.30 Å cut-off.

Amino acid/ Water molecule	*Tris molecule*	*Distance (Å)*
Asp 174 OD1	Tris N	2.34
Asp 174 OD1	Tris O3	2.80
Glu 200 OE1	Tris N	2.33
Glu 200 OE1	Tris O2	3.10
Glu 200 OE2	Tris N	2.99
Asp 264 OD2	Tris O3	2.93
Wat 1196	Tris O3	2.76
Wat 1203	Tris N	2.78
*(Arg 172 NH2	Tris O3	3.34)

Other examples of glycosyl hydrolases where Tris binding in the active site has been reported are: the three-dimensional structures of mutant Y48W glucoamylase from *Aspergillus niger* [36] and glucoamylase from *Saccharomycopsis fibuligera* [37].

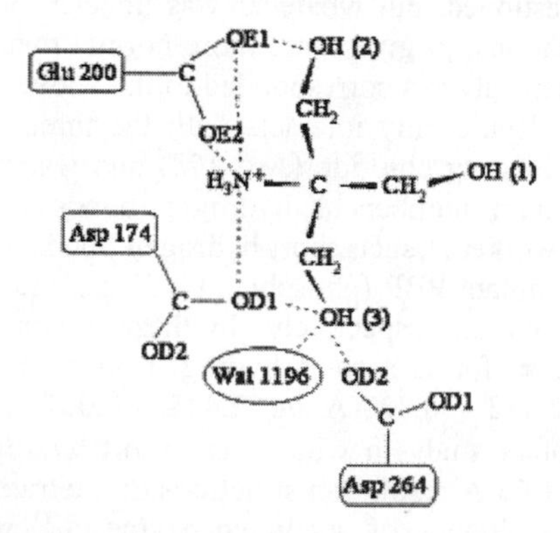

Figure 1 *Binding interactions between a Tris molecule and catalytic residues in the active site of psychrophilic AHA*

3 COMPLEXES WITH SUBSTRATE-ANALOGUE INHIBITORS

α-Amylases can serve as target molecules for studies of designed molecules with potential interest in diabete, hyperlipaemia and obesity treatments[38-43]. Some of the most commonly used oligosaccharide compounds for these studies and treatments origin from the trestatin family. Several three-dimensional studies of α-amylases in complex with oligosaccharides from this family have been carried out[44-48]. The most popular of these compounds, the pseudo-tetrasaccharide inhibitor acarbose, is now available in many countries on the pharmaceutical market for use in the treatment of diabetes. More than one reason causes the high interest of studying α-amylase complexes with acarbose. For example, in the light of mechanistic studies this tetrasaccharide is interesting not only because of its properties as inhibitor, but also because it has been shown that several amylolytic enzymes in the crystalline state, when in contact with acarbose seem to be able to catalyze transglycosylation reactions resulting in different reaction products, where the latter are also depending on the origin of the enzyme.

We have determined the crystal structures of two ternary complexes of AHA which simultaneously can bind two inhibitors[35]: a) one with acarbose and a Tris molecule, b) a second one, with Tris and the so-called component II (Figure 2), which corresponds to the pseudo-trisaccharide derived from acarbose by removing its glucose reducing unit. This study showed that both of the oligosaccharide inhibitors bind exclusively to the active site, and are able to push a pre-bound Tris molecule out of the active site, therefore probably acting as a competitive inhibitors. We find that the Tris molecule in fact stays in contact with both acarbose and component II in the respective complexes, leading in each case to a well defined ternary complex with the psychrophilic α-amylase. Moreover, in the presence of the tetrasaccharide substrate analogue acarbose, AHA performs a transglycosylation reaction resulting in a very well defined heptasaccharide product[35], which clearly displays increased inhibitory activity for the enzyme compared to that of the initial reacting acarbose molecule.

3.1 Binding of Component II to AHA: X-ray Analysis of the Complex

Pure *A. haloplanctis* α-amylase purified as described previously (1992) was kindly supplied by Ch. Gerday and G. Feller and Component II used for soaking experiments was a gift from Dr. H. Bischoff (Bayer Corp., Wuppertal, Germany). AHA was crystallized as described earlier[29] and soaking was performed with crystals in which the three dimensional structure showed the presence of a Tris molecule in the active site[20].

Diffraction data up to 1.8 Å resolution was collected using synchrotron radiation (λ=0.9828 Å) on the D2AM beamline, at the European Synchrotron Radiation Facility (ESRF, Grenoble, France). Difference electron density maps revealed very clearly, and before any refinement, the bound inhibitor and allowed fitting of side chains as well as inhibitor atoms into the electron density.

Electron density map interpretation followed by refinement showed the presence, in the AHA active site, of three rings consistent with the structure of component II (Figure 2). The valienamine part is unambiguously assigned to the sugar-binding subsite -1, whereas the adjacent unit, the 4,6-dideoxyglucose and the glucose rings occupy subsites +1 and +2, respectively (nomenclature according to ref 49). Binding of component II to AHA does not leave any doubt about either the nature or the direction of the sugar rings. It is also very clear that no density remains which could indicate the presence of other sugar binding sites. In fact the electron density terminates abruptly at both extremities, indicative of a well defined trisaccharide species. A great number of interactions occur between the enzyme and the inhibitor[35]. Indeed, among the 14 atoms of component II capable of hydrogen bonding, only 3 are not involved in such interactions with the enzyme environment. In total, 18 hydrogen bonds are formed between enzyme and inhibitor. The average B factors for the three glucosyl moieties (13.8, 15.3, and 22.7 $Å^2$ for sugars in subsites -1, +1, and +2, respectively) are similar and comparable to those of the surrounding protein residues, suggesting full occupancy of the three sugar binding sites. The mode of recognition of component II by the enzyme shows general features often observed in protein-carbohydrate interactions[50], including hydrogen bonding interactions mediated by water molecules. It should be mentioned here that the structure of the AHA/component II represents the first example of a complex between a glycosidase and such an inhibitor, the structure of the latter being still unknown in the free state.

3.2 The AHA Active Site Region Binds Two Inhibitors

The AHA complex with component II was prepared with AHA crystals obtained from an enzyme batch which had been purified in the presence of Tris buffer, and the later was shown to bind very tightly to the active site[20].The electron density resulting from component II binding to AHA showed that the Tris molecule is expelled from the active site by the incoming oligosaccharide inhibitor, but that it stays in the immediate environment, and in hydrogen bonding contact with the trisaccharide[35]. The new Tris binding site is completely different from its original position seen in the Tris inhibited AHA structure: in the AHA/component II complex, Tris is displaced by 7 Å and binds *via* its nitrogen atom to the non-reducing end (subsite -1) of the inhibitor. Therefore, oligosaccharide inhibitor binding to AHA induces large displacements of the Tris molecule, but remarkably, Tris remains bound to the incoming inhibitor. Despite its very strong electrostatic interactions with the AHA active site, Tris is pushed out by the new incoming substrate analogue. This can be easily understood when comparing the number and nature of interactions performed by each ligand and correlated to its specific affinity with AHA: including bonds with water molecules, Tris makes a total of 7 bonds with its environment, compared to 18 for component II. To our knowledge this is the first observation of competition between two inhibitors present simultaneously in an active site region of an enzyme in the crystalline state.

4 CONCLUSION

The crystal structures of the AHA/oligosaccharide complexes reveal both conserved and unique features in comparison with other α-amylases. They give the details of the binding modes, and illustrate inhibitor competition in the crystal as two inhibitors are simultaneously bound in the active site region. Common to the other α-amylases, the bacterial psychrophilic enzyme AHA displays an active site capable to bind at least a heptasaccharide (here only the trisaccharide binding is described) *via* an intricate pattern of interactions which appears highly similar to that seen in all known amylolytic enzymes, even as concerns a number of water molecules mediating the inhibitor/enzyme association. Surprisingly, in many aspects the resemblance of this bacterial psychrophilic α-amylase is greatest with its mesophilic, mammalian α-amylase counterparts, and the present AHA structures confirm the crucial role of some amino acids for recognition and substrate processing.

Component II is seen to bind exclusively to the active site of AHA, therefore enlarging the list of glycosidase inhibitors, some of which have already moved out of the laboratory and into the clinic as potential agents for the treatment of diseases including diabetes, cancer, hyperlipaemia.

Furthermore, the psychrophilic bacterial α-amylase like all enzymes from organisms adapted to cold environments function optimally at low temperature and have therefore considerable potential in biotechnological applications. Therefore, the present AHA models will prove very useful in engineering other active-site mutants, as well as mutants with various transglycosylation possibilities, and/or with enhanced stability without loss of catalytic efficiency in cold environments.
Finally, the structural basis for further understanding psychrophily and high enzymatic activity at low temperatures will now certainly increase rapidly with the increased number of high resolution crystal structures becoming available.

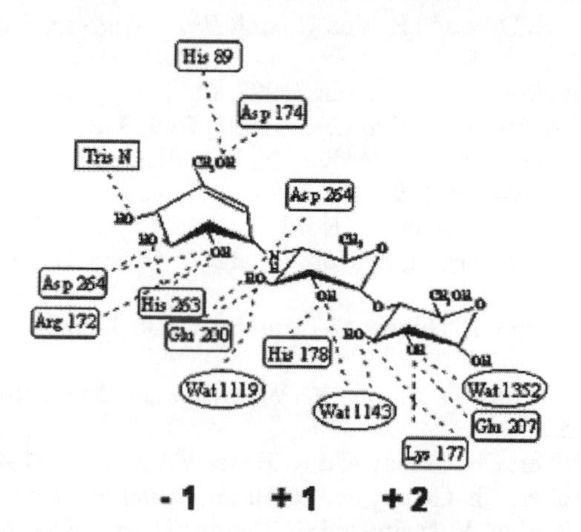

Figure 2 *Interactions between the pseudo-trisaccharide component II, AHA and water molecules*

Acknowledgments
We thank Dr. H. Bischoff (Bayer AG) for supplying us with acarbose and component II, and Prof. Ch. Gerday and Dr. Georges Feller for kindly providing us with pure psychrophilic α-amylase. The present work and N.A. were supported by the Human Capital and Mobility Network CHRX-CT94-0521 and the European Biotech program "Coldzyme".

References

1. D. J. Kushner, 'Aquatic Microbiology: an ecological approach', Blackwell Scientific Publications. Cambridge, 1993, p. 383.
2. S. Schmidt-Nielsen, *Centr. Bakteriol. Parasitenk,* 1902, **9**, 145
3. J. Forster, *Centr. Bakteriol. Parasitenk,* 1887, **2**, 337.
4. R. Y. Morita, *Bact. Rev*., 1975, **39**, 144.
5. J.L. Ingraham and J. L. Stokes, *Bacteriol. Rev*., 1959, **23**, 97.
6. J.L. Stokes, *Recent Prog. Microbiology*., 1963, **8**, 187.
7. J. A. Baross and R.Y. Morita, 'Microbial life in extreme environments', Academic Press, 1978, p. 9.
8. A. M. Gounot, *V Int. Kongr. Speläologie Stuttgart*. Abh. **Bd4**, 1969, B23/1.
9. A. M. Gounot, *Experienta*., 1986, **42**, 1192.
10. G. Feller and Ch. Gerday, *Cell. Mol. Life. Sci.,* 1997, **53**, 830.
11. G. Feller, E. Narinx, J.L. Arpigny, M. Aittaleb, E. Baise, S. Genicot and Ch. Gerday, *FEMS Microbiol. Rev*., 1996, **18**, 189.
12. A. A. Yayanos, A. S. Diez and R. Van Boxtel, *Proc. Natl. Acad. Sci. USA,* 1981, **78**, 5212.
13. N. J. Russell, *Adv. Bioch. Eng. Biotech.,* 1998, **61**, 1.
14. A. O. Smalås and A. Hordvik, *Acta Cryst*, 1993, **D49**, 318.
15. A. O. Smalås, E.S.Heimstad, A. Hordvik, N. P. Willassen and R. Male, *Prot. Struct. Func. Gen*, 1994, **20**, 149.
16. G. I. Berglund, A. O. Smalås and A. Hordvik, *Acta. Cryst*., 1995, **D51**, 725.
17. G. I. Berglund, N. P. Willassen, A. Hordvik and A. O. Smalås, *Acta. Cryst*., 1995, **D51**, 925.
18. I. Korndörfer, B. Steipe, R. Huber, A. Tomschy and R. Jaenicke, *J. Mol. Biol.*, 1995, **246**, 511.
19. M. Buehner, G. C. Ford, D. Moras, K. W. Olsen and M.G. Rossmann, *J. Mol. Biol*., 1974, **90**, 25.
20. N. Aghajari, G. Feller, Ch. Gerday and R. Haser, *Protein Sci*., 1998, **7**, 564.
21. N. Aghajari, G. Feller, Ch. Gerday and R. Haser, *Structure*, 1998, **6**, 1503.
22. M. Alvarez, J. P. Zeelen, V. Mainfroid, F. Rentier-Delrue, J. A. Martial, L. Wyns, R. K. Wierenga and D. Maes, *J. Biol. Chem*., 1998, **273**, 2199
23. R. J. M. Russell, U. Gerike, M. J. Danson, D. W. Hough and G. L. Taylor, *Structure*, 1998, **6**, 351.
24. A. M. Gounot, *J. Appl. Bact*., 1991, **71**, 386.
25. H. Kobori, C. W. Sullivan and H. Shizuya, *Proc. Natl. Acad. Sci. USA* , 1984, **81**, 6691.
26. R. Margesin and F. Schinner, *J. Biotech*., 1994, **33**, 1.
27. G. Feller, T. Lonhienne, C. Deroanne, C. Libioulle, J. Van Beeumen and Ch. Gerday, *J. Biol. Chem*., 1992, **267**, 5217.
28. G. Feller, F. Payan, F. Theys, M. Qian, R. Haser and Ch. Gerday, *Eur. J. Biochem*., 1994, **222**, 441.
29. N. Aghajari, G. Feller, Ch. Gerday and R. Haser, *Protein Sci*., 1996, **5**, 2128.
30. S. Strobl, K. Maskos, M. Betz, G. Wiegand, R. Huber, F. X. Gomis-Rüth and R. Glockshuber, *J. Mol. Biol*., 1998, **278**, 617.
31. G. Feller, O. le Bussy, C. Houssier and Ch. Gerday, *J. Biol. Chem.,* 1996, **271**, 23836.
32. Z. Wang, H. Luecke, N. Yao and F. A. Quiocho, *Nat. Struct. Biol*., 1997, **4**, 519.

33. K. C. Usher, J. Remington, D. P. Martin and D. G. Drueckhammer, *Biochemistry,*
34. B. Svensson, *Plant Mol. Biol.*, 1994, **25**, 141.
35. N. Aghajari, PhD Thesis, Université de la Méditerrannée Aix-Marseille II and Université de Liège, 1998.
36. B. Stoffer, T. P. Frandsen, B. Svensson and M. Gajhede, *1st Carbohydrate Bioengineering Meeting, Elsinore April 1995*, abstr. P16.
37. J. Sevcik, E. Hostinova, J. Gasperik and A. Solovicova, *Jahresbericht: Hamburger Synchrotronstrahlungslabor HASYLAB am Deutschen elektronen-synchrotron DESY*, 1997, 419.
38. D. D. Schmidt, W. Frommer, B. Junge, L. Müller, W. Wingender and E. Truscheit, *Naturwissenschaften*, 1977, **64**, 535.
39. D. D. Schmidt, W. Frommer, B. Junge, L. Müller, W. Wingender and E. Truscheit, 'First international symposium on Acarbose', Excerpta Medica, Amsterdam, 1981, p. 5.
40. E. W. Schmitt, 'A New Principle in the Treatment of Diabetes Mellitus. α-Glucosidase Inhibition of Acarbose', Leverkusen, 1987.
41. S. P. Clissold and C. Edwards *Drugs,* 1988, **35**, 214.
42. J. M. Brogard, B. Willemin, J. F. Blicklé, A. M. Lamalle and A. Stahl, *Rev. Med. Interne.*, 1989, **10**, 365.
43. H. Bischoff, H. J. Ahr, D. Schmidt and J. Stoltefuss, *Nachr. Chem. Tech. Lab.*, 1994, **42**, 1119.
44. M. Qian, R. Haser, G. Buisson, E. Duée and F. Payan, *Biochemistry*, 1994, **33**, 6284.
45. C. Gilles, J. P. Astier, G. Marchis-Mouren, C. Cambillau and F. Payan, *Eur. J. Biochem.*, 1996, **238**, 561.
46. M. Machius, L. Vértésy, R. Huber and G. Wiegand, *J. Mol. Biol.*, 1996, **260**, 409.
47. A. M. Brzozowski and G. J. Davies, *Biochemistry*, 1997, **36**, 10837.
48. A. Kadziola, M. Søgaard, B. Svensson and R. Haser, *J. Mol. Biol.*, 1998, **278**, 205.
49. G. Davies, K. S. Wilson and B. Henrissat, *Biochem. J.*, 1997, **321**, 557.
50. N. K. Vyas, *Curr. Opin. Struct. Biol.*, 1991, **1**, 723.

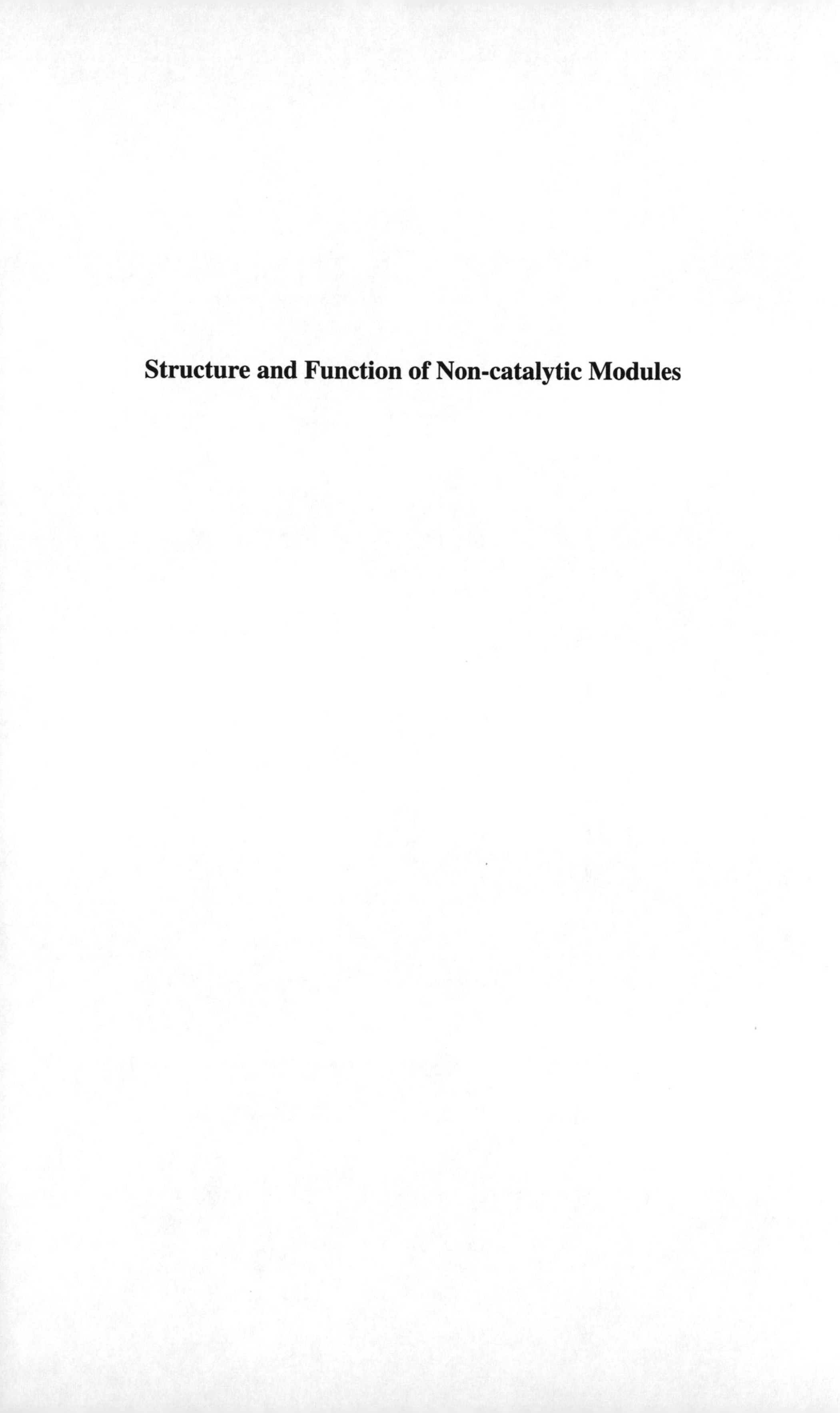

Structure and Function of Non-catalytic Modules

EMERGING PHYLOGENETICS OF CELLULOSOME STRUCTURE

Edward A. Bayer[1], Shi-You Ding[1,2], Adva Mechaly[1], Yuval Shoham[3], and Raphael Lamed[2]

[1]Department of Biological Chemistry, The Weizmann Institute of Science, Rehovot, Israel; [2]Department of Molecular Microbiology and Biotechnology, Tel-Aviv University, Ramat Aviv, Israel; and [3]Department of Food Engineering and Biotechnology, Technion — IIT, Haifa, Israel

1 SYNOPSIS

The cellulosome concept is one of the major paradigms by which microorganisms bring about the efficient enzymatic degradation of crystalline cellulosic substrates. The principal structural element of a cellulosome is a scaffoldin subunit which integrates the other enzymatic subunits into the cellulosome complex. In recent years, the three-dimensional structures of various functional modules, which characterize the typical cellulosome, and in some cases free cellulases as well, have been determined. Currently, phylogenetic analysis of cellulosome-related sequences is contributing novel evidence concerning the intra- and interspecies relationship of the functional modules that comprise a typical cellulosome.

2 INTRODUCTION

The cellulosome was originally described as a discrete multi-protein complex, consisting of cellulolytic and hemicellulolytic enzymes that occur mainly in anaerobic clostridia. [1] The different enzymes are organized into the complex by means of a unique type of cellulosome-integrating protein, called scaffoldin. Special cohesin domains on scaffoldin bind selectively and strongly to complementary dockerin domains on the enzyme subunits, thus incorporating them into the cellulosome complex. [2-6]

Four scaffoldin genes have thus far been reported from clostridal species: *Clostridium thermocellum* (*cipA*), [7] *C. cellulovorans* (*cbpA*), [8] *C. cellulolyticum* (*cipC*) [9] and *C. josui* ("*cipA*"). [10] All four contain multiple type-I cohesin domains, which integrate type-I dockerin-tagged enzymes into the cellulosome complex. In addition, a cellulose-binding domain (Family-IIIa CBD) in scaffoldin is responsible for the binding of the complex to its substrate, cellulose. [11] Another class of domain, a unique *C*-terminal dockerin domain (categorized as a type-II dockerin), has also been identified — until recently, only in the scaffoldin of *C. thermocellum*. [12,13] The type-II dockerin is involved in anchoring the latter scaffoldin to the bacterial cell wall, by interacting selectively with a type-II cohesin, borne by a series of cell-surface proteins. [14] Only 3 such anchoring proteins have yet been described, all in the latter bacterium. Finally, X2 modules of unknown function have been found in all four scaffoldin genes.

In addition to the above clostridial cellulosomes, another cellulosomal scaffoldin has recently been sequenced from *Acetivibrio cellulolyticus*.[15,16] Like the other cellulosomes, the *A. cellulolyticus* scaffoldin (termed CipV) contains a single Family-III CBD and numerous type-I cohesin domains. However, CipV is unique in that it differs from all other scaffoldins thus far described by the inclusion of a Family-9 catalytic domain as an integral part of its polypeptide chain. Nevertheless, the structural organization of the CipV scaffoldin resembles CipA from *C. thermocellum*, due to the arrangement of its CBD and the presence of a terminal type-II dockerin. The other three known scaffoldins collectively exhibit a different type of pattern and appear to form another group. Thus, the scaffoldins from *C. cellulovorans*, *C. cellulolyticum* and *C. josui* all contain an *N*-terminal CBD followed by a single Domain X and all lack a terminal type-II dockerin. Based on these observations, the currently known scaffoldins can be classified according to the provisional scheme in Figure 1.

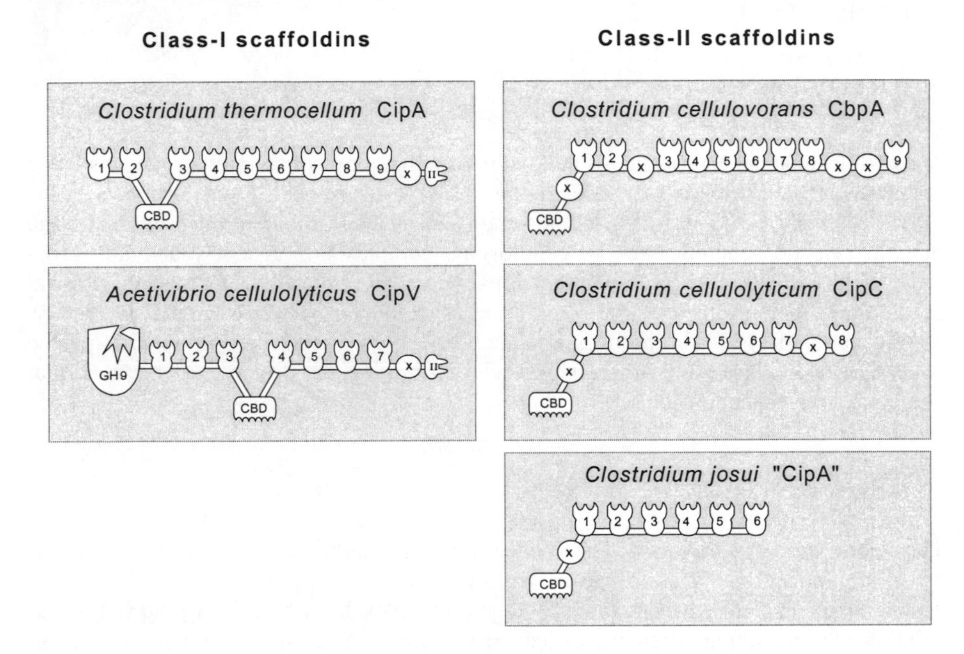

Figure 1 *Structural organization of currently sequenced scaffoldins.*

Additional cellulosome-related components have also been described. These include cellulosome-related oligosaccharides, the function of which is currently unknown, and cellulosome-anchoring proteins, through which the scaffoldin subunit, together with its complement of cellulosomal enzymes, is attached to the cell surface. The composition of cellulosomal oligosaccharides has been elucidated for only two species, *C. thermocellum* and *Bacteroides cellulosolvens*.[17,18] Three different cellulosome-anchoring proteins have been described to date — only in *C. thermocellum*.[19-21] The three known anchoring proteins contain one, two and four copies of a cohesin variant (type-II cohesins) that bind specifically to the type-II dockerin of the *C. thermocellum* scaffoldin.

Two other type-II cohesins (one complete and one partial) were sequenced from the *A. cellulolyticus* genome. [16] These cohesins were located tandemly and immediately downstream with respect to the scaffoldin subunit in same segment. These type-II cohesins presumably comprise part of a putative anchoring protein which would bind to the type-II dockerin of the scaffoldin subunit.

Finally, a 2.2-kb segment of an interesting cellulosome-related protein has recently been sequenced from *B. cellulosolvens*. [16,22] The segment contains a single CBD and five copies of type-II cohesin domains. To our knowledge, this is the first description of a CBD associated with type-II cohesins, thus raising the question as to the precise role of this protein. One possibility is that this segment represents part of a new kind of cellulosomal scaffoldin protein that contains type-II rather than type-I cohesins. Alternatively, the segment could be a novel CBD-containing, cell-surface anchoring protein which would also serve to bind the cell to the substrate.

In any case, it seems that the sequences of many cellulosome-related modules are either unique to or indicative of the presence of a cellulosome in a given organism. These sequences have been referred to as "cellulosome signature sequences". By subjecting numerous such sequences to phylogenetic analysis, the relationship of selected cellulosomal modules within a given species or among various species can be determined.

3 PHYLOGENETIC ANALYSIS OF CELLULOSOMAL MODULES

3.1 Cellulosomal CBDs

All currently known scaffoldin CBDs belong to Family-III, classified according to sequence alignment. [23] The CBDs of this family are separated into two major types: One type (which includes the Family-IIIa and -IIIb CBDs) binds strongly to crystalline cellulose. The other (the Family-IIIc CBDs, fused to a Family-9 glycosyl hydrolase) fails to bind crystalline cellulose but reportedly serves in a helper capacity by feeding a single incoming cellulose chain into the active site of the neighboring Family-9 catalytic module, pending hydrolysis. [24-26]

The phylogenetic relationship of the Family-III CBDs follows the general pattern of their function (Figure 2). Thus, all of the Family-IIIc CBDs form a distinct cluster on one side of the tree. On the opposite side of the weighted centroid are scattered the Family-IIIb CBDs. The Family-IIIa CBDs from the clostridial scaffoldins occupy a single branch, which emanates from an intermediary position among the Family-IIIb CBDs.

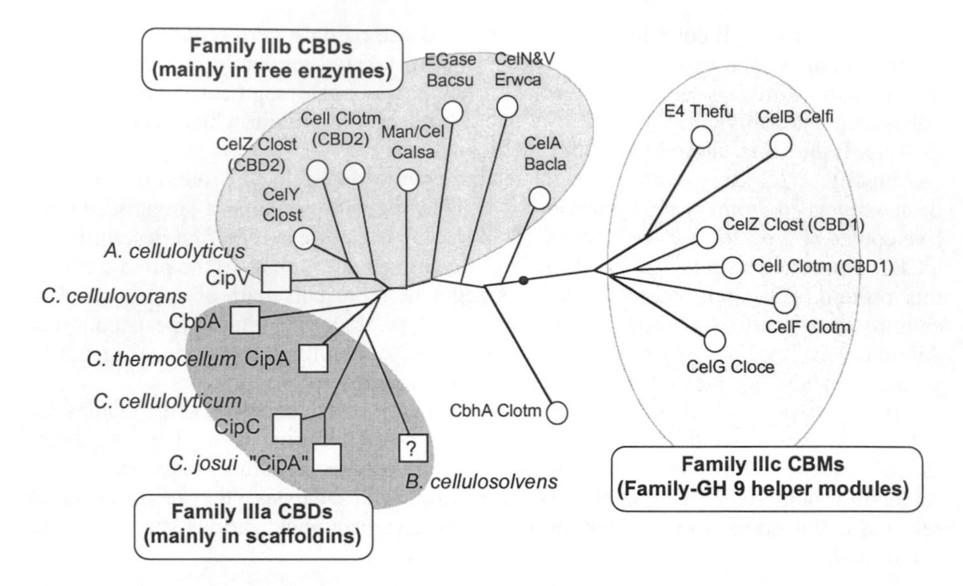

Figure 2 *Phylogenetic relationship of the Family-III CBDs. The scaffoldin CBDs are shown as squares and the CBDs from enzymes as circles. Species abbreviations (Clotm, Thefu, etc.) follow the SWISS-PROT standard.*

The CBDs of Families IIIa and IIIb were previously proposed to be distinguished by the nature of the parent protein, the Family-IIIa CBDs being component parts of cellulosomal scaffoldin subunits and the Family-IIIb CBDs being the targeting agent for free noncellulosomal enzymes. In this context, the Family-IIIa CBDs contain a 9-residue segment called the "scaffoldin loop"[27] that includes a proposed cellulose-binding tyrosine residue (Y67, numbered according to Tormo et al. [28]). This loop is missing in all of the Family-IIIb CBDs. The newly sequenced CBDs from the *A. cellulolyticus* scaffoldin and the *B. cellulosolvens* protein do not follow the Family-IIIa pattern. They both lack a scaffoldin loop and clearly occupy a distinct but related branch of the Family-IIIb CBDs, thus suggesting that the scaffoldin loop is not strictly definitive of the scaffoldins. It therefore follows that the relationship among the Family-IIIa and -IIIb CBDs is not necessarily a precise function of its parent protein, and further attempts at categorizing this particular family of CBD should await a larger collection of relevant sequences.

Nonetheless, in both Families IIIa and IIIb, a planar aromatic strip of amino acids is located on the cellulose-binding surface of the CBD, which is considered to play the major binding function of free cellulases or cellulosomes to crystalline cellulosic substrates. The Family-IIIc CBDs lack this aromatic strip and contain on the binding surface a different pattern of conserved polar amino acids which presumably interact with a single cellulose chain, freed from the crystalline surface of the substrate.

3.2 Cellulosomal Cohesin Domains

Multiple sequence alignment among the type-I cohesins has previously revealed a close interspecies relationship, which suggests that all of these cohesin modules would assume the same general structural fold. Indeed, two similar crystal structures have recently been described for different cohesins from the *C. thermocellum* cellulosomal scaffoldin subunit. [29,30] In both cases, an identical fold was demonstrated, i.e., a nine-stranded jelly-roll topology, which surprisingly matched that of the Family-IIIa CBD from the same subunit.

The type-II cohesins are somewhat larger than those of type-I. The first half of the two sequence types that encompasses the first four or five β strands is aligned quite closely, whereas the latter half tends to diverge. Nevertheless, secondary structure analysis of the type-II cohesin predicts an all β structure, and a very similar three-dimensional fold would be expected.

To date, information regarding the selectivity of the cohesins for the dockerins has been accumulated for only two species — *C. thermocellum* and *C. cellulolyticum.* [31,32] The results indicate that, in both species, each of the individual cohesin domains on the respective scaffoldin subunit recognizes the majority of the dockerins of the enzymatic subunits. Thus, within each of these two species, the recognition of the enzymes appears to be nonselective. In contrast, the cohesin-dockerin interaction between these two species appears to exhibit strict species specificity, such that a cohesin domain from *C. thermocellum* would not recognize a dockerin-bearing enzyme from *C. cellulolyticum* and *vice versa*. Likewise, in *C. thermocellum,* the type-I cohesins apparently fail to recognize the type-II dockerins, [13] and a similar fidelity thus seems to apply to the type-I versus the type-II cohesin-dockerin interaction. Hence, through an essentially identical mechanism, the cell can both integrate the enzymes to the scaffoldin subunit and, independently, attach the resultant cellulosome onto the cell surface in a highly selective manner.

An unrooted phylogenetic tree of the known cohesins is shown in Figure 3. The tree indicates that type-I and the type-II cohesins separate into two major clusters on opposite sides of the weighted centroid. The cellulosomal type-I cohesins from each species form a tight cluster, with the exception of the terminal cohesins of *C. cellulolyticum* and *C. josui* (cohesins 8 and 6, respectively). The latter two cohesins are clearly more similar to each other than to the other cohesins on the same scaffoldin subunit, although all of the cohesins from these two strains together form a tight cluster. The terminal scaffoldin cohesins of the other species also appear to be relatively distinct from their associates.

Interestingly, the type-I cohesins of the mesophilic bacteria are placed on a separate branch of the tree, while the cellulosomal cohesins of the thermophilic *C. thermocellum* occupy an opposing branch. The two known non-cellulosomal type-I cohesins (OlpA from *C. thermocellum* and OrfX from *C. cellulolyticum*) [9,33] maintain discrete positions on the tree, which radiate away from the other cohesin clusters. These two type-I cohesins appear together, on the same side of the weighted centroid, with those of the *C. thermocellum* scaffoldin.

The type-II cohesins of the *C. thermocellum* anchoring proteins appear to have diverged quite significantly compared to the type-I cohesins within a given species. With the exception of cohesins 2 and 3 from OlpB, the relationship of a type-II cohesin within the individual anchoring proteins is often closer to those of another protein. The *Acetivibrio* cohesin from the putative anchoring protein and the three highly related *Bacteroides* cohesins continue to map along the major branch of the type-II cohesins from the *C. thermocellum* anchoring proteins.

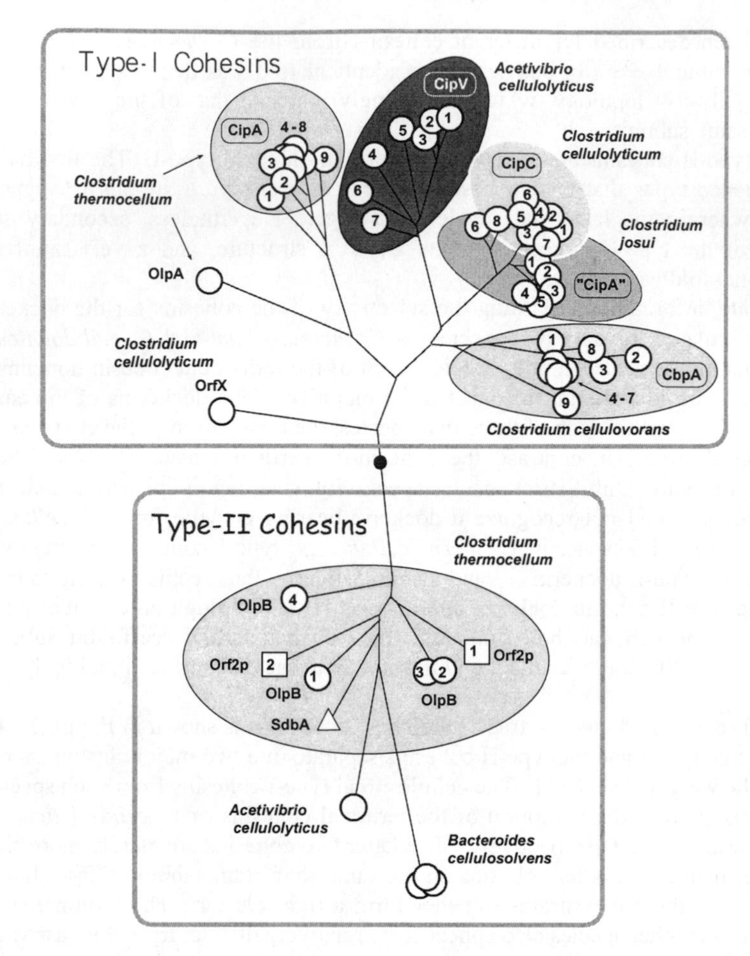

Figure 3 *Phylogenetic relationship of the type-I cohesin modules. The cohesins from a given scaffoldin are numbered according to their sequential arrangement with respect to the N-terminus as illustrated in Figure 1.*

3.3 The X2 Module: Hydrophilic Domains versus Domains X

This particular type of module [currently referred to as the X2 module according to B. Henrissat and P. Coutinho (URL: http://afmb.cnrs-mrs.fr/~pedro/DB/db.html)] was alternatively referred to as hydrophilic domain in *C. cellulovorans*[8] [as opposed to the hydrophobic cohesins], and to Domain X in *C. thermocellum.*[34] The function of this type of module was (and still remains) unknown. Initially, the connection between the respective modules from the scaffoldin subunits of the above-mentioned species escaped notice. However, careful analysis of the sequences revealed a clear correlation among the various modules of this type.[27] A more suitable name for the X2 module awaits the discovery of its structure or function. Secondary structure prediction, based on a collection of hydrophilic domains from *C. cellulovorans* and *C. cellulolyticum* and the Domain X from *C. thermocellum,* indicated an all β structure for this type of module.[27]

The connection between the hydrophilic domains and Domains X is verified by the unrooted phylogenetic tree of the known X2 modules, as shown in Figure 4. The tree shows two major clusters. Interestingly, the phylogenetic relationship of the homologous cellulosomal modules from the known scaffoldin subunits appears to reflect the overall modular organization of the parent scaffoldin (Figure 1). The sequences of the Domains X from the Group-I scaffoldins (i.e., *A. cellulolyticus* CipV and *C. thermocellum* CipA) are very similar, and the two form a small cluster on the right side of the phylogenetic tree (Figure 4). Likewise, the hydrophilic domains from the microorganisms of Group II are also clustered and map on opposite poles of the weighted centroid from those of Group I.

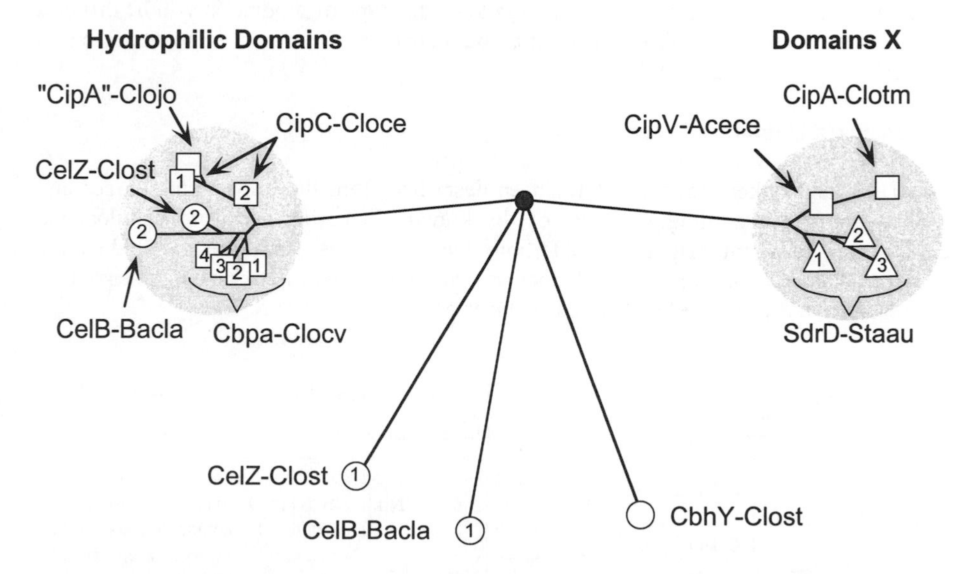

Figure 4 *Phylogenetic distribution of X2 modules. The scaffoldin-derived modules are shown as squares, enzyme-derived modules as circles and cell-wall associated protein from Staphylococcus aureus as triangles. Numbers indicate repeated modules, according to their sequential arrangement with respect to the N-terminus of the given protein. Species abbreviations (Clotm, Clocv, etc.) follow the SWISS-PROT standard.*

It is interesting to note that related X2 modules also occur in a few cellulases: CelZ from *C. stercorarium*[35] and CelB from *Bacillus lautus*[36] carry two tandem modules of this type. In both cases, the *C*-terminal copy joins the Group II scaffoldin cluster, whereas the *N*-terminal copy radiates in marked isolation from the weighted centroid. Another *C. stercorarium* cellulase (CbhY) [37] contains a single X2 module that also appears to diverge radically from a point at or near the weighted centroid.

Another set of newly described cell-wall-associated proteins from *Staphylococcus aureus* has recently been described, [38,39] that contain X2-module repeats (called B-motifs). The latter appear to be closely related to the modules of the Group-I scaffoldins. The B-motifs also include an unduplicated dockerin-like segment at its *N*-terminus, rather than the reiterated *C*-terminal dockerin domain, indicative of the Group-I scaffoldins.

It is difficult to speculate as to the role(s) the X2 module may play in nature. If function is related to sequence, then the similarity to known surface-associated proteins may suggest a possible anchoring role. In the case of the hydrophilic domains from the scaffoldins of the mesophilic clostridia, this would explain how their cellulosomes would be anchored to the cell surface. [40] This would also imply that the second X2 module of the cellulases from *C. stercorarium* and *Bacillus lautus* might also play such a role, perhaps in contrast to the undetermined role of the other copy in the same two enzymes. However, the surface proteins from *Staphylococcus aureus* map on the other side of the centroid, together with the Domains X from *C. thermocellum* and *A. cellulolyticus*, both of which appear to contain other types of anchoring proteins for this purpose.

Direct biochemical evidence concerning the biological roles(s) of the X2 module is lacking. Until we know the precise function(s) of this type of module, it will be difficult to assess the biological and/or structural consequences of the observed sequence-based clustering shown in Figure 4.

3.4 Dockerin Domains

Thus far, dockerin domains have been described from three different sources and can be grouped accordingly. Most of the known dockerins are from cellulosomal enzymes of bacterial origin. In addition, two dockerins from the scaffoldins are currently known. A third type of dockerin has been identified in fungal enzymes. Representative examples of dockerin sequences are shown in Figure 5.

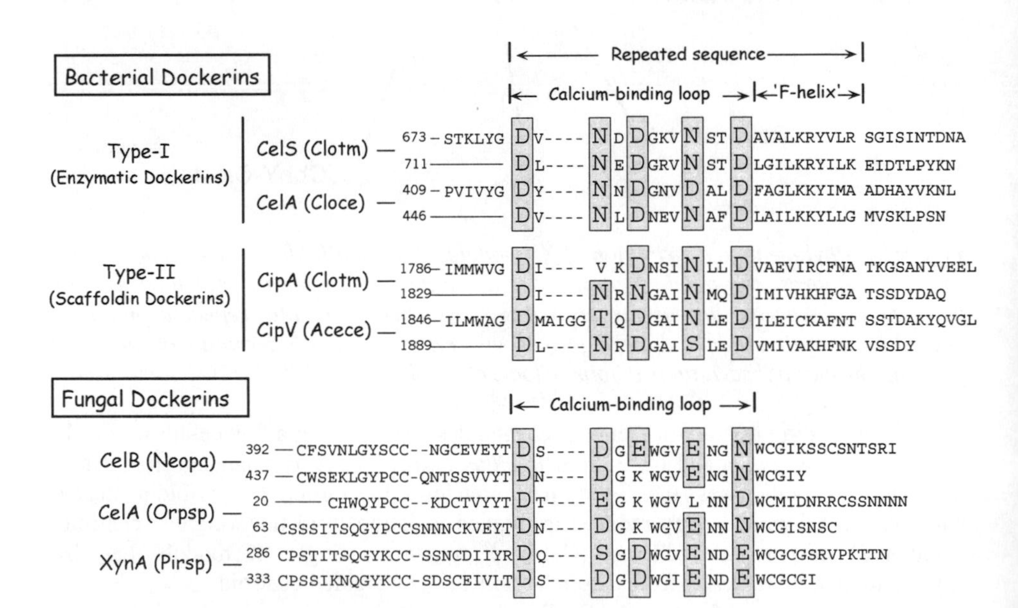

Figure 5 *Deduced amino acid sequence alignment of the C-terminal type-II dockerin domain from CipV with that of* C. thermocellum, *and their relationship to selected type-I dockerins from various cellulosomal enzyme subunits.*

The dockerins characteristically include a duplicated sequence that contains a 12-residue calcium-binding loop, which is essentially synonymous with that of the EF-hand motif of various calcium-binding proteins, such as calmodulin and troponin C. [41] The designated calcium-binding residues usually involve conserved aspartic and/or glutamic acids, asparagines, and sometimes hydroxyamino acids (serines or threonines). In isolated cases, putative dockerins were found to have either a single or a triple calcium-binding motif, although a duplicated motif is usually the rule. In some cases, one or more of the calcium-binding residues may be substituted with a residue or residues which would not form interactions *via* their side chains. This would not necessarily indicate a lack of calcium binding but may suggest a weaker overall interaction between the dockerin domain and calcium. Moreover, alternative interactions, e.g., via mainchain atoms, may take place and compensate for the lack of an appropriate sidechain at a given position.

3.4.1 Bacterial Type-I Dockerin Domains. The bacterial dockerin sequences are organized in a seven-division arrangement that includes the duplicated calcium-binding loop and "F-helix" (but apparently not the E-helix) as described previously. [32] These dockerins generally appear at the *C*-terrminus of the protein although both *N*-terminal and internal dockerins have also been reported. The internal dockerins often occur in cellulosomal enzymes, particularly xylanases, which contain multiple catalytic domains.

The type-I cohesin-dockerin interaction defines the cellulosome structure. Unfortunately, we only have three-dimensional structures for the cohesin component; structures for the dockerin alone or the cohesin-dockerin complex are still wanting, and the molecular mechanism of recognition and binding of the enzymatic subunits into the complex is still elusive. Nevertheless, the abundance of known cohesin and dockerin sequences from the cellulosomes of different microorganisms has enabled us to speculate as to the amino acid residues on each type of domain which may contribute towards their mutual recognition. By employing correlation analyses among dockerins and cohesins of distinct specificities, putative recognition determinants were predicted. [32] The validity of this approach is currently being examined by site-directed mutation and biochemical analysis. For the time being, homology modeling of the F-helix theme provides a provisional model for the dockerin structure, based on the known crystal structure for the EF-hand motif of the above-mentioned calcium-binding proteins. [27] Ultimately, the determination of a solution or crystal structure for the cohesin-dockerin complex will define unequivocally the interaction on the molecular level.

3.4.2 Bacterial Type-II Dockerin Domains. As mentioned above, only two such domains have thus far been described. The CipA scaffoldin from *C. thermocellum* exhibits a *C*-terminal type-II dockerin domain, which immediately flanks the Domain X of the protein. A type-II dockerin also appears to occur at the C-terminus of the CipV scaffoldin from *A. cellulolyticus*. It presumably interacts with a type-II cohesin of a putative anchoring protein. Indeed, the type-II cohesins of the incomplete ORF that appears immediately downstream from the scaffoldin gene signifies a potential anchoring protein in this organism. Both of the known type-II dockerins appear in tandem with a Domain X, and this modular dyad may represent a functional theme for tethering scaffoldins onto the cell surface via selective binding to specialized anchoring proteins.

Like the bacterial type-I dockerins, the type-II sequences also appear to be organized in the same type of seven-division arrangement, [32] that includes the duplicated calcium-binding motif. However, the sequence of the first duplicated calcium-binding motif appears to digress in both of the known type-II dockerins (Figure 5). In *A.*

cellulolyticus CipV, a 4-residue insert is apparent between the presumed first and second calcium-binding amino acid. In CipA from *C. thermocellum*, the second calcium-binding residues is replaced by a valine. Digressions from the canonical EF-hand calcium-binding loop are infrequent but have also been observed in some of the type-I dockerins. Such replacements or inserts could either result in reduced binding affinity for calcium or alternative components (e.g., backbone atoms) may compensate for the substituted side chain. Reduced affinities may be a feature of the type-II cohesin-dockerin interaction and may provide one of the mechanisms by which the release of the cellulosome from the cell surface may be regulated.

3.4.3 Fungal Dockerin Domains. Early biochemical evidence presaged the presence of cellulosome-like complexes in anaerobic rumen fungi. [42] More recently, the formation of cellulosomes in *Neocallimastix*, *Piromyces* and *Orpinomyces* has been corroborated by molecular biology studies. [43-45] In this context, many cellulase and hemicellulase genes from these organisms were shown to include dockerin-like duplicated sequences. Unlike the bacterial dockerins, the fungal sequences were distinguished by numerous cysteines and otherwise lacked a clear similarity with their putative bacterial cognates. Nonetheless, the fungal dockerins clearly include a 12-residue stretch that contains the necessary residues in suitable postions for functioning as a calcium-binding loop, synonymous to that of the bacterial dockerins (see Figure 5). In several places, some of the presumed calcium-binding residues appear to be substituted, and it is unclear how this might affect their presumed calcium-binding properties and specificity for supposed cohesin. In any case, the fungal sequences have been shown to interact with protein components of the cellulosome-like complexes, — the inference being that the latter proteins serve as scaffoldins and contain cohesin domains that would bear specificity for the putative dockerins.

The cellulosome status in the anaerobic fungi awaits final verification. Scaffoldins that bear cohesin-like domains have yet to be identified. However, these microbes produce high-molecular-weight cellulase/hemicellulase complexes, the enzymes of which contain calcium-binding sequences within duplicated domains. The resemblance to thebacterial cellulosome prototype is unmistakable.

4 VIEWPOINT

It has long been noted that the cellulases and hemicellulases from evolutionarily diverse species can be very similar. Likewise, the results presented in this communication indicate that the phylogeny of the various cellulosomal modules does not necessarily reflect the phylogenetic relationship of the microorganisms themselves. It seems that the individual cellulosomal modules have evolved from a restricted number of common ancestral sequences.

The accumulated sequence evidence presents an elegant case for a variety of mechanisms, including horizontal gene transfer, gene duplication and domain shuffling, that could collectively account for the observed phylogenetic distribution of the different modules. For example, horizontal gene transfer could account for the correlation of cellulosome organization among diversified species which may, however, share the same ecosystem. The apparent functional separation of the Family-III CBDs into the observed subfamilies may also be indicative of such a process. On the other hand, gene duplication could account for the multiplicity of the type-I cohesins and their close intraspecies relatedness. The process of gene shuffling could explain the internal distribution of a given type of repeated module (e.g., the cohesins). The same mechanism could also explain interspecies permutation of the arrangement of individual functional modules (e.g., a CBD) relative to others (cohesins). Future accumulation of more sequence data will undoubtedly refine our views of the phylogenetics of cellulosome structure.

Acknowledgements

This work was supported by a contract from the European Commission (Biotechnology Programme, BIO4-97-2303) and by grants from the Israel Science Foundation (administered by the Israel Academy of Sciences and Humanities, Jerusalem) and from the Minerva Foundation (Germany).

References

1. R. Lamed, E. Setter, R. Kenig and E. A. Bayer, *Biotechnol. Bioeng. Symp.*, 1983, **13,** 163.
2. C. R. Felix and L. G. Ljungdahl, *Annu. Rev. Microbiol.*, 1993, **47,** 791.
3. E. A. Bayer, E. Morag and R. Lamed, *Trends Biotechnol.*, 1994, **12,** 378.
4. P. B guin and M. Lemaire, *Crit. Rev. Biochem. Molec. Biol.*, 1996, **31,** 201.
5. E. A. Bayer, H. Chanzy, R. Lamed and Y. Shoham, *Curr. Opin. Struct. Biol.*, 1998, **8,** 548.
6. E. A. Bayer, L. J. W. Shimon, R. Lamed and Y. Shoham, *J. Struct. Biol.*, 1998, **124,** 221.
7. U. T. Gerngross, M. P. M. Romaniec, T. Kobayashi, N. S. Huskisson and A. L. Demain, *Mol. Microbiol.*, 1993, **8,** 325.
8. O. Shoseyov, M. Takagi, M. A. Goldstein and R. H. Doi, *Proc. Natl. Acad. Sci. USA*, 1992, **89,** 3483.
9. S. Pag s, A. Belaich, H.-P. Fierobe, C. Tardif, C. Gaudin and J.-P. Belaich, *J. Bacteriol.*, 1999, **181,** 1801.
10. M. Kakiuchi, A. Isui, K. Suzuki, T. Fujino, E. Fujino, T. Kimura, S. Karita, K. Sakka and K. Ohmiya, *J. Bacteriol.*, 1998, **180,** 4303.
11. D. M. Poole, E. Morag, R. Lamed, E. A. Bayer, G. P. Hazlewood and H. J. Gilbert, *FEMS Microbiol. Lett.*, 1992, **99,** 181.
12. S. Salamitou, K. Tokatlidis, P. B guin and J.-P. Aubert, *FEBS Lett.*, 1992, **304,** 89.
13. S. Salamitou, O. Raynaud, M. Lemaire, M. Coughlan, P. B guin and J.-P. Aubert, *J. Bacteriol.*, 1994, **176,** 2822.
14. E. Leibovitz and P. B guin, *J. Bacteriol.*, 1996, **178,** 3077.
15. S. Y. Ding, D. Steiner, R. Kenig, S. Yaron, E. Morag, Y. Shoham, E. A. Bayer and R. Lamed, (1998) *in* Genetics, Biochemistry and Ecology of Cellulose Degradation (Conference abstract) (K. Ohmiya and K. Hayashi, Eds.), p. 47, Suzuka, Japan.
16. E. A. Bayer, S. Y. Ding, Y. Shoham and R. Lamed, *in* 'Genetics, biochemistry and ecology of lignocellulose degradation' (K. Ohmiya and K. Hayashi, Eds.), in press.
17. G. Gerwig, P. de Waard, J. P. Kamerling, J. F. G. Vliegenthart, E. Morgenstern, R. Lamed and E. A. Bayer, *J. Biol. Chem.*, 1989, **264,** 1027.
18. G. Gerwig, J. P. Kamerling, J. F. G. Vliegenthart, E. Morag (Morgenstern), R. Lamed and E. A. Bayer, *Eur. J. Biochem.*, 1992, **205,** 799.

19. T. Fujino, P. B guin and J.-P. Aubert, *J. Bacteriol.*, 1993, **175,** 1891.
20. E. Leibovitz, H. Ohayon, P. Gounon and P. B guin, *J. Bacteriol.*, 1997, **179,** 2519.
21. M. Lemaire, H. Ohayon, P. Gounon, T. Fujino and P. B guin, *J. Bacteriol.*, 1995, **177,** 2451.
22. S. Y. Ding, D. Steiner, R. Kenig, S. Yaron, Y. Shoham, E. Morag, E. A. Bayer and R. Lamed, (1998) *in* Genetics, Biochemistry and Ecology of Cellulose Degradation (Conference abstract) (K. Ohmiya and K. Hayashi, Eds.), p. 49, Suzuka, Japan.
23. P. Tomme, R. A. J. Warren, R. C. Miller, D. G. Kilburn and N. R. Gilkes, *in* 'Enzymatic degradation of insoluble polysaccharides' (J. M. Saddler and M. H. Penner, Eds.), p. 142, American Chemical Society, Washington, D.C., 1995.
24. J. Sakon, D. Irwin, D. B. Wilson and P. A. Karplus, *Nature Struct. Biol.*, 1997, **4,** 810.
25. D. Irwin, D.-H. Shin, S. Zhang, B. K. Barr, J. Sakon, P. A. Karplus and D. B. Wilson, *J. Bacteriol.*, 1998, **180,** 1709.
26. L. Gal, C. Gaudin, A. Belaich, S. Pag s, C. Tardif and J.-P. Belaich, *J. Bacteriol.*, 1997, **179,** 6595.
27. E. A. Bayer, E. Morag, R. Lamed, S. Yaron and Y. Shoham, *in* 'Carbohydrases from *Trichoderma reesei* and other microorganisms' (M. Claeyssens, W. Nerinckx and K. Piens, Eds.), p. 39, The Royal Society of Chemistry, London, 1998.
28. J. Tormo, R. Lamed, A. J. Chirino, E. Morag, E. A. Bayer, Y. Shoham and T. A. Steitz, *EMBO J.*, 1996, **15,** 5739.
29. L. J. W. Shimon, E. A. Bayer, E. Morag, R. Lamed, S. Yaron, Y. Shoham and F. Frolow, *Structure*, 1997, **5,** 381.
30. G. A. Tavares, P. B guin and P. M. Alzari, *J. Mol. Biol.*, 1997, **273,** 701.
31. S. Yaron, E. Morag, E. A. Bayer, R. Lamed and Y. Shoham, *FEBS Lett.*, 1995, **360,** 121.
32. S. Pag s, A. Belaich, J.-P. Belaich, E. Morag, R. Lamed, Y. Shoham and E. A. Bayer, *Proteins*, 1997, **29,** 517.
33. S. Salamitou, M. Lemaire, T. Fujino, H. Ohayon, P. Gounon, P. B guin and J.-P. Aubert, *J. Bacteriol.*, 1994, **176,** 2828.
34. R. Lamed and E. A. Bayer, *in* 'Genetics, biochemistry and ecology of lignocellulose degradation' (K. Shimada, S. Hoshino, K. Ohmiya, K. Sakka, Y. Kobayashi and S. Karita, Eds.), p. 1, Uni Publishers Co., Ltd., Tokyo, Japan, 1993.
35. S. Jauris, K. P. RŸcknagel, W. H. Schwarz, P. Kratzsch, K. Bronnenmeier and W. L. Staudenbauer, *Mol. Gen. Genet.*, 1990, **223,** 258.
36. P. L. Jorgensen and C. K. Hansen, *Gene*, 1990, **93,** 55.
37. K. Bronnenmeier, K. Kundt, K. Riedel, W. Schwarz and W. Staudenbauer, *Microbiology*, 1997, **143,** 891.
38. E. Josefsson, D. O'Connell, T. J. Foster, I. Durussel and J. Cox, *J. Biol. Chem.*, 1998, **273,** 31145.
39. E. Josefsson, K. W. McCrea, D. Ni Eidhin, D. O'Connell, J. Cox, M. Hook and T. J. Foster, *Microbiology*, 1998, **144,** 3387.
40. R. H. Doi, J. S. Park, C. C. Liu, L. M. Malburg, Y. Tamaru, A. Ichiishi and A. Ibrahim, *Extremophiles*, 1998, **2,** 53.
41. S. Chauvaux, P. B guin, J.-P. Aubert, K. M. Bhat, L. A. Gow, T. M. Wood and A. Bairoch, *Biochem. J.*, 1990, **265,** 261.
42. C. Wilson and C. M. Wood, *Appl. Microbiol. Biotechnol.*, 1992, **37,** 125.
43. C. Fanutti, T. Ponyi, G. W. Black, G. P. Hazlewood and H. J. Gilbert, *J. Biol. Chem.*, 1995, **270,** 29314.

44. X. Li, H. Chen and L. Ljungdahl, *Appl. Environ. Microbiol.*, 1997, **63,** 4721.
45. X. Li, H. Chen and L. Ljungdahl, *Appl. Environ. Microbiol.*, 1997, **63,** 628.

CARBOHYDRATE-BINDING MODULES: DIVERSITY OF STRUCTURE AND FUNCTION

A.B. Boraston[1,2], B.W. McLean[1,2], J.M. Kormos[1], M. Alam[1,2], N.R. Gilkes[1,2], C.A. Haynes[1,3], P. Tomme[1,2], D.G. Kilburn[1,2] and R.A.J. Warren[2]

Protein Engineering Network of Centres of Excellence
Biotechnology Laboratory[1], Microbiology & Immunology[2], Chemical & Bio-Resource Engineering[3], University of British Columbia, Vancouver Canada V6T 1Z3

1 INTRODUCTION

The structural polysaccharides of plant cell walls make up the bulk of biopolymers on this planet. Numerous organisms have evolved diverse polysaccharolytic enzyme systems to exploit this abundant energy source. Many of these enzymes are modular in organisation and often include distinct substrate binding modules. The cellulose binding domains (CBDs) were the first such substrate-binding modules to be identified and characterised in detail [1]. Since then many binding modules with differing carbohydrate binding properties have been identified. Some of the features of these binding modules are discussed in this paper in addition to a new paradigm of grouping them structurally and functionally. Also described is an updated family classification of the carbohydrate binding modules (CBM) and a novel, systematic method of naming these CBMs.

2 CARBOHYDRATE-BINDING MODULES

Since their discovery, the substrate binding modules of polysaccharidases and cellulosome scaffolding proteins have been referred to as cellulose-binding domains (CBDs) based on their preference for cellulose as a ligand. Despite the emergence of similar substrate binding moieties with preferences for carbohydrate ligands other than cellulose, the general label of CBD has persisted. We propose the use of "carbohydrate-binding module" (CBM) to encompass the expanded specificity of these modules. We define a module as a contiguous polypeptide sequence that folds into a discrete functional unit. In contrast, a domain may be composed of non-contiguous polypeptide segments.

2.1 CBM Family Classification

Three methods are commonly used to identify CBMs: gene fragment deletion experiments, limited proteolysis of the intact enzymes, and amino acid sequence similarity. The functionality of unique CBMs is often demonstrated by deletion experiments where DNA corresponding to all or portions of the CBM encoding region are subcloned or deleted and the engineered protein produced recombinantly and characterised [3,4]. Alternatively, as with the first identified CBMs, the CBMs can

Table 1 *Carbohydrate Binding Module Families*

Family	Representative Source Enzyme	Amino Acids	Binding Module Ligand
I	*T. reesei* Cel7A	~33	insoluble cellulose
IIa*	*C. fimi* Xyn10A	~110	insoluble cellulose
IIb	*C. fimi* Xyn11A	~90	xylan
IIIa	*C. cellulovorans* CbpA	~135	insoluble cellulose
IIIb	*T. fusca* Cel9A	~135	insoluble cellulose
IV	*C. fimi* Cel9B	~150	cellulose and cello-oligosaccharides
V	*E. chrysanthemi* Cel5A	63	insoluble cellulose
VI	*C. stercorarium* Xyn11A	~100	cellulose, cello-oligosaccharides, xylan
VII	Entry Deleted		
VIII	*D. discoideum* Cel9A	150	insoluble cellulose
IX	*T. maritima* Xyn10A	~180	cellulose, mono-, di-, and oligosaccharides
X	*P. fluorescens* Cel5A	~55	insoluble cellulose
XI	*F. succinogenes* CBP	~170	insoluble cellulose
XII	*B. agaradherens* Cel5A	~55	insoluble cellulose
XIII	*S. lividans* Xyn10A	~130	hemicellulose, mono-, di-, and oligosaccharides
XIV	*P. parisitica* CBEL	~114	insoluble cellulose
XV	*S. reticuli* AbpS	?	insoluble cellulose
XVI	*R. marinus* Xyn10A	~165	cellulose, xylan
XVII	*C. cellulovorans* Cel5A	~200	cellulose
XVIII	*P. chrysosporium* CDH	~50	insoluble cellulose
XIX	*C. fimi* Man26A	~180	mannan
XX	*T. polysaccharolyticum* Man5A	~120	insoluble cellulose

* *A Roman numeral followed by a letter indicates a sub-family with significant but lower amino acid similarity.*

occasionally be separated from the catalytic module by limited proteolysis [1]. The boundaries of carbohydrate-binding modules can also be approximated by amino acid sequence alignments if similar CBMs have been previously identified. Such methods have been used to identify over 200 amino acid sequences as CBMs or putative CBMs. These polypeptides were previously grouped into thirteen families based on amino acid similarity [5]. In the four years since this publication the CBM classification has grown to include twenty families (Table 1). At least one member of each family has been demonstrated to have carbohydrate-binding activity.

2.2 Nomenclature

The designation of the CBMs (or CBDs) has traditionally been linked to the name of its parent enzyme. The family II CBM from the *Cellulomonas fimi* mixed function exoglucanase/xylanase, Xyn10A (formerly Cex) has been referred to as 'CBD_{Cex}'; the

Table 2 *Designations of CBMs from a selected number of families and organisms*

Family	Organism	Current Designation	New Designation
I	*T. reesei*	CBD_{CBHI}	TrCel7Abm
IIa	*C. fimi*	CBD_{CenA}	CfCel6Abm
		CBD_{CenD}	CfCel5Abm
		CBD_{CbhB}	CfCel48Abm
		CBD_{Cex}	CfXyn10Abm
	P. fluorescens	$XylA_{CBD}$	PfXyn10Abm
IIIa	*C. cellulolyticum*	CbpA	CcCbpAbm
	C. thermocellum	Cip-CBD	CtCipCbm
IV	*C. fimi*	CBD_{N1}	CfCel9Bbm1
		CBD_{N2}	CfCel9Bbm2
		CBD_{N1-N2}	CfCel9Bbm1.2
V	*E. chrysanthemi*	CBD_{EgZ}	EcCel5Abm

family I CBM from *the Trichoderma reesei* endoglucanase I (Cel7B, formerly EG I), has been called 'CBD_{EGI}'. There have been exceptions: the family IV CBM closest to the N-terminus of the endoglucanase Cel9B (formerly CenC) was named 'CBD_{N1}' and the second binding module of the tandem called 'CBD_{N2}'.

As the number of newly identified CBMs grows it has become imperative to establish a naming convention to uniquely identify each binding module. We suggest a unified system of CBM nomenclature consistent with the glycosyl hydrolase family naming scheme proposed by Henrissat et al. (1998) [2]. We propose that the binding modules be designated by the following: a two letter abbreviation of the source organism, the enzyme designation, the binding module abbreviation 'bm' followed by the module number, if there is more than one binding module present in the enzyme. For example, the two binding modules of the *C. fimi* endoglucanase Cel9B would now be referred to CfCel9Bbm1.2; other examples are included for reference (Table 2)

This system of nomenclature uniquely identifies the source of each of the binding module and identifies the particular module (or combination of modules). In the instance where the CBMs are subunits of non-catalytic proteins, such as the scaffoldin of *Clostridium thermocellum* (CipC) or cellulose binding protein of *C. cellulovorans* (CbpA), the enzyme-like abbreviations designating these modules have been adapted for use with this system.

3 STRUCTURE AND FUNCTION OF CBMs

Quiocho (1986) [6] classified carbohydrate-binding proteins into two groups based on structural features of the binding sites and binding affinities. The group I carbohydrate binding proteins are characterised by high association constants (>10^6 M^{-1}) and substrate binding sites that enclose the ligand and sequester it from the solvent. The catalytic sites of glycosyl hydrolases and the substrate binding sites of carbohydrate transport proteins are examples of this type of carbohydrate binding protein. The group II carbohydrate-binding proteins have open, relatively shallow substrate binding sites. The association constants of this type of binding domain are generally less than 10^6 M^{-1}. The affinities and structural properties of the currently characterised binding modules found in polysaccharidases classify them as group II carbohydrate binding proteins. However, as more CBMs are characterised both structurally and biochemically, it is becoming evident that this classification can be further elaborated. Based on structural and functional features, we suggest that CBMs can be organised into three types: type A, insoluble polysaccharide or "surface" binders; type B, soluble polysaccharide or "chain" binders; and type C, small sugar binding CBMs. It should be noted that this concept is based on structural and biochemical properties; it is independent of the CBM family classification, which is based on amino acid sequence similarity.

3.1 Type A Binding Modules

Type A is comprised of binding modules that bind insoluble cellulose or similar β-1,4 glycans such as chitin. Families I, IIa, IIIa, V and possibly families XII, and XV are assigned to this group. Families I to III have relatively high affinities for insoluble matrices that range from $1x10^5$ M^{-1} to approximately $8x10^7$ M^{-1} (Table 3). They may show some interaction with or weak affinity for cello-oligosaccharides [7,8]. The biological significance of this interaction is not known.

Salient structural features of these modules are a β-sheet topology and a platform or ridge of highly conserved solvent-exposed aromatic amino acid residues. This is evident as a series of tyrosines of the family I TrCel7Abm (replaced by tryptophan or phenylalanine at some positions in other members); tryptophans of family IIa CfXyn10Abm; tyrosine, tryptophan and histidine of the family IIIa CtCipCbm; the tryptophans and tyrosine of the *Erwinia chrysanthemi* EcCel5Abm (Fig 1). Aromatic residues are established mediators of protein-carbohydrate interactions via van der Waals interactions and hydrogen bonding. The surface aromatics have been shown to be essential determinants of substrate affinity for many of these members. Other residues on the binding face with hydrogen bonding potential likely contribute to further interactions with the glycan surface.

The most extensively studied modules are the family I TrCel7Abm and the family IIa CfXyn10Abm. The family I module from *T. reesei* is small, composed of only 36 amino acids, and binds reversibly to BMCC [10]. Mutation of the surface tyrosines reduces binding affinity for insoluble cellulose without significant structural perturbation [11].

A number of tryptophans are highly conserved among the members of family IIa. From the structure determined by multi-dimensional NMR, three of these tryptophans of

Table 3 *The association affinity constants of representatives of three families of CBMs for the insoluble β-1,4 glycans bacterial microcrystalline cellulose (BMCC), Avicel, phosphoric acid swollen Avicel (PASA), and chitin. N/A indicates that values are not available.*

		Association Constant (x10^6 M^{-1})			
Family	**CBM**	**BMCC**	**Avicel**	**PASA**	**Chitin**
I	TrCel7Abm	0.1	0.2	N/A	N/A
IIa	CfCel5Abm	2.2	N/A	1.9	0.8
	CfCel6Abm	1.1	0.6	0.7	0.4
	CfCel48A bm	1.8	N/A	1.9	N/A
	Pf Xyn10A bm	1.8	N/A	1.9	N/A
	Cf Xyn10Abm	3.2	1	1.5	0.4
IIIa	CcCbpAbm	1.5	2.5	1.2	0.2
	CtCipCbm	29	77	67	N/A

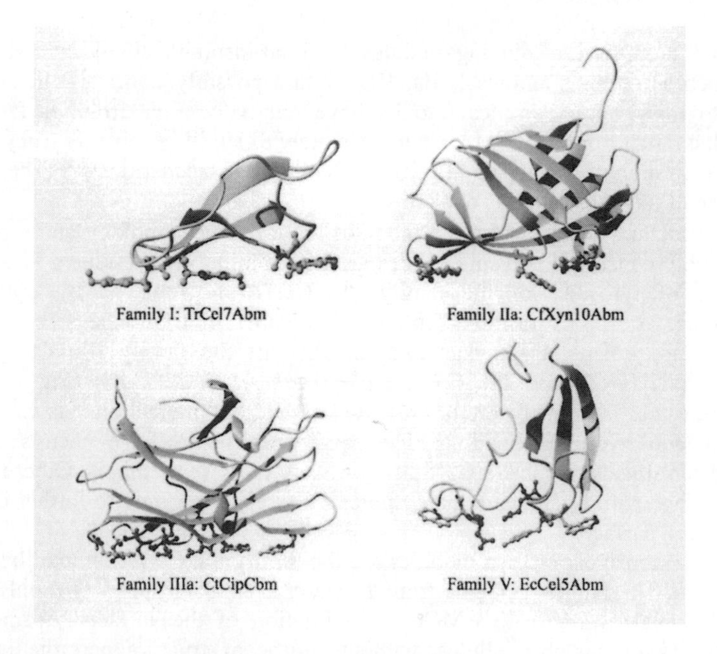

Figure 1 *Representation of the three-dimensional structures of example type A binding modules. Residues on the proposed binding faces of each of the modules thought to be involved with ligand interaction are shown in a ball and stick representation. This figure prepared with the program MOLMOL* [9]

CfXyn10Abm are aligned on one face of the molecule forming a hydrophobic ridge. Mutation (unpublished results) or chemical modification[12] of these residues decreases or abrogates affinity for cellulose. Unlike the family I, this module binds cellulose irreversibly. It is the first CBM to be mobile on cellulose surfaces surface [13]. Thermodynamically, the binding to bacterial microcrystalline cellulose is entropically driven and is presumable by the dehydration of the CBM's hydrophobic tryptophan ridge and the cellulose surface [14]. It is likely that other group A binding modules have a similar binding mode. The common structural arrangement of these functional residues of the type A binding modules suggests the evolutionary convergence of a structural motif specifying affinity for "flat" insoluble glycan surfaces.

3.2 Type B Binding Modules

Type B binding modules differ from type A binding modules in that they bind soluble polysaccharides. They may also bind insoluble substrates, however, they do not bind mono- or disaccharides. The type B binding modules can be considered as "chain binders" because their binding proficiency is in part determined by the degree of polymerisation of the ligand. The type B binding modules currently include members of families IIb, IV and possibly XVI.

The family IIb binding modules, despite their amino acid sequence similarity to the family IIa binding modules [5], appear to be specific for soluble and insoluble xylan. SlXyn10Bbm, the binding module from *Streptomyces lividans* xylanase Xyn10B, and CfXyn11Abm2, the C-terminal module of the two family IIb binding modules from *C. fimi* xylanase Xyn11B, bind to soluble birchwood xylan with association constants determined by affinity electrophoresis of $2x10^3$ per mole of xylan polymer. No binding to mono- or disaccharides could be detected by competition affinity electrophoresis or fluorescence spectroscopy (data not shown). The mode of substrate binding by these binding modules is currently not known; however, the ability to bind soluble polysaccharides suggests a significant rearrangement of the substrate combining site when compared with the family IIa binding modules.

The best-characterised example of a type B binding module is CfCel9Bbm1, a family IV CBM from the endoglucanase Cel9B found in *Cellulomonas fimi*. It binds with association constants of 10^4-10^5 M^{-1} to amorphous cellulose, β-glucan, soluble cellulose derivatives and cello-oligosaccharides of greater than three glucose units in length, but has no apparent affinity for crystalline cellulose [15]. The solution structure of CfCel9Bbm1, determined by NMR spectroscopy, showed a substrate-binding groove on one face of the polypeptide sufficient in length and width to accommodate an oligosaccharide five glucose units in length [16] (Figure 2). Hydrophobic residues run the length of this groove with hydrophilic hydrogen bond donors and acceptors flanking the groove. The postulated mechanism of substrate binding involves hydrogen bonding of the oligosaccharide hydroxyls with the flanking hydrophilic residues and van der Waals interactions between the glucopyranosyl rings of the oligosaccharide and the hydrophobic amino acid strip in the binding groove [16,17]. This binding mechanism is analogous to the protein-carbohydrate interactions observed for lectins and other carbohydrate binding proteins [18]. Binding affinity increases with the number of glucose residues in the cello-oligosaccharide beginning with a minimum of three to a maximum of five or more. A relatively long binding groove is required to accommodate the lengths of the ligands

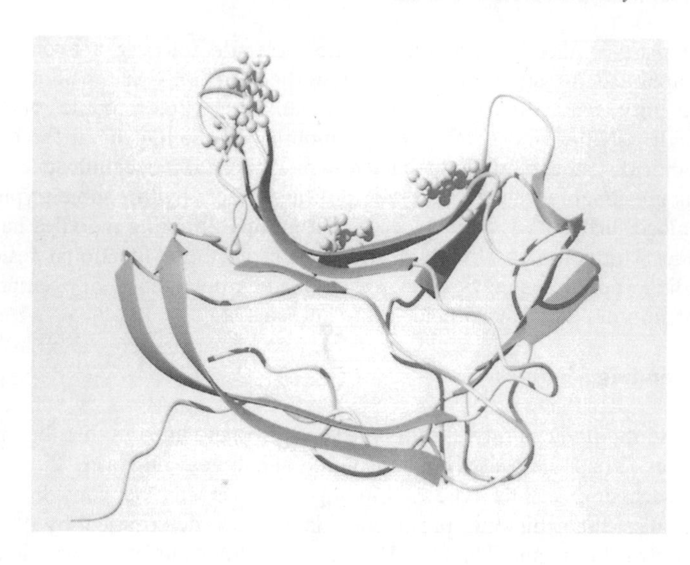

Figure 2 *Schematic representation of the three dimensional structure of CBDN1. Illustrated in this view is the binding cleft, which can accommodate single sugar strands. This figure prepared with the program MOLMOL* [9]

required for sufficient binding to the type B binding modules. Thus, as other CBM families with type B binding specificities are structurally characterised, such a binding groove will conceivably be seen as common structural feature of this type of binding module.

Xylanase Xyn10A from *Rhodothermus marinus* has tandem binding modules, which have been classified as family XVI modules, at the N-terminus of the protein [19]. These binding modules show a small degree of amino acid similarity to CfCel9Bbm1 and CfCel9Bbm2 suggesting that family XVI may be a sub-family of family IV. This sequence similarity, in addition to specificity for soluble polysaccharides, tentatively classifies RmXyn10Abm1 and RmXyn10Abm2 as type B binding modules. Further biochemical and structural characterisation of family XVI binding modules will clarify their status as type B binding modules.

3.3 Type C Binding Modules

The type C carbohydrate-binding modules are the least characterised type of CBM. They may bind soluble or insoluble polysaccharides and are currently differentiated from type A and B CBMs by their ability to bind mono- and disaccharides with relatively high affinity. Families VI, IX, XIII, and XVII currently meet these criteria.

The sources of family XIII CBMs are the most diverse of all the CBM families. These sources include animals, crustaceans, plants, and microbes. Notable members of this family are the B-chains of the toxic ribosome-inactivating proteins (RIPs) ricin, abrin, sambucus nigra lectin, and mistletoe lectin. These well-characterised proteins are galactose specific carbohydrate binding modules which are thought to mediate the attachment of RIPs to cell surface glycans. The binding modules of *Streptomyces lividans* xylanase Xyn10A and arabinofuranosidase AbfB show significant sequence identity with the RIPs though clearly have a different biological function with their specificity for xylan [20]. Based on the conservation of amino acids involved in the binding of ligand to the RIP lectins it is likely that SlXyn10Abm and SlAbfBbm also have an affinity for monosaccharides.

The family VI binding modules of Xyn11A from *Clostridium stercorarium*, the tandem family IX binding modules of Xyn10A from *Thermotoga maritima*, and the family XVII binding module of Cel5A from *Clostridium cellulovorans* have all been reported to be desorbed from cellulose by cellobiose [3,21,22]. The interaction with cellobiose is specific in the cases of the *C. stercorarium* family VI CBMs and TmXyn10Abm2 (unpublished results). We have found that TmXyn10Abm2 binds to a variety of other mono- and disaccharides. Unfortunately no structural information and little biochemical information is available for these binding modules. As this information becomes available the details of the modes of substrate binding will more accurately define how the type C binding modules differ from the type A and B binding modules.

4 BIOLOGICAL ROLES

The complete biological role of CBMs is not fully understood. The only property, consistently verified through studies with different polysaccharidases is that CBMs potentiate the hydrolysis of insoluble cellulose. The removal of a CBM impairs cellulase activity on insoluble substrates, but has little effect on the hydrolysis of soluble substrates [23-25]. It is postulated that the CBM targets the enzyme to the substrate. This results in increased local polysaccharolytic activity. Some CBMs may also disrupt cellulose fibres by nonhydrolytic means. The disruption of surface fibres was originally observed in electron micrographs of cellulose treated with one of two family IIa CBMs [26,27]. It has not been reported for other CBMs to date, including other family IIa CBMs [8]. The mechanism of cellulose fibre disruption by a CBM is currently unclear.

CBMs are not limited to cellulolytic enzymes. They are also found in other polysaccharidases including β1,3-glucanases, xylanases, chitinases, mannanases, and arabino-furanosidases. Though some of these enzymes contain binding modules with affinities for substrates other than cellulose, many of the CBMs present in these hydrolases do bind exclusively to cellulose. This is likely related to the biological role of these enzymes in degrading plant cell wall material, which is predominantly cellulose with hemi-celluloses associated with it. Targeting a hemi-cellulase to the cellulose in plant cell walls still brings the enzyme in proximity with the enzymes preferred substrate [28]. In this way, organisms can maximise the energy yield from plant biomass by ensuring the complete degradation of all carbohydrates present.

5 BIOTECHNOLOGICAL APPLICATIONS

CBMs are well suited as affinity tags for the purification and/or immobilisation of fusion proteins. For immobilisation applications the type A CBMs, particularly family IIa and IIIa CBMs, are effective immobilisation tags since they bind crystalline cellulose irreversibly. The type B CBMs, represented by the family IV CBM CfCel9Bbm1, also show potential as affinity tags in phase separation systems involving soluble cellulosic polymers [29]. For biopurification in chromatographic applications the type C CBMs, notably members of family VI and IX, are emerging as particularly useful affinity tags since they can be eluted from insoluble cellulose with inexpensive competitive sugars [21,30].

Perhaps the biggest advantage of using CBMs as opposed to other affinity tags is the use of cellulose as an affinity matrix. Often referred to as the world's most abundant polymer, cellulose represents an extremely cheap, inert, and safe compound, upon which speciality products can then be made [29]. It is commercially available in numerous physico-chemical forms, including cloth, powder, filters, membranes, tubing, fibres, beads, and soluble polymers with various degrees of polymerisation, substitution and viscosity. Because of its molecular organisation and chemical properties [31-34], cellulose offers outstanding structural integrity with low non-specific protein adsorption. The chromatography industry has benefited from such properties. Many commercial resins for ion exchange, HIC, or gel filtration applications are prepared from cellulose. Porous cellulose beads can have solid contents as low as 2%, yet still maintain enough structural integrity for chromatographic applications. They offer exceptional hydrodynamic properties, unrestricted mass transfer, and large surface areas. These properties are ideally suited for purification and immobilisation applications that utilise CBMs. In addition, these applications only require the raw cellulose bead, eliminating the cost of chemically modifying the cellulose for speciality resins.

There is clear potential in the use of CBM/cellulose systems for protein purification and immobilisation. Through continued research, new CBMs with novel properties are being discovered, ultimately extending the list of potential applications based on CBM technology.

Reference List

1. N.R. Gilkes, R.A. Warren, R.C. Miller Jr., D.G. Kilburn, *J.Biol.Chem.*, 1988, **263**, 10401
2. B. Henrissat, T.T. Teeri, R.A. Warren, *FEBS Lett.*, 1998, **425**, 352
3. A. Ishi, S. Sheweita, R.H. Doi, *Appl.Environ.Microbiol.*, 1998, **64**, 1086
4. R. Ramalingam, J.E. Blume, H.L. Ennis, *J.Bacteriol.*, 1992, **174**, 7834
5. P. Tomme, R.J. Warren, R.C. Miller Jr., D.G. Kilburn, N.R. Gilkes, 'Enzymatic Degradation of Insoluble Carbohydrates', American Chemical Society, Washington, DC, 1995, p. 142.
6. F.A. Quiocho, *Annu.Rev.Biochem.*, 1986, **55**, 287
7. G.Y. Xu, E. Ong, N.R. Gilkes, D.G. Kilburn, D.R. Muhandiram, M. Harris-Brandts, J.P. Carver, L.E. Kay, T.S. Harvey, *Biochemistry,* 1995, **34**, 6993
8. D.N. Bolam, A. Ciruela, S. McQueen-Mason, P. Simpson, M.P. Williamson, J.E. Rixon, A. Boraston, G.P. Hazlewood, H.J. Gilbert, *Biochem.J.*, 1998, **331**, 775
9. R. Koradi, M. Billeter, K. Wüthrich, *J.Mol.Graphics,* 1996, **14**, 51
10. M. Linder, T.T. Teeri, *Proc.Natl.Acad.Sci.U.S.A.*, 1996, **93**, 12251
11. T. Reinikainen, L. Ruohonen, T. Nevanen, L. Laaksonen, P. Kraulis, T.A. Jones, J.K. Knowles, T.T. Teeri, *Proteins,* 1992, **14**, 475
12. M.R. Bray, P.E. Johnson, N.R. Gilkes, L.P. McIntosh, D.G. Kilburn, R.A. Warren, *Protein Sci.,* 1996, **5**, 2311
13. E.J. Jervis, C.A. Haynes, D.G. Kilburn, *J.Biol.Chem.*, 1997, **272**, 24016
14. A.L. Creagh, E. Ong, E. Jervis, D.G. Kilburn, C.A. Haynes, *Proc.Natl.Acad.Sci.U.S.A.,* 1996, **93**, 12229
15. P. Tomme, A.L. Creagh, D.G. Kilburn, C.A. Haynes, *Biochemistry,* 1996, **35**, 13885

16. P.E. Johnson, M.D. Joshi, P. Tomme, D.G. Kilburn, L.P. McIntosh, *Biochemistry,* 1996, **35** , 14381
17. P.E. Johnson, P. Tomme, M.D. Joshi, L.P. McIntosh, *Biochemistry,* 1996, **35**, 13895
18. H. Lis, N. Sharon, *Chemical Reviews,* 1998, **98**, 637
19. E.N. Karlsson, E. Bartonek-Roxa, O. Holst, *FEMS Microbiol.Lett.,* 1998, **168**, 1
20. C. Dupont, M. Roberge, F. Shareck, R. Morosoli, D. Kluepfel, *Biochem.J.,* 1998, **330**, 41
21. C. Winterhalter, P. Heinrich, A. Candussio, G. Wich, W. Liebl, *Mol.Microbiol.,* 1995, **15** , 431
22. K. Sakka, G. Takada, S. Karita, K. Ohmiya, *Ann.N.Y.Acad.Sci.,* 1996, **782**, 241
23. P. Tomme, H. Van Tilbeurgh, G. Pettersson, J. Van Damme, J. Vandekerckhove, J. Knowles, T. Teeri, M. Claeyssens, *Eur.J.Biochem.,* 1988, **170**, 575
24. T. Reinikainen, O. Teleman, T.T. Teeri, *Proteins,* 1995, **22**, 392
25. J. Hall, G.W. Black, L.M. Ferreira, S.J. Millward-Sadler, B.R. Ali, G.P. Hazlewood, H.J. Gilbert, *Biochem.J.,* 1995, **309**, 749
26. N. Din, I.J. Forsythe, L.D. Burtnick, N.R. Gilkes, R.C. Miller Jr., R.A.J. Warren, D.G. Kilburn, *Biotechnol.,* 1991, **9**, 1096
27. E. Ong, N.R. Gilkes, R.C. Miller Jr., R.A.J. Warren, D.G. Kilburn, *Biotechnol.Bioeng.,* 1993, **42**, 401
28. L.M. Ferreira, A.J. Durrant, J. Hall, G.P. Hazlewood, H.J. Gilbert, *Biochem.J.,* 1990, **269**, 261
29. P. Tomme, A. Boraston, B. McLean, J. Kormos, A.L. Creagh, K. Sturch, N.R. Gilkes, C.A. Haynes, R.A.J. Warren, D.G. Kilburn, *J.Chromatogr.B,* 1998, **715**, 283
30. K. Sakka, S. Karita, T. Kimura, K. Ohmiya, *Ann.N.Y.Acad.Sci.,* 1998, **864**, 485
31. F.J. Kolpak, M. Weih, J. Blackwell, *Polymer,* 1999, **19**, 123
32. R. Nardin, M. Vincendron, *Makromol.Chem.,* 1988, **189**, 153
33. Sarko, A., 'Wood and cellulosics: industrial utilization, biotechnology, structure and properties', Ellis Horwood, New York, 1986, p. 55.
34. R.H. Attala, '*Trichoderma resei* Cellulases and Other Hydrolases', Foundation for Biotechnological and Industrial Fermentation, Helsinki, 1993, p. 25.

HOW THE *N*-TERMINAL XYLAN-BINDING DOMAIN FROM *C. FIMI* XYLANASE D RECOGNISES XYLAN

Michael P Williamson,[1] Peter J Simpson,[1] David N Bolam,[2] Geoffrey P Hazlewood,[3,5] Antonio Ciruela,[3] Alan Cooper[4] and Harry J Gilbert[2]

[1]Department of Molecular Biology and Biotechnology, The Krebs Institute, University of Sheffield, Sheffield S10 2TN, UK, [2]Department of Biological and Nutritional Sciences, University of Newcastle upon Tyne, Newcastle upon Tyne NE1 7RU, UK, [3]Laboratory of Molecular Enzymology, The Babraham Institute, Babraham, Cambridge CB2 4AT, UK and [4]Department of Chemistry, Joseph Black Building, University of Glasgow, Glasgow G12 8QQ, UK. [5]Current address: Finnfeeds International Ltd., PO Box 777, Marlborough SN8 1XN, UK.

1 INTRODUCTION

Plant cell walls are complex structures, and their function requires them to be strong and resistant to degradation. The structural core of the cell wall is the microfibril, which is made of cellulose (Figure 1a) and is considered to be reasonably crystalline, although with amorphous regions. The microfibrils are crosslinked together by lignin and by a mesh of polysaccharide strands known collectively as hemicelluloses, and consisting of xylans, arabinans and mannans. In contrast to the cellulose microfibrils, hemicellulose is amorphous and often consists of single or weakly associated strands rather than the crystalline bundles of strands seen in microfibrils. Xylan is a polymer of xylose, which is similar to glucose but lacks the C_6 carbon (Figure 1b). In contrast to cellulose, hemicelluloses are often structurally heterogeneous. For example, xylans may be quite heavily derivatised by acetylation or glycosylation. The degree and nature of derivatisation is species-specific.

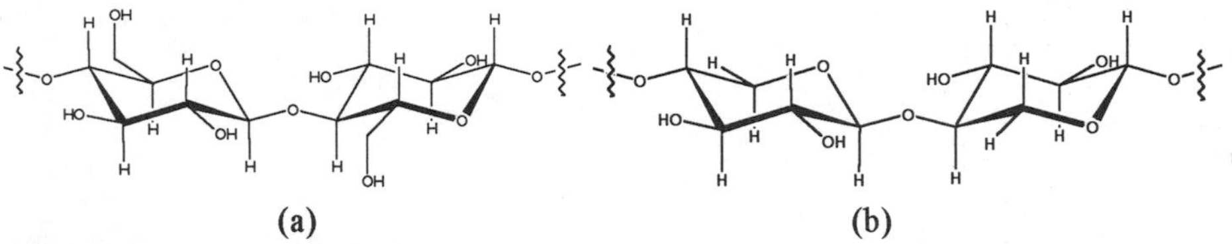

(a) (b)

Figure 1 *The covalent structure of (a) cellulose (b) xylan*

Many microorganisms, both bacteria and fungi, live on plants, and have to degrade the cell walls. They do this by secreting several different digestive enzymes, which act in concert.[1] Almost always this group of enzymes will include an endocellulase and an exocellulase (which usually releases the glucose dimer, cellobiose) together with a β 1,4-glucosidase to cleave cellobiose and therefore remove product inhibition. Many microorganisms also secrete additional cellulases and hemicellulases, some of which have been shown to act synergistically in degrading the cell wall.

1.1 Binding Domains

All enzyme-catalysed reactions require the substrate to bind at the active site of the enzyme. This creates a potential problem for cellulases and hemicellulases, because their substrate is polymeric and usually insoluble, which implies that the rate of reaction can become limited by the rate at which substrate is delivered to the active site. The majority of cellulases and hemicellulases therefore contain additional domains that act as polysaccharide-binding domains (BDs), and are linked to the catalytic domain by flexible linkers.[1,2] These domains increase the catalytic rate against insoluble polymeric substrates but have no effect against oligosaccharides.[3] In the case of cellulases these are always cellulose-binding domains (CBDs). Hemicellulases sometimes contain specific binding domains (eg xylanases containing xylan-binding domains), but a surprisingly high proportion of hemicellulases contain CBDs.[2] It has been speculated that the reason for this is that it is much easier to recognise the structurally homogeneous cellulose than the heterogeneous hemicelluloses.[4] Furthermore, hemicelluloses are always closely associated with cellulose, implying that a domain that binds to cellulose will still locate the enzyme close to hemicelluloses.

Polysaccharide-binding domains have been classified into a number of families based on sequence comparisons, of which the most abundant are families I and II.[5] Family II has been further subdivided into two sub-families, IIa and IIb, which have roughly 30% sequence similarity (Figure 2). There are structures of representatives from a number of these families, which so far all consist entirely of β-sheet, and have characteristic surface-exposed aromatic residues that act as the primary binding sites for the ligands. For example, the family IIa CBD from *Cellulomonas fimi* exoglucanase Cex has three aromatic rings in a linear, coplanar arrangement that have been identified as the binding site for crystalline cellulose.[6] It has been suggested that family IIa bind to cellulose, while family IIb bind xylan.[7] Here we describe the structure of a family IIb xylan-binding domain (XBD) from *C. fimi* xylanase D (XYLD), and discuss how it recognises xylan, while the closely related CBD_{Cex} recognises cellulose. Some of this material has been published elsewhere.[8]

```
Cex   PAGCQVLWGV NQ-WNTGFTA NVTVKNTSSA PVDGWTLTFS FPSGQQVTQA.WSSTV
CenA  APGCRVDYAV TNQWPGGFGA NVTITNLGD- PVSSWKLDWT YTAGQRIQQL.WNGTA
XBD1  STGCSVTATR AEEWSDRFNV TYSVS-GSS- ---AWTVNLA LNGSQTIQAS.WNANV
XBD2  TGSCSVSAVR GEEWADRFNV TYSVS-GSS- ---SWVVTLG LNGGQSVQSS.WNAAL

Cex   TQSGS.AVTVRNAPWN.GSIPAGGTAQ.FGFNGSHTGT.NAAPTAFSLN.GTPCTVG
CenA  STNGG.QVSVTSLPWN.GSIPTGGTAS.FGFNGSWAGS.NPTPASFSLN.GTTCTGT
XBD1  T-GSG.STRTVTPN--.----------.-GSGNTFGVT.VMKNGSSTTP.AATCAGS
XBD2  T-GSS.GTVTARPN--.----------.-GSGNSFGVT.FYKNGSSATP.GATCATG
```

Figure 2 *Sequence alignment of family II polysaccharide binding domains. The first 2 sequences are family IIa CBDs, while the other 2 are family IIb XBDs. All sequences are from* C. fimi *enzymes. The first residue listed for XBD1 corresponds to residue 246 in the intact enzyme.*

2 THE LIGAND SPECIFICITY OF THE *C. FIMI* XYLD BINDING DOMAIN

XYLD contains an N-terminal family 11 catalytic domain, a family IIb BD (XBD1), a domain that exhibits sequence similarity to *Rhizobium* NodB and catalyses the hydrolysis of ester bonds in acetylated xylan, and a C-terminal domain (XBD2) that exhibits 70% sequence identity with XBD1.[9,10] The ligand specificity of XBD1 was evaluated by non-denaturing affinity gel electrophoresis, using a fusion protein of XBD1 with glutathione *S*-transferase (GST). The data, presented in Figure 3, show that the electrophoretic migration of GST-XBD1 is retarded by the presence of barley β-glucan or xylans derived from wheat, rye or oat spelt. In contrast, no retardation of the fusion protein was observed in gels containing arabinan, galactan (data not shown), hydroxyethyl cellulose or carboxymethyl cellulose. GST alone was not retarded by the presence of any polysaccharides in the acrylamide gels. These data therefore suggest that XBD1 binds to both soluble and insoluble xylans. It also exhibits some affinity for β-glucan, but does not bind to either soluble or insoluble forms of cellulose, other hemicelluloses or pectins. Isothermal titration calorimetry and NMR spectroscopy confirmed these binding specificities, and showed that the binding site accommodates a maximum of 6 xylose units (Table 1). The slightly weaker binding to oat spelt xylan compared to xylohexaose listed in Table 1 may be rationalised by the heterogeneous and more highly derivatised nature of oat spelt xylan. The $\Delta H°$ and $T\Delta S$ of xylohexaose binding were *ca.* -12 ± 0.6 kcal mol^{-1} and -6.3 ± 0.02 kcal mol^{-1}, respectively.

3 THE STRUCTURE AND XYLAN BINDING SITE OF XBD1

The structure of XBD1 has been determined using NMR spectroscopy. Using ^{15}N-labelled protein, the spectrum was almost completely assigned, and a list of 1162 distance restraints (296 intraresidue, 222 sequential, 77 medium-range, 310 long-range and 205 ambiguous NOEs, and 26 pairs of H-bond restraints) and 110 angle restraints (60 ϕ, 49 χ^1 and 1 χ^2) was compiled. 50 structures were calculated using hybrid distance geometry/simulated

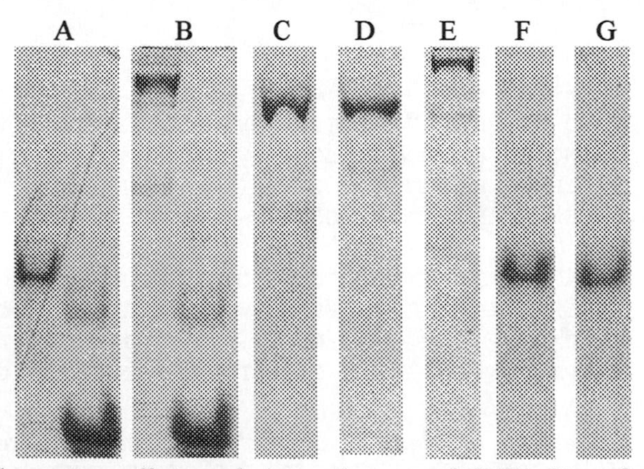

Figure 3 *Non-denaturing affinity gel electrophoreses of XBD1. Ligands: A, none; B, rye xylan; C, wheat xylan; D, barley glucan; E, oat xylan; F, carboxymethyl cellulose; G, hydroxyethyl cellolose. All panels show the result for a GST fusion, except for the right-hand side of the first two panels, which show the result for GST alone.*

Table 1 *The affinity of XBD1 for oligosaccharides and polysaccharides*

Ligand	*Dissociation constant*/mM	
	K_d[a]	K_d[b]
Xylobiose	> 20[c]	ND[d]
Cellohexaose	> 20[c]	≥ 10
Xylotriose	22 ± 6.5	ND
Xylotetraose	1.19 ± 0.04	ND
Xylopentaose	0.41 ± 0.04	ND
Xylohexaose	0.19 ± 0.03	0.29
Oat spelt xylan	0.41 ± 0.01	ND
Barley β-glucan	8.55 ± 0.80	ND

[a]Ligand affinity determined by ITC; pH 7.5, 25 °C, Tris buffer.
[b]Ligand affinity estimated by 1D NMR; pH 7.0, 30 °C, sodium phosphate buffer.
[c]No binding could be detected by ITC.
[d]ND: not determined

annealing in X-PLOR starting from random co-ordinates and subsequently refined. The 38 best structures had no NOE violations greater than 0.25 Å, and are shown in Figure 4. The atomic RMSD for backbone and all heavy atoms when superimposed by the regions of secondary structure are 0.39 ± 0.08 Å and 0.72 ± 0.06 Å respectively. When all residues between the N- and C-terminal disulfide bridge inclusive are used for the superimposition, the atomic RMSD for backbone and all heavy atoms goes up to 0.60 ± 0.13 and 0.83 ± 0.10 Å respectively. 98.8% of residues are in the allowed regions of the Ramachandran plot, as measured using PROCHECK.[11] These results imply that the structure is of high quality.

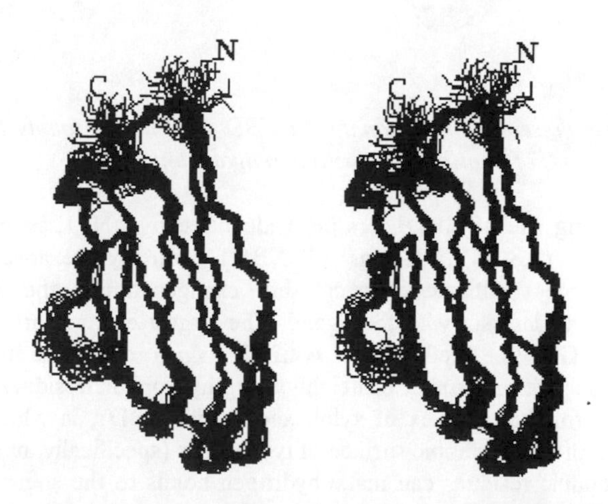

Figure 4 *Superposition of 38 XBD1 structures shown as a stereo diagram. The superposition used the backbone atoms (N, C^α, C') of residues in regular secondary structure elements (residues 249-259, 262-270, 275-281, 285-291, 293-297, 301-306, 311-319 and 326-330).*

The structure of XBD1 consists of two, 4-stranded antiparallel β-sheets which form a twisted 'β-sandwich' motif about an extensive hydrophobic core. The topology of the secondary structure is very similar to that reported for the family IIa cellulose binding domain CBD_{Cex} (Figure 5).[6] Figure 5 highlights the three tryptophan residues that have been identified in CBD_{Cex} as forming the cellulose binding site. One of these is not present in XBD1, because in CBD_{Cex} it is in a long loop, which is truncated in XBD1. Therefore in XBD1, only two of these tryptophans are present.

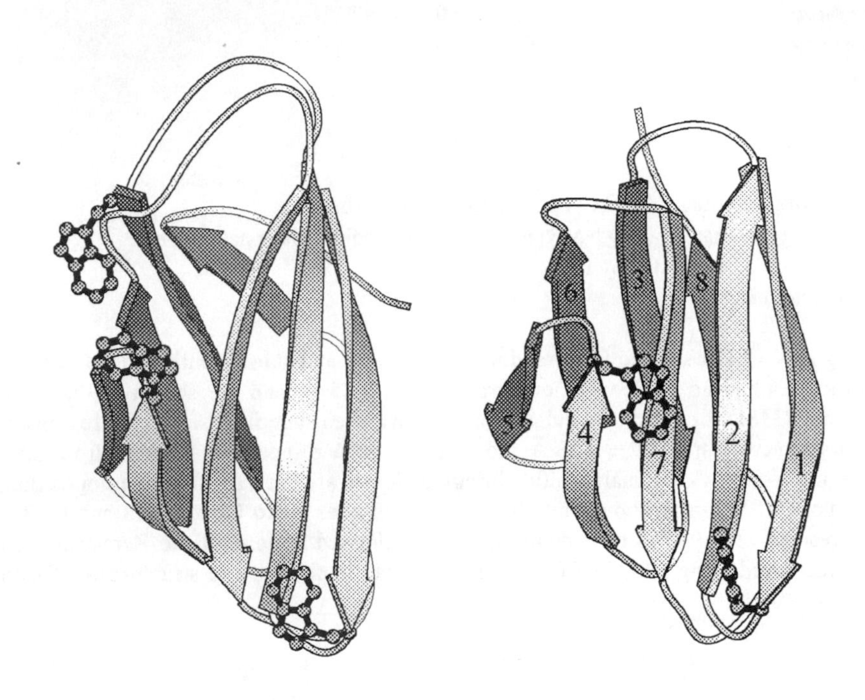

Figure 5 *Ribbon representations of family IIa CBD_{Cex} (left) and family IIb XBD1 (right), illustrating the positions of tryptophans involved in ligand binding*

The xylan binding site of XBD1 has been identified by ^{15}N-^{1}H heteronuclear single-quantum coherence (HSQC) titrations of XBD1 with xylohexaose. A number of resonances experience significant chemical shift changes during the titration, and are therefore presumed to interact with the ligand. The largest changes are seen for residues Trp259, Asn264, Gln288, Trp291 and Asn292, which constitute the two exposed tryptophans together with a number of neighbouring hydrophilic residues. It is possible to construct a model for the complex of xylohexaose with XBD1, in which the tryptophan rings stack against the hydrophobic surface of two xylans (specifically, monomers i and i + 2), and the hydrophilic residues can make hydrogen bonds to the sugar hydroxyl groups (Figure 6). The xylan chain sits on the convex surface of the protein at one end, but in the region of the tryptophan residues it is held in a shallow groove.

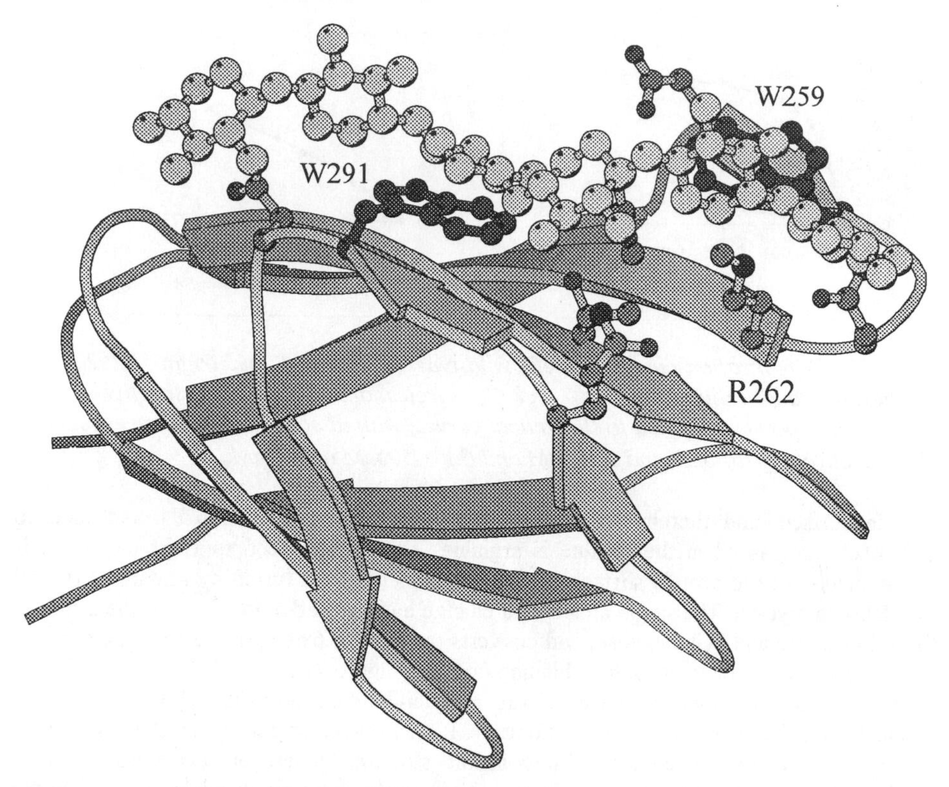

Figure 6 *A ribbon representation of XBD1 with xylohexaose docked into the binding site, showing the complementarity of the twisted binding site and xylohexaose. Residues whose sidechains are implicated in binding are highlighted: Trp259, Trp291, Glu257, Asp261, Arg262, Asn264, Gln288, Asn292 and Thr316.*

4 DISCUSSION

The structure of XBD1 is similar to that already found for CBD_{Cex} (Figure 5), as expected from their approximately 30% sequence similarity. The structure of the model permits a rationalisation of many of the unusual features of XBD1 binding. In particular, the origin of the difference in binding specificity comes from the orientation of Trp259, which lies parallel to the protein surface in CBD_{Cex}, but is twisted by approximately 90° in XBD1. This difference in the protein surface matches the difference in the secondary structure of the polysaccharide ligand: cellulose forms flat sheets, with an approximately 180° rotation between one monomer and the next,[1] whereas xylan forms helices, with an approximately 120° rotation between monomers.[12]

The sequence similarities between families IIa and IIb suggest that the change in ligand specificity has arisen by divergent evolution. On comparing the sequences (Figure 2), one consistent difference is that residue 262 is an arginine in family IIb but a glycine in family IIa. Inspection of the structures of XBD1 and CBD_{Cex} suggests that this is a crucial difference: when the residue is glycine, it allows the nearby Trp259 to lie flat against the

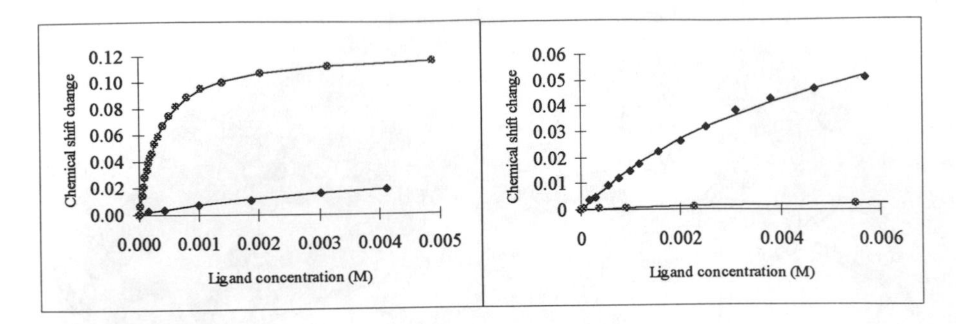

Figure 7 *NMR titration curves of (left) wild-type (WT) XBD1 and (right) R262G with xylohexaose and cellohexaose (X_6 and C_6: circles and diamonds respectively). The experimental points are fitted to theoretical curves, with dissociation constants to X_6 and C_6 respectively of (a) 0.29 and ≥ 10 mM and (b) > 50 mM and 5 mM.*

protein surface, and thereby form a coplanar arrangement with Trp291 and bind to cellulose, whereas when the residue is arginine, Trp259 is forced to twist so that it is perpendicular to the protein surface. This hypothesis has been tested by mutating Arg262 of XBD1 to glycine. The single amino acid change has a remarkable effect on the affinities for xylohexaose and cellohexaose, and converts the protein from being essentially a xylan-binding domain to being a cellulose-binding domain (Figure 7).

It is interesting that xylanase D has a xylan-binding domain, while many other hemicellulases have cellulose-binding domains. It was suggested above that the reason may lie in the fact that cellulose has a constant structure wherever it occurs, whereas hemicelluloses can be much more variable, both in conformation and in substitution.[4] In the model of the XBD1/xylan complex (Figure 6), most of the xylose 2′ and 3′ hydroxyl groups are pointing either away from the protein surface or parallel to it. The model can therefore accommodate a high level of xylan substitution at these positions, in line with the experimental results (Table 1). This may explain why an XBD has proved useful to this enzyme, as opposed to the more common CBDs.

4.1 Do Binding Domains have any Additional Function?

There has been considerable debate over the function of binding domains. It is clear that all binding domains attach the enzyme to the substrate surface, and thus increase the local substrate concentration at the active site. In addition, it has been shown that some CBDs, when added *in trans* (ie, as the free binding domain, not attached to the catalytic domain), can increase the cellulolytic activity of the catalytic domain.[13] This has been interpreted as showing that the CBDs are disrupting the crystalline surface of cellulose. A similar observation has also been made for the starch-binding domain of *Aspergillus niger* glucoamylase;[14] in this case, there is a clear rationale for how the domain achieves the rate enhancement, because the binding domain binds two independent starch strands and apparently twists them apart.[15] In the case of CBDs, only a single binding site has ever been observed. By contrast, several other CBDs cause no detectable rate enhancement.[16] A possible resolution of this difficulty can be found by consideration of the energetics of binding.

It was noted above that the enthalpy for binding to xylohexaose is favourable, while the entropy is unfavourable. A similar observation was also made for the CBD from *C. fimi* CenC, which binds to amorphous cellulose.[17] By contrast, domains that bind to crystalline cellulose (such as CBD_{Cex}) have favourable entropy but unfavourable enthalpy.[18] The difference is readily explicable, in that mobile ligands, such as xylan or amorphous cellulose, lose considerable rotational entropy on binding, whereas crystalline cellulose loses virtually none. Thus, the entropic cost of binding to mobile ligands is certain to be higher. However, it is not so easy to make a similar argument for the enthalpic change, which comes largely from hydrogen bonding interactions and van der Waals attraction, and so should be similar for all binding domains. The observation that the enthalpy of binding is actually unfavourable for domains that bind to crystalline cellulose suggests that the maximum binding free energy is not being used, and therefore that there is an advantage for the enzyme in having submaximal binding energy. One could suggest several explanations for this: for example, perhaps binding domains should not bind too tightly, because otherwise they could never detach from the polysaccharide. However, an attractive alternative explanation is similar to that described in detail by Jencks and named the Circe effect:[19] that some of the free energy of binding is being used to distort the substrate and therefore make the subsequent catalytic step easier. In other words, it is possible that some of the hydrogen bonding interactions made by domains that bind to crystalline cellulose serve not to attach the CBD to cellulose but largely to disrupt the cellulose surface and increase the availability of substrate.

References

1. P. Tomme, R. A. J. Warren and N. R. Gilkes, *Adv. Microb. Physiol.*, 1995, **37,** 1.
2. N. R. Gilkes, B. Henrissat, D. G. Kilburn, R. C. Miller Jr and R. A. J. Warren, *Microb. Rev.*, 1991, **55,** 303.
3. G. W. Black, J. E. Rixon, J. H. Clarke, G. P. Hazlewood, M. K. Theoderou, P. Morris and H. J. Gilbert, *Biochem. J.*, 1996, **319,** 515.
4. L. M. A. Ferreira, A. J. Durrant, J. Hall, G. P. Hazlewood and H. J. Gilbert, *Biochem. J.*, 1990, **269,** 261.
5. P. Tomme, R. A. J. Warren, R. C. Miller Jr, D. G. Kilburn and N. R. Gilkes, in 'Enzymic Degradation of Insoluble Polysaccharides' (eds. J. N. Saddler and M. Penner), American Chemical Society, Washington DC, 1995, p. 142.
6. G. Y. Xu, E. Ong, N. R. Gilkes, D. G. Kilburn, D. R. Muhandiram, M. Harris-Brandts, J. P. Carver, L. E. Kay and T. S. Harvey, *Biochemistry*, 1995, **34,** 6993.
7. C. Dupont, M. Roberge, F. Shareck, R. Morosoli and F. Kluepfel, *Biochem. J.*, 1998, **330**, 41.
8. P. J. Simpson, D. N. Bolam, A. Cooper, A. Ciruela, G. P. Hazlewood, H. J. Gilbert and M. P. Williamson, *Structure*, 1999, **7**, in press.
9. S. J. Millward-Sadler, D. M. Poole, B. Henrissat, G. P. Hazlewood, J. H. Clarke and H. J. Gilbert, *Mol. Microbiol.*, 1994, **11,** 375.
10. J. I. Laurie, J. H. Clarke, A. Ciruela, C. B. Faulds, G. Williamson, H. J. Gilbert, J. E. Rixon, J. Millward-Sadler and G. P. Hazlewood, *FEMS Microbiology Lett.*, 1997, **148**, 261.
11. R. A. Laskowski, J. A. C. Rullmann, M. W. MacArthur, R. Kaptein and J. M. Thornton, *J. Biomol. NMR*, 1996, **8**, 477.

12. E. D. T. Atkins, in 'Xylan and Xylanases: Progress in Biotechnology' (eds. J. Visser, G. Beldman, v. S. Kusters and A. G. L. Voragen), Elsevier, Amsterdam, 1992, Vol. 7, p. 39.
13. N. Din, H. G. Damude, N. R. Gilkes, R. C. Miller Jr, R. A. J. Warren and D. G. Kilburn, *Proc. Natl. Acad. Sci. USA*,1994, **91**, 11383.
14. S. M. Southall, P. J. Simpson, H. J. Gilfbert, G. Williamson and M. P. Williamson, *FEBS Letts*, 1999, **447**, 58.
15. K. Sorimachi, M.-F. Le Gal-Coëffet, G. Williamson, D. B. Archer and M. P. Williamson, *Structure*, 1997, **5**, 647.
16. D. N. Bolam, A. Ciruela, S. McQueen-Mason, P. J. Simpson, M. P. Williamson, J. E. Rixon, A. Boraston, G. P. Hazlewood and H. J. Gilbert, *Biochem. J.*, 1998, **331**, 775.
17. P. Tomme, A. L. Creagh, D. G. Kilburn and C. A. Haynes, C.A., *Biochemistry*, 1996, **35**, 13885.
18. A. L. Creagh, E. Ong, E. Jervis, D. G. Kilburn and C. A. Haynes, *Proc. Natl. Acad. Sci. USA*, 1996, **93,** 12229.
19. W. P. Jencks, *Adv. Enzymol.*, 1975, **43**, 219.

ANALYSIS OF THE CELL WALL-ANCHORED AVICEL-BINDING PROTEIN (ABPS) FROM *STREPTOMYCES RETICULI*

S. Walter, M. Machner and H. Schrempf

University of Osnabrück
FB5/ Angewandte Genetik der Mikroorganismen
Barbarastr. 11
49069 Osnabrück
Germany

1 INTRODUCTION

Streptomycetes are Gram-positive bacteria, which are highly abundant in soil[1]. They are optimally adapted to their natural environment. Thus they exhibit a differentiated life cycle which closes with the production of spores resistant to heat, dryness and cold. Many streptomycetes are able to degrade biopolymers, the abundant carbon sources in soil[9]. Starch, xylan, chitin and cellulose are efficiently hydrolyzed due to the action of the corresponding extracellular enzymes. A number of these enzymes have been well characterized biochemically, and their genes been identified.

Streptomyces reticuli, the strain used for our studies, is able to utilize crystalline cellulose (Avicel), due to the production of an exoglucanase (Cel1, Avicelase)[11,12]. Physiological studies showed that Avicel, to which *S. reticuli* strongly adheres during cultivation, is the only known inducing carbon source. A low-molecular weight inducer, such as the breakdown products of cellulose (glucose, cellobiose, -triose, -tetraose or -pentaose), able to enter the mycelia could be excluded[15]. On the molecular level, the regulation of the *S. reticuli cel1* gene is realized by both transcriptional activation and repression[16]. Contact of bacteria and surfaces represents an additional signal for stimulation of intracellular responses. Thus *Pseudomonas aeruginosa* cells have been shown to produce an extracellular alginate matrix. The transcription of the *alg*C gene (encodes for essential enzymes in the synthesis of the alginate matrix) is only activated by contact of the cells with Teflon or glass[5,14]. *Vibrio parahaemolysis* synthesizes accessory lateral flagella only when cultivated on agar surfaces. In liquid media, the transcription of the flagellum-encoding gene *laf* is repressed[2].

2 RESULTS AND DISCUSSION

2.1 Identification and purification of the Avicel-binding protein (AbpS) from *Streptomyces reticuli*

Using the total proteins of the crude extract prepared from *Streptomyces reticuli* mycelia, only a 35 kDa protein could be purified by its high affinity to crystalline cellulose (Avicel). It was named AbpS (stands for Avicel-binding protein of

Streptomyces reticuli). Having sequenced internal peptides, the corresponding gene was identified by reverse genetics (accession number Z97071 EMBL)[17].

2.2 Characterization of the binding specificity

AbpS interacts strongly with crystalline forms of cellulose (Avicel, bacterial microcrystalline cellulose and tunicin cellulose); other polysaccharides are recognized only weakly (valonia cellulose or chitin) or not at all (starch, xylan or agar-agar). The binding of AbpS to Avicel could not be inhibited by soluble forms of cellulose (carboxymethyl- or hydroxyethylcellulose) or by the breakdown products of cellulose (cellodextrins, cellobiose, glucose). The Avicel-AbpS interaction occurred very rapidly; after 1 min all protein was bound. Non-ionic or ionic detergents, salts in high concentrations (NaCl or KCl), and the chelating agent EDTA did not induce a release of Avicel-associated AbpS. In contrast, the interaction could be completely inhibited by the addition of 5 M guanidium hydrochloride, 5 M urea or buffer with a pH of less than 4.

2.3 Computer-supported analysis of the AbpS sequence

When we investigated the hydrophobicity of the deduced AbpS sequence, one putative transmembrane segment was predicted at the C-terminus. By computer-supported analysis of the secondary structure, a large centrally located α-helix which has weak homology to the tropomyosin family and the M-proteins was found (Figure 1)[4].

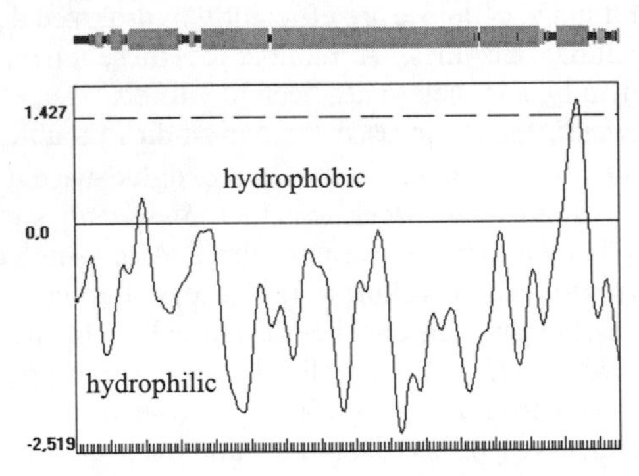

Figure 1 *Prediction of the secondary structure and the hydrophobicity of AbpS. The predicted model of the secondary structure is given in the top, whereas α-helical parts are drawn in grey boxes.*

2.4 Localization of AbpS

After purification of antibodies, which are highly specific to the N-terminal part of AbpS, the localization and orientation of AbpS were analyzed using the techniques of fluorescence, transmission electron or immunofield scanning electron microscopy (Figure 2).

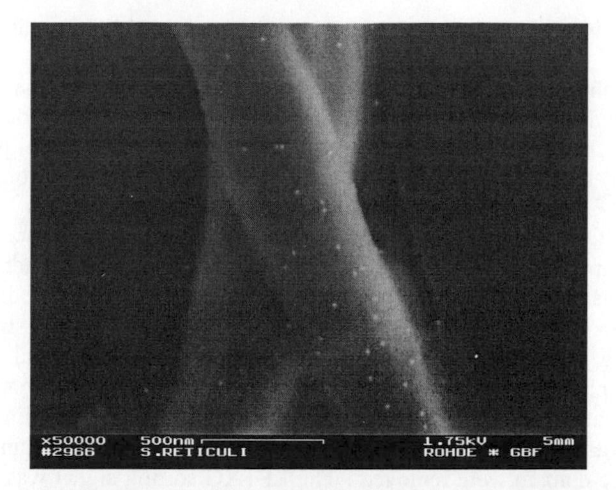

Figure 2 *Immunofield scanning electron microscopy. S. reticuli hyphae were labelled with antibodies specific to the N-terminal part of AbpS and subsequently with proteinA-gold complexes (10 nm). The immunofield electron microscopy was performed by M. Rohde, GBF Braunschweig, Germany.*

Additionally native mycelia were treated with proteinaseK for varying time periods, and the presence of AbpS was immunologically determined. Contrary to the control (without protease), a major portion of AbpS was found to be truncated by approximately 4 kDa.

Combining the results of the above-described methods, we clearly demonstrated that AbpS is located on the surface of the hyphae, that the N-terminus protrudes from the murein layer and the major portion of AbpS is covered by the surrounding envelope murein structure[18].

In contrast, the largest portion of AbpS was found to be associated with protoplasts generated from *S. reticuli* hyphae by lysozyme treatment[18]. The protein was only released by subjecting the protoplasts to osmotic shock, sonification or detergents, whereas small quantities could be isolated in association with the membranes. The predicted COOH-terminally located hydrophobic domain, which shows all characteristics of a transmembrane segment, might anchor AbpS to the membrane of *S. reticuli*. This mechanism of anchoring is well-known from proteins of eukaryotic cells lacking the polyglucane layer, i.e. the T-cell-associated IgGs, the class I and II MHC proteins, the CD4 and CD8 receptors of T-lymphocytes or the peripheral myelin proteins L1 and 0[3].

2.5 Interaction of AbpS with the murein layer

Inspection of ultrathin sections of the murein layer by transmission electron microscopy revealed a linkage between the polyglycane and AbpS. After isolation of the murein layer AbpS could be released from the insoluble polyglycane by sonification or by the action of lysozyme. Neither salts, detergents, buffers with varying pH nor heating with reducing agents dissolved the binding of the protein to murein. Therefore a covalent linkage is proposed.

Incubating the isolated murein-AbpS complexes with proteinaseK and subsequently with lysozyme AbpS was found to be converted into an approximately 21 kDa truncated protein. In contrast soluble AbpS is completely degradable by the protease.

Covalently anchored surface proteins were identified among other bacteria: The immunoglobulin binding protein A from *Staphylococcus* spec. was found to be covalently linked to the pentaglycine of the cell wall, and the NH_2-terminal part of the protein was shown to protrude from the murein. Internalin (InlA) from *Listeria monocytogenes* (a Gram-positive, facultative intracellular parasite) is covalently anchored to its COOH-terminal part, whereas the NH_2-terminus is exposed to the extracellular environment. InlA is a virulence factor, as it assists the adherence to surfaces and the invasion of the eukaryotic host cells[6,7]. ‚Similarly the Ig-binding P- and M-proteins from several streptococci were shown to be cell wall-associated[10]. In all given examples, anchoring requires a COOH-terminally located sorting signal (LPTXG) followed by a hydrophobic domain and a tail of charged amino acids[13]. After secretion, an extracellularly located sortase was shown to modify the surface proteins. First the side chain of threonine was covalently linked to the pentaglycine, second the glycine together with the hydrophobic domain were removed[8]. The LPTXG sorting signal was not found in the deduced amino acid sequence of AbpS. This result indicates a different anchoring mechanism of AbpS.

2.6 Multimerization of AbpS

With the help of immobilized AbpS proteins differing in their sizes, AbpS could be specifically extracted from the *S. reticuli* crude extract, indicating an intermolecular interaction of the AbpS molecules. Combining the different immobilized target proteins and their binding ability the forefront part of the centrally located α-helix seems to be essential for multimerization. Crosslinking experiments with AbpS which was bound to Avicel indicate the formation of homotetramers (Figure 3).

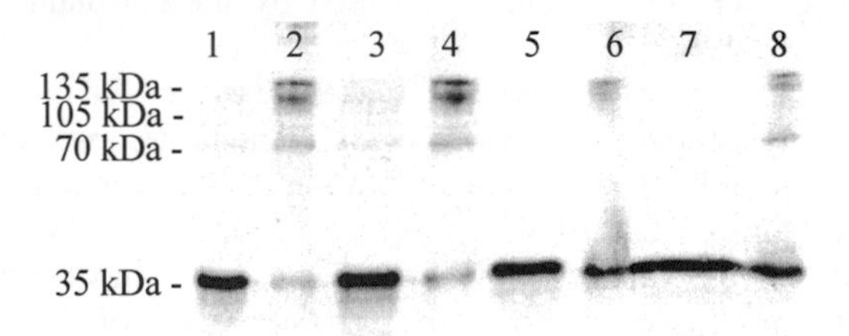

Figure 3 *Multimerization of AbpS. AbpS in different amounts (lane 1+2 : 1 ug; lane 3+4 700 ng; lane 5+6: 400 ng; lane 7+8: 100 ng) was bound to Avicel and treated with the crosslinker DSP(di-thio-bis-succinimydyl-propinate). Subsequently the proteins were released by heating using SDS (lane 2,4,6 and 8) or SDS and mercaptoethanol (lane 1,3,5 and 7), which dissolved the covalent crosslinking, and subjected to SDS-PAGE.*

2.7 Current model of AbpS

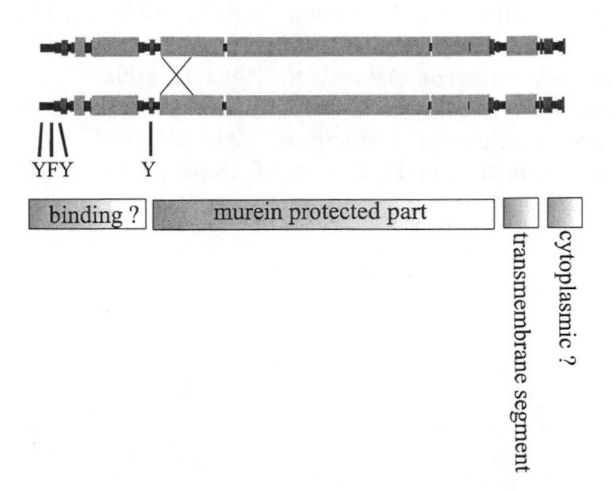

The predicted secondary structure of AbpS with the postulated area responsible for the formation of homomultimers is shown. Deduced from the results of several proteinaseK experiments a part of AbpS which is centrally located was protected from proteolytical digestion by the murein structure. In the postulated N-terminal binding domain several aromatic amino acids were found, which could participate in the interaction of AbpS with Avicel. The short segment following the transmembrane spanning α-helix is predicted to be located intracellular.

References

1. M. Alexander, 'Introduction to Soil Microbiology', John Wiley & Sons, New York, Chichester, Brisbane, Toronto, 1977.
2. R. Belas, M. Simson, and M. Silverman, *J. Bacteriol.*, 1986, **167,** 210.
3. C. Chothia and E.Y. Jones., *Annu. Rev. Biochem.*, 1997, **66,** 823.
4. P. Cleary and D. Retnoningrum, *Trends Microbiol.*, 1994, **2,** 131.
5. D.G. Davies and G.G. Geesey, *Appl. Environ. Microbiol.*, 1995, **61,** 860.
6. S. Dramsi, I. Biswas, E. Maguin, L. Braun, P. Mastroeni, and P. Cossart, *Mol. Microbiol.*, 1995, **16,** 251.
7. M. Lebrun, J. Mengaud, H. Ohayon, F. Nato, and P. Cossart, *Mol. Microbiol.*, 1996, **1,** 579.
8. W.W. Navarre and O. Schneewind, *Mol. Microbiol.*, 1994, **14,** 115.
9. W. Peczynska-Czoch and M. Mordarski.. 'Actinomycete enzymes', pp.219-283. *In* M. Goodfellow, S.T. Williams, and M. Mordarski (eds.), Actinomycetes in Biotechnology. Academic Press, London, 1988.
10. G.N. Phillips Jr., P.F. Flicker, C. Cohen, B.N. Manjula, and V.A. Fischetti, *Proc. Natl. Acad. Sci. USA*, 1981, **78,** 4689.
11. A. Schlochtermeier, F. Niemeyer, and H. Schrempf, *Appl. Environ. Microbiol.* 1992, **58,** 3240.

12. A. Schlochtermeier, S. Walter, J. Schröder, M. Moormann, and H. Schrempf, *Mol. Microbiol.*, 1992, **6**, 3611.
13. O. Schneewind, D. Mihaylova-Petkov, and P. Model, *EMBO J.*, 1993, **12**, 4803.
14. P. Vandevivere and D.L. Kirchman, *Appl. Environ. Microbiol.*, 1993, **59**, 3280.
15. S. Walter and H. Schrempf, *Appl. Environ. Microbiol.*, 1996, **62**, 1065.
16. S. Walter and H. Schrempf, *Mol. Gen. Genet*. 1996, **251**, 186.
17. S. Walter, E. Wellmann, and H. Schrempf, *J. Bacteriol*. 1998, **180**, 1647.
18. S. Walter, M. Rohde, M. Machner, and H. Schrempf, *Appl. Environ. Microbiol.* 1999, **65**, 886.

CHARACTERISTICS OF CHITIN-BINDING PROTEINS FROM STREPTOMYCETES

H. Schrempf, S. Kolbe, A. Becirevic, A. Zeltins

FB Biologie/Chemie, Universität Osnabrück, Barbarastraße 11, D-49069 Osnabrück Germany; Fax: +49 541 969-2804, Phone: +49 541 969-2895
e-mail: schrempf@sfbbio1.biologie.uni-osnabrueck.de

1 INTRODUCTION

Streptomycetes are mycelia- and spore-forming Gram-positive bacteria abundant in soil. They produce a wide range of antibacterial and antifungal compounds. In addition, they hydrolyze a number of macromolecules, including chitin. Some chitinases were enriched from *Streptomyces antibioticus*,[7] *Streptomyces griseus*,[22] *Streptomyces plicatus*,[16] *Streptomyces erythraeus* (now *Saccharopolyspora erythraea*),[6] and *Streptomyces lividans*.[9] *Streptomyces olivaceoviridis* was shown to degrade chitin very efficiently [1] and several of its chitinases were purified.[17]

Our recent studies have revealed that the chitinolytic systems of *S. olivaceoviridis*, *S. reticuli*, and several other investigated streptomycetes comprise the synthesis of a formerly unknown type of extracellular small proteins targeting specifically α-chitin. The characteristics of these chitin-binding proteins (CHBs) have been investigated in more detail.

1.1 Detection of Chitin-Binding Proteins (CHBs)

During growth with chitin, *Streptomyces olivaceoviridis* secretes an 18.7 kDa protein which was shown to adhere strongly to the insoluble substrate and was hence named CHB1. The protein can be released from the substrate by up to 1 M of guanidine hydrochloride,[19] and be purified to homogeneity by consecutive chromatography. Using antibodies raised against CHB1, the crossreacting proteins can frequently be found among streptomycetes, when grown in the presence of chitin-containing substrates.

Several other *Streptomyces* species, including strains of *S. albus*, *S. canescens*, *S. citrofluorescens*, *S. coelicolor* A3(2), *S. coelicolor* Müller, *S. lividans*, *S. parvulus*, *S. vinaceus*, *S. rimosus*, and *S. tendae* secrete homologues of CHB1 and CHB2 during cultivation with insoluble chitin. We also detected proteins crossreacting with *anti*-CHB1 from several other chitinolytic bacteria of different genera.

1.2 Characteristics of *chb* Genes and Deduced Proteins

The genes encoding CHBs of two *Streptomyces* species (*S. olivaceoviridis* and *S. reticuli*) have been identified, cloned, sequenced and overexpressed by reverse genetics.

The deduced CHB1 and CHB2 contain 200 and 201 amino acids, respectively. The relative positions of all aromatic residues, including five tryptophans, match in both proteins.

It is interesting that the CBP21 *Serratia* protein deduced from the corresponding gene shares 45.3 % amino acid identity with the *Streptomyces* CHB1. Thus the relative positions of several short motifs, of aromatic amino acid residues, as well as of several threonin, proline, histidine and lysine residues are conserved among the *Streptomyces* CHB1 and the *Serratia* protein.

1.3 Properties of CHB1

Both CHBs do not display any catalytic or antifungal activity. The proteins interact strongly with various types of α-chitin, including crab shell powder. Weaker adsorption of CHBs has been found with colloidal chitin; this is derived from native crab shell α-chitin after treatment with acids, and its structure is probably predominantly amorphous. The proteins neither adhere to ß-chitin from various sources (i.e. diatom spike, squid pen, ciliates), nor to carboxychitin or chitosan. CHBs do not interact with crystalline bacterial cellulose (chains arranged in parallel orientation [cellulose I]), plant-derived cellulose (Avicel, consisting of crystalline [type I] and amorphous regions), or mureine from *E. coli*, either.

From these results we concluded that both proteins recognize and interact highly specifically with α-chitin. More detailed binding studies revealed that CHB1 has a binding capacity of 0.063 μM (i.e. 1.2 mg protein/mg purified crab shell powder). CHB2 has about half (i.e. 0.037 μM) of this capacity. The dissociation constants (K_d) for CHB1 and CHB2 have been determined to be 0.11 μM and 0.27 μM, respectively.[8,25]

Recently a 21 kDa protein (CBP21) was found to be secreted by *Serratia marcescens* 2170. It binds most strongly to ß-chitin (from squid), followed by colloidal and regenerated chitin.[21]

1.4 Targeting α-Chitin: the Role of Individual Tryptophan Residues

Our spectroscopical investigations suggested an involvement of tryptophan residues in the interaction of CHB1 with α-chitin.[19] In order to explore the role of tryptophan (W) residues within CHB1 in more detail (Fig. 1), they were individually changed to leucine, in some cases to tyrosine residues. Based on the fluorescence spectra, the presence of three buried W-residues can be calculated for CHB1 and the mutant protein W57. Contrary to CHB1, the protein W57 lacks one exposed W-residue. The CD-spectra and the affinity studies with antibodies suggest that there is comparatively little conformational change of the mutant protein W57L, compared to CHB1. The significant reduction of the binding efficiency of W57L to α-chitin can be primarily attributed to the lack of one exposed W-residue. As the replacement of W57 by a tyrosine (Y) residue also leads to the formation of the mutant protein W57Y, whose affinity to α-chitin is even slightly lower than that of W57L, it can be concluded that the W57-residue is essential and cannot be replaced by the other aromatic Y-residue. The CD spectrum suggests that the secondary structure of the mutant protein W134L diverges from that of CHB1. The data obtained by fluorimetry indicate that W134L has two buried and two exposed W-residues. However, according to the hydrophobicity plot of the deduced protein, the mutant protein retains one exposed and three buried W-residues. Thus the secondary structure of the mutant W134L protein must significantly differ from that of

CHB1; this is also indicated by its considerably reduced affinity to *anti*-CHB1 antibodies. Consequently no conclusion can be drawn about the contribution of the W134 residue to the binding affinity of CHB1 to α-chitin. The CD- and fluorescence spectra, as well as the slightly lower affinity to *anti*-CHB1 antibodies suggest that the conformational change in W184 is less pronounced than that in W134L. Therefore it can be assumed that also the W184 residue plays a part in the specific interaction of CHB1 with α-chitin.[25]

Number and relative position of the tryptophan residues within CHB2 correspond to those of CHB1.[8] Therefore their role may correspond to that of the CHB1 residues, too. No common amino acid motifs were identified among the deduced CHBs and chitin-binding domains from several chitinases, including a chitin-binding domain from the *Streptomyces olivaceoviridis* exochitinase.[2] A clustering of cysteine and glycine residues typical of WGA (wheat germ agglutinin) is also missing in the deduced CHBs. The novel CHBs do not share amino acid identities with a new *Anopheles gambiae* type I peritrophic, glycosylated matrix protein (30 kDa) binding to chitin, or with a chitin-binding lectin (39 kDa) secreted by human macrophages.[15]

1.5 The Biological Role of CHBs

Detailed investigations revealed that *Streptomyces* strains secrete the CHBs only when grown with ground crab shells, or with living or autoclaved mycelia from chitin-containing fungi (i.e. *Aspergillus proliferans*, *Neurospora crassa*), but not in the presence of chitobiose or glucose. The *Streptomyces* hyphae adhere closely to the chitin-containing substrates, and the secreted CHBs seem to act like a glue. We showed by confocal microscopy (using CHB1 labelled with fluorescence dye, see below) that, in the course of the close interactions, CHBs invade (in contrast to wheat germ agglutinin [WGA], see below) deeper layers of crab chitin or of the fungal cell wall containing α-chitin.

In their natural habitat, streptomycetes encounter different α-chitin types, which vary as to the length of their *N*-acetylglucosamine chains and the degree of crystallinity, and in dependence on the presence of accessory inorganic compounds or proteins. As a consequence, the consistency of the chitinous layers in organisms like arthropods (comprising crustaceans and insects), molluscs, nematodes, worms, and fungi differs considerably.[12] Within fungal cell walls chitin is embedded in several types of other polysaccharides, including glucans. It is thus assumed that binding constants of CHB1 and CHB2 and of the different chitin types vary and consequently affect the contact to CHB-producing *Streptomyces* strains.

The marine bacterium *Vibrio parahemolyticus* secretes a large lectin (134 kDa) named chitovibrin which shows a high affinity to swollen α-chitin prepared from crab chitin, as well as longer chitooligomers (> dp 9). It has been speculated that chitovibrin plays a part in the adhesion to chitin.[4] Within *Vibrio harveyi*, cell-associated proteins (55 kDa and 150 kDa) lacking catalytic activities were discovered, which maybe mediate the specific attachment of *V. harveyi* to chitin.[11]

1.6 The Use of CHBs for the *in situ* Localization of α-Chitin

Fluorescent dyes, such as Calcofluor,[13] Congo Red,[23] and Primulin [18] interact with chitin and other polysaccharides. Calcofluor, the most commonly used one, was found to bind to many ß-linked polysaccharides,[13] including cellulose and exopolysaccharides from *Rhizobium* strains.[5] Thus the detection of Calcofluor-binding material does not prove the presence of chitin.

WGA (wheat germ agglutinin) is a plant lectin consisting of 171 amino acids. It recognizes and binds to N-acetyl-glucosamine (GlcNAc) and exhibits an especially high affinity for ß(1→4)-linked GlcNAc-oligomers which are composed of three or more residues.[14] Sectioned cells can be treated with gold-conjugated WGA, and the location of the bound electron-dense particles be determined by transmission electron microscopy.[20] Although this method provides the highest resolution currently obtainable, it has several drawbacks. First it requires considerable technical skill and time. Second, WGA is not absolutely specific for GlcNAc, it also interacts weakly with N-acetyl-neuraminic acid, N-acetyl-galactosamine, and other compounds containing GlcNAc. To achieve a more rapid assay, fluorescein-isothiocyanate (FITC) or rhodamine can be coupled with WGA and used for binding assays.[10]

In order to quantify chitin, it can be treated with a chitinase, and the released chitobiose can be enzymatically cleaved to GlcNAc. Although this is a highly reliable method to definitely identify chitin, its sensitivity is relatively low and it cannot be applied to ascertain the location of chitin within biological material.[3]

None of the above described methods can discriminate between polymer chains arranged in a parallel (ß) and in an antiparallel (α) fashion; this is only possible by X-ray diffraction spectra of purified chitin.[12] However, FITC-labelled CHB1 is most suitable to detect the relative position of α-chitin in native samples (including native α-chitin-containing fungi, shells or cuticles of arthropods, insects and crustaceans) by fluorescence microscopy.[19,24] Thus the novel chitin-binding protein has proven to be superior to any of the other known methods to rapidly detect the relative location of crystalline α-chitin. It is also expected that a gold-conjugated CHB1 could be utilized for an analysis of sections through cells. With confocal laser microscopy we found that FITC-labelled WGA binds more strongly to the surface of chitin, whereas FITC-marked CHB1 invades its deeper layers.[24] The CHB1-labelled protein is therefore well-suited, too, to estimate the width of chitin layers. Since CHB1 binds neither to chitotriose or chitooligomers, nor to chitin of low crystallinity (situated at the tips of fungi), it can in future be used to study the crystallization process of chitin within various organisms.

Acknowledgements

I am grateful to M. Lemme for supporting the writing of the manuscript and to D. Müller for taking the photographs. The work was financed by the Deutsche Forschungsgemeinschaft (DFG).

References

1. M. Beyer and H. Diekmann, *Appl. Microbiol. Biotechnol.*, 1985, **23,** 140.
2. H. Blaak and H. Schrempf, *Eur. J. Biochem.*, 1995, **229,** 132.
3. C. E. Bulawa, M. Slater, E. Cabib, J. Au-Young, A. Sburlati, W. A. Lee, Jr. and P. W. Robbins, *Cell*, 1986, **46,** 213.
4. O. S. Gildemeister, B. C. R. Zhu and R. A. Laine, *Glycoconjugate Journal*, 1994, **11,** 526.
5. J. Glazebrook and G. C. Walker, *Cell* , 1989, **56,** 661.

6. S. Hara, Y. Yamamura, Y. Fujii, T. Mega and T. Ikenada, *J. Biochem.*, 1989, **105,** 484.
7. C. Jeuniaux, *Methods Enzymol.*, 1966, **8,** 644.
8. S. Kolbe, S. Fischer, A. Becirevic, P. Hinz and H. Schrempf, *Microbiology*, 1998, **144,** 1291.
9. K. Miyashita, T. Fujii and Y. Sawada, *J. Gen. Microbiol.*, 1991, **137,** 2065.
10. J. Molano, B. Bowers and E. Cabib, *J. Cell Biol.*, 1980, **85,** 199.
11. M. T. Montgomery and D. L. Kirchman, *Appl. Environ. Microbiol.*, 1994, **60,** 4284.
12. R. A. A. Muzzarelli. 'Chitin', Pergamon Press, Oxford, 1977, p.
13. J. R. Pringle, A. E. M. Adams, D. G. Drubin and B. K. Haarer. In 'Guide to Yeast Genetics and Molecular Biology', eds. C. Guthrie and G. R. Fink, Academic Press, San Diego, 1991, p. 565.
14. N. V. Raikhel, H.-I. Lee and W. F. Broekaert, *Annu. Rev. Plant Physiol. Plant Mol. Biol.*, 1993, **44,** 591.
15. G. H. Renkema, R. G. Boot, F. L. Au, W. E. Donker-Koopman, A. Strijland, A. O. Muijsers, M. Hrebicek and J. M. F. G. Aerts, *Eur. J. Biochem.*, 1998, **251,** 509.
16. P. W. Robbins, C. Albright and B. Benfield, *J. Biol. Chem.*, 1988, **263,** 443.
17. A. Romaguera, U. Menge, R. Breves and H. Diekmann, *J. Bacteriol.*, 1992, **174,** 3450.
18. R. Schekman and V. Brawley, *Proc. Natl. Acad. Sci. USA*, 1979, **76,** 645.
19. J. Schnellmann, A. Zeltins, H. Blaak and H. Schrempf, *Mol. Microbiol.*, 1994, **13,** 807.
20. J. A. Shaw, P. C. Mol, B. Bowers, S. J. Silverman, M. H. Valdivieso, A. Durán and E. Cabib, *J. Cell. Biol.*, 1991, **114,** 111.
21. K. Suzuki, M. Suzuki, M. Taiyoji, N. Nikaidou and T. Watanabe, *Biosci. Biotechnol. Biochem.*, 1998, **62,** 128.
22. A. L. Tarentino and F. Maley, *J. Biol. Chem.*, 1974, **249,** 811.
23. G. L. Vannini, F. Poli, A. Donini and S. Pancaldi, *Plant Sci. Lett.*, 1983, **31,** 9.
24. A. Zeltins and H. Schrempf, *Anal. Biochem.*, 1995, **231,** 287.
25. A. Zeltins and H. Schrempf, *Eur. J. Biochem.*, 1997, **246,** 557.

Industrial Exploitation of Carbohydrate Modifying Enzymes

AMYLOSUCRASE : A GLUCOSYLTRANSFERASE THAT USES SUCROSE AND BELONGS TO THE α-AMYLASE SUPER FAMILY

G. Potocki de Montalk[a], P. Sarçabal[a], M. Remaud-Simeon[a], R. M. Willemot[a], V. Planchot[b] and P. Monsan[a]
[a]Centre de Bioingénierie Gilbert Durand, U. M. R. C.N.R.S. 5504, U.R. I.N.R.A. 792, D.G.B.A., I.N.S.A., Complexe Scientifique de Rangueil, 31 077 Toulouse CEDEX, France
[b]INRA URPOI, 44 316 Nantes CEDEX 3, France

1 INTRODUCTION

Amylosucrase is a glucosyltransferase (E.C.2.1.4) which catalyses the transfer of the D-glucopyranosyl unit from sucrose onto an α-glucan primer, and synthesises an insoluble polymer. This reaction does not require any activated substrates like α-D-glucosyl-nucleoside-di-phosphate (13, 14, 15, 18, 19, 31), unlike glycogen-synthase.

Amylosucrase was first discovered in 1946 from *Neisseria perflava* by Hehre et al (6). In 1974, *Neisseria polysaccharea* was isolated from the throats of healthy children in Europe and Africa (24). This strain was shown to possess an extracellular amylosucrase that synthesises an extracellular polymer from sucrose, constituted by α-(1→4) linked glucopyranosyle units (23, 25), having strong similarities with amylose (4).

The gene encoding amylosucrase from *N. polysaccharea* has been cloned and sequenced (21). This report describes the specific comparison of the deduced amino acid sequence with those of enzymes from the α-amylase family. This work allowed us to localise crucial regions conserved in related enzyme sequences. The recombinant enzyme expressed by *E. coli* was then purified by affinity chromatography using fusion protein to a degree compatible with crystallisation assays and X-ray diffraction studies. The catalytic mechanism of the enzyme was then investigated and the synthesised polymer characterised.

2 RESULTS

2.1 Sequence comparison of amylosucrase with related enzymes

The gene encoding amylosucrase from *Neisseria polysaccharea* (ATCC 43768) consists of 1911 nucleotides, encoding a protein of 636 amino acids.

A characteristic $(\beta/\alpha)_8$-barrel fold was predicted by the Pfam Software, interrupted by the presence of a separate folding module (70 amino acids) homologous to a calcium-binding domain, protruding between β strand 3 and α helix 3 (11). This domain could correspond to a part of domain B, conserved in amylolytic enzymes (8).

This study also revealed a C-terminal β-barrel, which confirms that amylosucrase is a member of the α-amylase family.

Moreover, the sequence of amylosucrase contains seven of the eight highly conserved regions in amylolytic enzymes (7). The multiple alignment corresponding to these sites is presented in figure 1. Sites II, V, VI and VII are the four best-conserved regions in the active site of the enzymes belonging to the α-amylase family. Sites III is not sufficiently similar to the amylosucrase sequence to be localised.

Some of the highly conserved residues have been shown to play a role in amylosucrase activity. That is the case of amino acids D-294, D-401 and E-336. They have been individually replaced by N, N and Q, respectively, using site-directed mutagenesis. The three mutants obtained were totally inactive. These amino acids correspond to the catalytic triad involved in glucosidic bond cleavage in α-amylases. They are equivalent to D-206, D-297 and E-230 of α-amylase from *A. oryzae*. Consequently, they could be involved in the formation of a glucosyl-enzyme intermediate (29, 30).

Two other invariant amino acids (H-195 and H-400) have also been submitted to site-directed mutagenesis. Their replacement in Q and N, respectively, led to a strong decrease of amylosucrase activity. As for α-amylase from *A. oryzae,* they could be involved in stabilisation of the substrate-binding transition state in α-amylase from *A. oryzae* (9,16).

	I	II	III	IV	V	VI	VII	VIII
	β2	β3	L3	L3	β4	β5	β7	β8
AS	134-GLTYLHLM-P-142	190-DFIFNH-195		262-QWDLN-266	290-ILRMDAVAF-298	336-EAIV-339	396-YVR--SHD-401	488-GLPLIYLGD-496
AMY	36-GFGGVQVS-P-44	96-DAVINH-101	150-SYND-153	165-LLDLA-169	193-GFRIDASKH-201	233-EVID-236	295-FVD--NHD-300	334-GFTRVMSSY-342
OGL	44-GIDVIWLS-P-52	98-DLVVNH-103	150-QYDE-153	167-QPDLN-171	195-GFRMDVINF-203	255-EMPG-258	324-YWN--NHD-329	360-GTPYIYQGE-368
AGL	52-GVDAIWVC-P-60	106-DLVINH-111	164-TFDE-167	181-QVDLN-185	210-GFRIDTAGL-218	276-EVAH-279	344-YIE--NHD-349	381-GTLYVYQGQ-389
PUL	210-GVTHVELL-P-218	281-DVVYNH-286	308-AYGN-311	319-GNDIA-323	348-GFRFDLMGI-356	381-EGWD-384	464-YVE--SHD-469	505-GIPFLHSGQ-513
APU	435-GISVIYLN-P-443	488-DGVFNH-493	527-PYGD-530	565-WADFI-569	593-GWRLDVANE-601	626-ELWG-629	698-LLG--SHD-703	745-GMPSIYYGD-753
CMD	187-GVNALYFN-P-195	240-DAVFNH-245	284-TYDT-287	294-MPKLN-298	323-GWRLDVANE-331	356-EIMH-359	418-LLG--SHD-423	450-GTPCIYYGD-458
MTH	50-GFSAIWMPVP-59	112-DVVPNH-117	145-NYPN-148	160-ESDLN-164	189-GFRFDFVRG-197	219-ELWK-222	289-FVD--NHD-294	327-GTPVVYWSH-335
ISA	217-GVTAVEFL-P-225	291-DVVYNH-296	328-TSGN-331	341-GANFN-345	370-GFRFDLASV-378	416-EFTV-419	502-FID--VHD-507	570-GTPLMQGGD-578
DGL	44-GVMAIWLS-P-52	98-DLVVNH-103	145-QYDD-148	162-QPDLN-166	190-GFRMDVIDM-198	236-ETWG-239	308-FWN--NHD-313	344-GTPYIYQGE-352
MHH	38-GITAVWIP-P-46	102-DVVMNH-107	166-DWDQ-169	203-YADID-207	232-GFRIDAVKH-240	266-EFWK-269	328-FVD--NHD-333	362-GYPSVFYGD-370
NPU	188-GINGIYLT-P-196	241-DAVFNH-246	284-NYDT-287	294-MPKLN-298	323-GWRLDVANE-331	356-EVWH-359	418-LLG--SHD-423	450-GTPCIYYGD-458
BRE	280-GFTHLELL-P-288	335-DWVPGH-340	356-LYEH-359	367-HQDWN-371	401-ALRVDAVAS-409	458-EEST-461	519-FVLPLSHD-526	555-GWMWAFPGK-563
CGT	70-GVTALWISQP-79	135-DFAPNH-140	185-SLEN-188	197-LADFN-201	225-GIRVDAVKH-233	257-EWFL-260	323-FID--NHD-328	354-GVPAIYYGT-362
GDE	137-GYNMIHFT-P-145	198-DVVYNH-203	248-KYKE-251	451-LRNFA-455	505-GVRLDNCHS-513	538-ELFT-541	604-FMD-ITHD-610	642-GYDELVPHQ-650
TAK	56-GFTAIWIT-P-64	117-DVVANH-122	155-YEDQ-158	173-LPDLD-177	202-GLRIDTVKH-210	230-EVLD-233	292-FVE--NHD-297	323-GIPIIYAGQ-331

	I	II	IV	V	VI	VII	VIII
AS	134-GLTYLHLM-P-142	190-DFIFNH-195	262-QWDLN-266	290-ILRMDAVAF-298	336-EAIV-339	396-YVR--SHD-401	488-GLPLIYLGD-496
DSRB	939-GITSFQLA-P-947	1004-DWVPDQ-1009	491-ANDVD-495	529-GIRVDAVDN-537	571-EDWS-574	639-FVR--AHD-644	709-TVPRVYYGD-717
DSRA	699-GITSFEMA-P-707	764-DWVPDQ-769	244-ANDVD-248	282-GYRVDAVDN-290	324-QFWK-327	392-FIR--AHD-397	464-TIPRVYYGD-472
GTFD	894-GVTSFEMA-P-902	959-DWVPDQ-964	423-ANDID-427	461-GVRVDAVDN-469	503-EAWS-506	579-FIR--AHD-584	649-SITRLYYGD-657
GTFK	893-GITSFEIA-P-901	958-DWVPDQ-963	417-SNDID-421	457-GIRVDAVDN-465	499-EAWS-502	577-FVR--AHD-582	647-AATRVYYGD-655
GTFS	849-GITQFEMA-P-857	914-DLVPNQ-919	495-ANDVD-499	433-GVRVDAVDN-441	475-EAWS-478	542-FIR--AHD-547	614-TVTRVYYGD-622
GTFI	860-GITDFEMA-P-868	926-DWVPDQ-931	405-ANDVD-409	443-SIRVDAVDN-451	485-EAWS-488	553-FAR--AHD-558	623-SIPRVYYGD-631
GTFC	889-GVTDFEMA-P-897	954-DWVPDQ-959	435-ANDVD-439	473-SIRVDAVDN-481	515-EAWS-518	583-FIR--AHD-588	653-SVPRVYYGD-661
GTFB	862-GVTDFEMA-P-870	927-DWVPDQ-932	409-ANDVD-413	447-SIRVDAVDN-455	489-EAWS-492	557-FIR--AHD-562	627-SVPRVYYGD-635
DSRS	958-GVTSFQLA-P-966	1023-DWVPDQ-1028	509-ANDVD-513	547-GIRVDAVDN-555	589-EDWS-592	657-FVR--AHD-662	727-TVPRVYYGD-735

Figure 1 *Conserved sequence stretches (roman numbers) in amylosucrase, in the α-amylase superfamily and in glucosyltransferases. The second line denotes the elements of secondary structure, as determined for pig pancreatic α-amylase. The enzymes are numbered from the N-terminal end. The invariable residues are in bold type. Enzyme sources: AS, amylosucrase (*Neisseria polysaccharea*); AMY, α-amylase (pig pancreatic); OGL, oligo-1,6-glucosidase (*Bacillus cereus*); AGL, α-glucosidase (*Saccharomyces cerevisiae*); PUL, pullulanase (*Bacillus stearothermophilus*); APU, amylopullulanase (*Clostridium thermohydrosulfuricum*); CMD, cyclomaltodextrinase (*Bacillus sphaericus*); MTH, maltotetraohydrolase (Pseudomonas saccharophila); ISA, isoamylase (*Pseudomonas amyloderamosa*); DGL, dextran-glucosidase (*Streptococcus mutans*); MHH, maltohexaohydrolase (*Bacillus sp. strain 707*); NPU, neopullulanase*

*(*Bacillus stearothermophilus*); BRE, branching enzyme (*Escherichia coli*); CGT, cyclodextrin-glycosyltransferase (*Bacillus circulans*); GDE, glycogen debranching enzyme (*Human muscle*); TAK, α-amylase (*Aspergillus oryzae*); AS, amylosucrase (*Neisseria polysaccharea*); DSRB (*Leuconostoc mesenteroides *NRRL B-1299); DSRA (*Leuconostoc mesenteroides *NRRL B-1299); GTFD (*Streptococcus mutans *GS5);* GTFK (Streptococcus salivarius *ATCC 25975); GTFS (*Streptococcus downei *Mfe 28); GFTI (*Streptococcus sobrinus *OMZ176 Serotype D); GTFC (*Streptococcus mutans *GS5); GTFB (*Streptococcus mutans *GS5); DSRS (*Leuconostoc mesenteroides *NRRL B-512F).*

These consensus sequence regions were also conserved in other glucosyltransferases consuming sucrose to produce α-glucan polymers, like dextransucrases and glucosyltransferases from *Leuconostoc* sp. and *Streptococcus* sp.. The multiple alignment shown in Figure 1, supports the studies predicting that glucosyltransferases are also members of α-amylase family (5, 12), since they possess a circularly permuted $(\beta/\alpha)_8$ barrel (12). The catalytic triad of α-amylases is also conserved in glucosyltransferases, and has been shown to be crucial for activity (5, 10).

An unrooted evolutionary tree was constructed using ClustalW. It was based on the conserved sites I, II, IV, V, VI, VII and VIII of amylolytic enzymes, glucosyltransferases and amylosucrase. In spite of the fact that amylosucrase is itself a glucosyltransferase, it is situated on one of the longest branches of the tree, showing that it is very far from dextransucrases and amylolytic enzymes, from an evolutionary point of view.

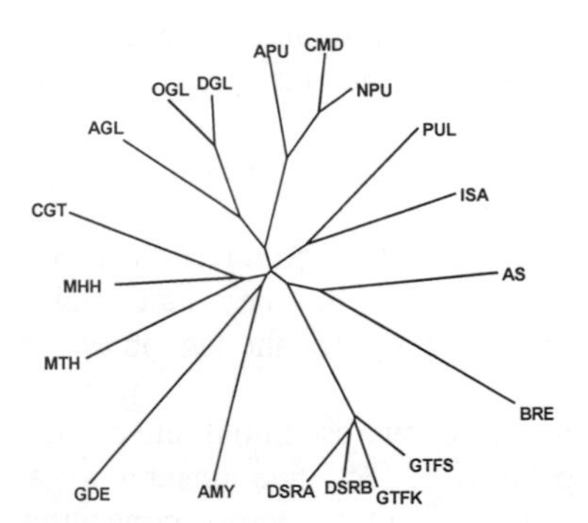

Figure 2 *Evolutionary tree of amylolytic enzymes, glucosyltransferases, and amylosucrase, based on the stretches I, II, IV, V, VI, VII, and VIII. The abbreviations of enzyme sources are given in the legends to Figure 1. The branch lengths are proportional to the sequence divergency.*

2.2 Amylosucrase purification

Amylosucrase was purified by affinity chromatography between fusion protein Glutathion-S-Transferase/Amylosucrase and Glutathion-Sepharose-4-B. The cloning of the amylosucrase gene fused to glutathione-S-transferase was carried out using pGEX-6-P-3, as it was described by Potocki *et al* (21). The fusion protein was expressed in *E. coli* strain BL21. The activity of the glutathione-S-transferase/amylosucrase fusion protein and the purification fractions was measured in the presence of 50 g/l sucrose and 0.1 g/l glycogen.

The purification process enabled to obtain a pure amylosucrase, with a specific activity of 5657 U/g in the assay conditions. The total yield of purification was 58 %. The total purification factor was 7.6. The molecular mass and the isoelectric point of amylosucrase are 70 ± 2 kDa and 4.9 ± 0.2, respectively.

In order to understand the catalytic mechanism of amylosucrase, and the features shared with both amylolytic enzymes and glucosyltransferases, the pure enzyme has been submitted to crystallisation and X-ray diffraction studies. Experiments are in progress.

2.3 Kinetic characterisation

Catalytic mechanism has been investigated using kinetic studies, in the presence of sucrose alone, and using glycogen as a glucan primer. The amylosucrase assay was carried out at 30 °C in 50 mM Tris-HCl buffer, pH 7, supplemented with 0.6 to 600 mM sucrose and 0 to 30 g/l oyster glycogen, with 39 mg/l of pure enzyme. One unit of amylosucrase corresponds to the amount of enzyme that catalyses the production of one μmole of fructose per min in the assay conditions. Sucrose, glucose and fructose concentrations were measured by ion-exchange chromatography with an Aminex HPX87H column (Biorad Chemical Division, Richmond, CA).

2.3.1 Reactions in the presence of sucrose. When acting on sucrose, amylosucrase displays several catalytic activities:
- hydrolysis, which is measured by the rate of glucose and fructose produced
- maltose and maltotriose synthesis, by transfer of glucopyranosyl units from sucrose onto glucose moieties released in the medium
- α-glucan polymer growth, which is exclusively composed of α-(1→4)-glucosidic bonds.

During initial activity measure, fructose was produced without any visible lag phase. This result is not in accordance with the preliminary data obtained with a crude preparation of recombinant enzyme where a significant lag phase occurred. It was probably due to interfering activities.

The Lineweaver-Burk representation of sucrose initial rate consumption does not correspond to a Michaelis-Menten pattern. In fact, two linear curves with different slopes are obtained before and after 20 mM initial sucrose concentration. Values of apparent Vm and Km obtained from the curve corresponding to sucrose initial concentration lower than 20 mM are 510 U/g amylosucrase and 2 mM, respectively. For initial sucrose concentrations higher than 20 mM, values are 906 and 26 mM, respectively, as if amylosucrase was activated at high sucrose concentrations.

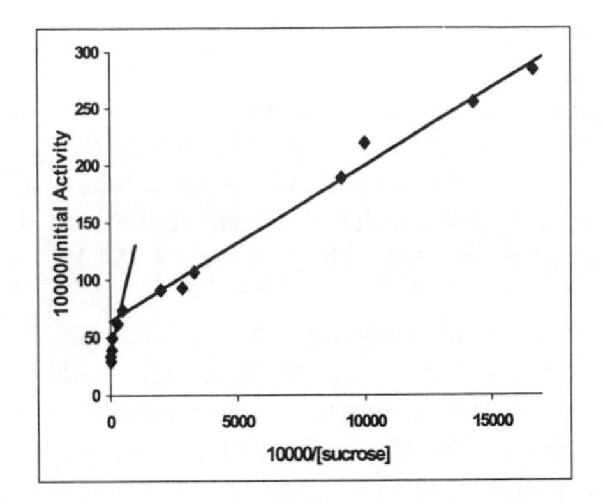

Figure 3 *Lineweaver-Burk curve without any primer in the medium*

2.3.2 Kinetic characterisation in the presence of sucrose and glycogen. Amylosucrase catalyses the synthesis of α-1→4-glucan from sucrose without any primer (4). Nevertheless, studies on crude extracts of recombinant amylosucrase showed that the polymer-synthesis rate greatly increased in the presence of a glucan primer, and particularly of glycogen (23).

The role of glycogen was investigated, by studying both sucrose and glycogen concentration effects on initial sucrose consumption rate. The results are presented in fig 4. For a same sucrose concentration, the initial rate of sucrose consumption by amylosucrase strongly increases with glycogen concentration. At 105 mM sucrose, the initial activity increases 98 times when glycogen concentration increases from 0 to 30 g/l. Nevertheless, the activator effect of glycogen decreases with high sucrose concentrations. The optimum sucrose concentration increases with glycogen concentration, suggesting a competition between sucrose and glycogen.

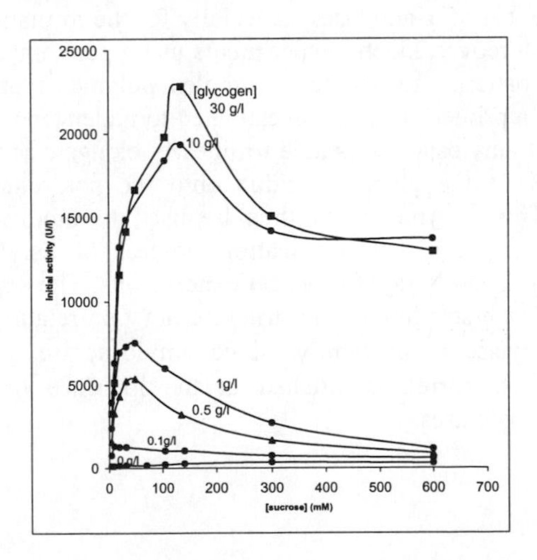

Figure 4 *Initial activity of amylosucrase versus sucrose concentration for variable glycogen concentrations*

2.4 Polymer characterisation

Glycogen exerts an activating effect on sucrose consumption depending on its concentration. By using a ratio sucrose/glycogen primer of 50 w/w, it was possible to produce an insoluble polysaccharide. Average molecular weight values for native glycogen and synthesised polysaccharide were based on high-performance size-exclusion chromatography results. They were $9.954.10^6$ and $15.27.10^6$ g/mol, respectively.

Deramification of the initial glycogen with isoamylase led to a broad population of DP 12. For the synthesised polysaccharide, a major peak was found at the same average DP and a white precipitate appeared, corresponding to longer chains located in the dissolving gap described by Aberle et al (1). β-amylolysis was 67 % and 77 % respectively for native and modified glycogens. Lastly, iodine-complex of native glycogen presents a λmax at 435 nm, in contrast to 605 nm for the modified glycogen. λmax at 435 nm is in agreement with values usually obtained for short chains, while λmax at 605 nm is characteristic of linear chains with an average DP of around 75 glucosyl units.

According to these results, amylosucrase have produced the insoluble polysaccharide by extension of some external chains of the native glycogen from the non-reducing ends, from 12 to 75 glucosyl units.

3 CONCLUSION

From these studies, it can be concluded that amylosucrase is a member of the α-amylase superfamily, and that some of the invariant residues are involved in amylosucrase activity. Particularly, multiple alignments and site-directed mutagenesis experiments allowed to localise the catalytic triad conserved and previously described both in amylolytic enzymes and glucosyltransferases. The catalytic mechanism of this enzyme may therefore resemble that of α-amylases, especially for the formation of the glucosyl enzyme intermediate. Moreover, kinetic experiments using pure amylosucrase revealed non-Michaelis-Menten patterns, related to a complex polymer synthesis mechanism. The use of glycogen as a primer has been investigated to understand its role in reaction activation, showing that amylosucrase is able to linearly elongate some of the polymer branchings, by transfer of the glucose residue onto the non-reducing ends of the glycogen molecules. This enzyme could thus be used to modify glucan polymer structure. Lastly, the very efficient purification process of amylosucrase allowed initiation of crystallisation and X-ray diffraction experiments. The expected results will be crucial for a deeper understanding of the structure/function relationships in enzymes belonging to the α-amylase super family, since amylosucrase, possessing a non-circularly-permuted $(\beta/\alpha)_8$ barrel, is situated at the interface between amylolytic enzymes and glucosyltransferases.

References

1. Th. Aberle, W. Burchard, W. Vorwerg and S. Radosta, *Starch/Stärke*, 1994, **46**, 329.
2. A. Bello-Perez, P. Roger, B. Baud, P. Colonna, *J. Cereal Sci.*, in press.
3. M.M. Bradford, *Anal. Biochem.*, 1976, **72**, 248.
4. V. Büttcher, T. Welsh, L. Willmitzer and J. Kossmann, *J. Bacteriol.*, **1997**, 179, 3324.
5. K.S. Devulapalle, S.D. Goodman, Q. Gao, A. Hemsley and G. Mooser, *Protein Sci.*, 1997, **6**, 2489.
6. E.J. Hehre, D.M. Hamilton and A.S. Carlson, *J. Biol. Chem.*, 1949, **177**, 267.
7. S. Janecek, *FEBS Lett.*, 1994, **353**, 119.
8. S. Janecek, B. Svensson, and B. Henrissat, *J. Mol. Evol.*, 1997, **45**,322.
9 H.M. Jespersen, E.A. MacGregor, B. Henrissat, M.R. Sierks and B. Svensson, *J. Protein Chem.*, 1993, **12**, 791.
10. C. Kato, Y. Nakano, M. Lis and H.K. Kuramitsu, *Biochem. Biophys. Res. Co.*, 1992, **189**, 1184.
11. S.B. Larson, A. Greenwood, D. Cascio, J. Day, and A. McPherson, *J. Mol. Biol.*, 1994, **235**, 1560.
12 E.A. Mac Gregor, H.M. Jespersen, B. Svensson, *FEBS Lett.*, 1996, **378**,263.
13. C.R. Mac Kenzie, K.G. Johnson and I.J. McDonald, *Can. J. Microbiol.*, 1977, **23**, 1303.
14. C.R. Mac Kenzie, I.J. McDonald and K.G. Johnson, *Can. J. Microbiol.*, 1978, **24**, 357.
15. C.R. Mac Kenzie, M.B. Perry, I.J. McDonald and K.G. Johnson, *Can. J. Microbiol.*, 1978, **24**, 1419.
16. Y. Matsuura, M. Kusunoki, W. Harada and M. Kakudo, *A. J. Biochem.*, 1984, **95**, 697.
17. H. Nelson, *J. Biol. Chem.*, 1944, **153**, 375.
18. G. Okada and E.J. Hehre, *Carbohyd. Res.*, 1973, **26**, 240.
19. G. Okada and E.J. Hehre, *J. Biol. Chem.*, 1974, **249**, 126.
20. V. Planchot, P. Colonna and L. Saulnier, *In* B. Godon and W. Loisel (ed.), Guide Pratique d'Analyses dans les Industries des Céreales Publication Lavoisier, Paris, France, 1996, p.11.
21. G. Potocki de Montalk, M. Remaud-Simeon, R.M. Willemot, V. Planchot and P. Monsan, *J. Bacteriol.*, 1999, **181**, 375.
22. J. Preiss, J.L. Ozbun, J.S. Hawker, E. Greenberg and C. Lammel, *Ann. N. Y. Acad. Sci.*, 1973, **210**, 265.
23. M. Remaud-Simeon, F. Albaret, B. Canard, I. Varlet, P. Colonna, R.M. Willemot and P. Monsan, *In* S. B. Petersen, B. Svensson, and S. Pedersen (ed.), Carbohydrate bioengineering, Elsevier Science B. V., Amsterdam, The Nederlands, 1995, p.313.
24. J.Y. Riou, M. Guibourdenche and M.Y. Popoff, *Ann. Microbiol. (Inst. Pasteur)*, 1983, **134B**, 257.
25 J.Y. Riou and M. Guibourdenche, M.B. Perry, L.L. MacLean and D.W. Griffith, *Can. J. Microbiol.*, 1986, **32**, 909.

26. J. Sambrook, E.F. Fritsch and T. Maniatis, 'Molecular cloning: a laboratory manual', 2nd ed. Cold Spring Harbor Laboratory Press, Cold Spring Harbor, N. Y., 1989.
27. F. Sanger, S. Nicklen and A.R. Coulson, 'DNA sequencing with chain-terminating inhibitors', Proc. Natl. Acad. Sci. USA, 1979, 74,5463.
28. J.B. Sumner, S.F. Howell, *J. Biol. Chem.*, 1935, **108**, 51.
29 B. Svensson, *Plant Mol. Biol.*, 1994, **25**, 141.
30 B. Svensson, T.P. Frandsen, I. Matsui, N. Juge, H.P. Fierobe, B. Stoffer and K.W. Rodenburg, *In* S.B. Petersen, B. Svensson and S. Pedersen (ed.), Carbohydrate bioengineering, Elsevier Science B. V., Amsterdam, The Nederlands, 1995, p.125.
31 B.Y. Tao, P.J. Reilly and J.F. Robyt, *Carbohyd. Res.*, 1988, **181**, 163.

α-1,4-GLUCAN LYASE, MOLECULAR FEATURES AND ITS USE FOR PRODUCTION OF 1,5-ANHYDRO-D-FRUCTOSE FROM STARCH

Shukun Yu and Jan Marcussen

Danisco Biotechnology, Danisco A/S,
Langebrogade 1, PO Box 17, DK 1001,
Copenhagen K, Denmark.

1 INTRODUCTION

It is well known that among polysaccharides, starch and glycogen, and their degradation are most studied as starch is the main photosynthetic product after the cell wall material cellulose and lignin, and is used for both food and non-food purposes. Polysaccharides are made up of sugar residues or anhydro-sugar units, and in the case of starch it is 1,4-anhydroglucose. The degradation of polysaccharides is catalyzed by two groups of enzymes: hydrolases, which result in the formation of the constituent sugars of the polysaccharides, and lyases, which yield unsaturated sugars. There are more than 100 lyases known today that are active toward polysaccharides, such as pectin-, pectate, heparin-, and alginate lyase, which are represented with the EC number of 4.2.2.1 to EC4.2.2.12. However, the discovery of a lyase active toward starch and glycogen is only a recent event. The starch lyase, i,e. α-1,4-glucan lyase (EC 4.2.2. 13) is further found to be an exo-lyase and releases a 1,2-enol sugar that tautomerizes to 1,5-anhydro-D-fructose (1,5AnFru, 1,5-Anhydro-D-*arabino*-hex-2-ulose), representing the only polysaccharide lyase known today that acts on homopolymers and produces a monosugar as the major product. Here is presented some of the work performed at our laboratory in the last several years.[1-6]

2 MECHANISTIC CONSIDERATIONS

Scheme 1 shows the reaction catalyzed by α-1,4-glucan lyase (1) and its comparison with those by other polysaccharide lyases exemplified by pectate lyase (2), and α-glucosidase (3). The glucan lyase cleaves *carbon-oxygen* bond and introduces a double bond between C1 and C2, yielding the1,2-enol 1,5AnFru (Scheme 1). In contrast, the other polysaccharide lyases reported up to now cleave *oxygen-carbon* bond and introduce a double bond between C4 and C5, resulting in an oligosaccharide with 4-deoxy 4-enuronosyl group on the non-reducing end (Scheme 1). It is therefore proposed that these two groups of lyases are substantially different in their catalysis mechanisms. Our studies indicate that the catalysis mechanism of the glucan lyase is similar largely to α-glucosidase. This explains their shared substrate and inhibitor specificity. Such hypothesis is further supported by the fact that α-glucosidase catalyzes the protonation of D-glucal, an analogue of the1,2-enol form of 1,5AnFru (Scheme 1), which is virtually the reverse reaction of the glucan lyase.

Scheme 1 A comparison of the reactions catalyzed by α-1,4-glucan lyase, pectate lyase and α-glucosidase

1). The cleavage of C—O by α-1,4-glucan lyase

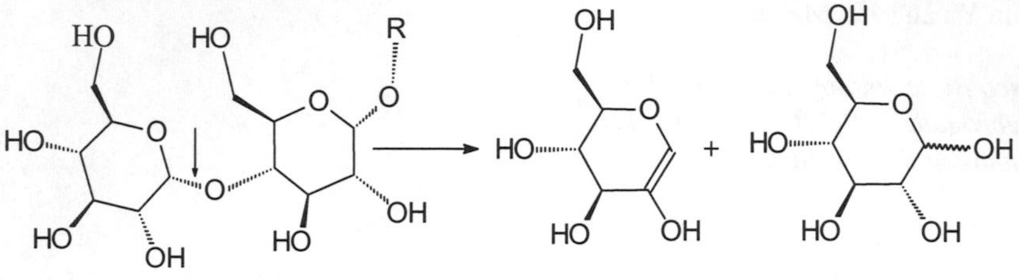

Linear α-1,4-glucan (n residues) → 1,5AnFru (n-1) + Glc (1)

2). The cleavage of O—C by pectate lyase

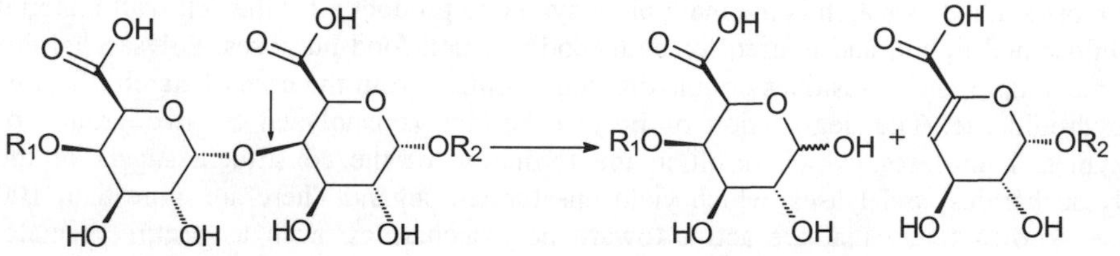

Poly(α-1,4-D)galacturonate → Oligogalacturonate + 4-Deoxy 4-enuronoside

3). The protonation of D-glucal by α-glucosidase

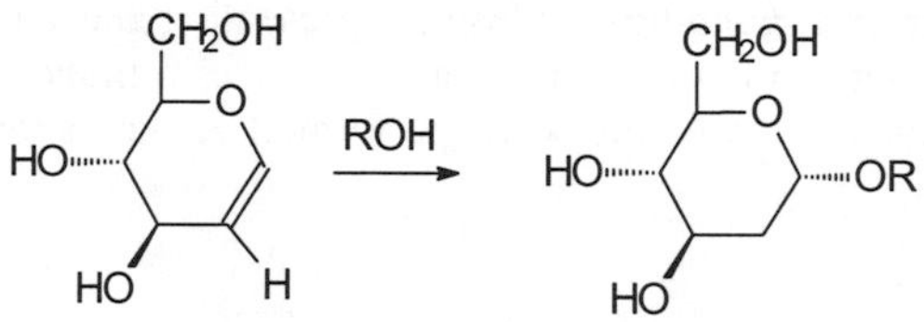

D-Glucal —ROH→ 2-Deoxy-α-D-glucoside

3 THE MOLECULAR AND ENZYMATIC FEATURES OF THE GLUCAN LYASE

3.1 Lyase source and their molecular features

In our laboratory we isolated and studied seven lyases from red seaweeds and three from fungi (Table 1), which represent α-1,4-glcuan lyases that have up to now been characterized. This work includes cloning and heterologous expression in *Aspergillus niger* and *Pichia pastoris*. We believe, however, the wide occurrence of the lyase. There are, for instance, clear indications for the occurrence of the lyase in *E. coli* and mammals, but their purification and characterization remain to be achieved.

Table 1 *Lyase sources and the number of lyase isozymes isolated from each source.*

Lyase sources	Number of isozymes described	Their names
Gracilariopsis lemaneiformis subspecies from Qingdao (Shandong, China)	3	GLq1-, GLq2-, GLq3-lyase
G. lemaneiformis subspecies from Santa Cruz, (California, USA)	2	GLs1-, GLs2-lyase
G. lemaneiformis subspecies from Araya Peninsula (Sucre, Venezuela)	1	GLa1-lyase
Gracilaria verrucosa from Araya Peninsula (Sucre,Venezuela)	1	GLv1-lyase
Morchella costata	1	MC-lyase
Morchella vulgaris	1	MV-lyase
Peziza ostracoderma	1	PO-lyase

Table 2 *A summary of the identities and similarities of α-1,4-glucan lyases as compared to the algal GLq1-lyase; the fungal lyases are further compared to MC-lyase.*

Lyases	No. of amino acids used for the comparison	Identity (%)[a]	Similarity (%)[b]
GLq1-lyase	1088	100	-
GLq2-lyase	1091	78	6
GLq3-lyases	316	83	5
GLs1-lyase	1092	75	8
GLs2-lyase	570	79	8
Glal-lyase	181	82	7
GLmc-lyase	1066	21	14
GLmv-lyase	1070	20	14
Glpo-lyase	164	24	15
GLmc-lyase	1066	100	-
GLmv-lyase	1070	86	5
Glpo-lyase	164	76	7

[a] Calculated by using the PCgene program; [b] Amino acids said to be 'similar' are: A,S,T; D,E; N,Q; R,K; I,L,M,V; F,Y,W.

From Table 2 it is clearly seen that the glucan lyases are diverged into two subfamilies, as high identity exists within both the algal and fungal lyases, but the identity between the algal and fungal lyases is only 20%.

Table 3 shows the molecular masses determined by mass spectrometry MALDI-TOF, which are close to those by cloning, indicating the lyases are not highly glycosylated if any. This is in agreement with that the lyase is not able to bind to ConA-Sepharose and the negative staining on sugars after SDS-PAGE. It further indicates that in contrast to the fungal lyases, the algal lyases possess transit polypeptides of 50 amino acid long.[5]

3.2 Substrate and inhibitor specificity

α-1,4-Glucan lyases from both red algae and fungi are highly specific for α-1,4-glucosidic bonds and show little or no activity toward other types of glucosidic bonds. The lyase is an exo-lyase and releases1,5AnFru from the non-reducing end of an α-1,4-

Table 3 *The molecular masses (Da) of the lyases determined by cloning and mass spectrometry.*

Lyases	Molecular mass of the coding region by cloning [a]	Molecular mass of the mature lyases by cloning [a]	Molecular mass determined by MALDI-TOF
GLq1-lyase	122,337	117,035	116, 876
GLq2-lyase	123,254	117,663	-[b]
GLsc1-lyase	123,169	117,796	117,206
GLmc-lyase	121,535	121,535	122,421
GLmv-lyase	121,924	121,924	119,714

[a] Calculated by using the PCgene program; [b] Not determined.

glucan. The smallest substrate for the lyase is p-nitrophenyl α-glucoside. Several inhibitors that are inhibitory to starch hydrolases are also found inhibiting both the algal and fungal lyases. These inhibitors include 1-deoxynojirimycin, acarbose, bromoconduritol and p-chloromercuribenzoic acid (PCMB).

According the mechanism, Glc is one of the products and maltose is an intermediate product during the complete degradation of linear α-1,4-glucans. It is therefore interesting to see if they have any effect on the lyase, for example, feedback inhibition. At a substrate concentration of 1% (w/v), the activity of the fungal MC-lyase decreased by 19.3 % in the presence of 0.1 M Glc when amylopectin was used as substrate (Table 4); the activity was not affected when glycogen was used as substrate. In the presence of 0.1 M of maltose the activity decreased by 48.8 % and 73.4% for glycogen and amylopectin, respectively.

Table 4 *Inhibition of the fungal MC-lyase by 0.1 M of glucose and maltose in two substrate concentrations.*

Substrates	Inhibitors	
	Glucose	Maltose
Amylopectin 1% (2%)	19.3%(7%)	73.4% (67.2%)
Glycogen 1% (2%)	0.0 (0.0)	48.8% (49.7%)

3.3 The pH and temperature optima

The pH optimum of the algal lyase is around pH 4, whereas that of the fungal lyases is at pH 6.5 (Table 5). There are therefore more than 2 pH units difference between the two subfamilies of the glucan lyases. The temperature optima for both groups of lyases vary between 37 and 50 °C, depending on which type of substrate is used (Table 5). There are indications that at higher temperatures glycogen stabilizes the lyase more efficient than amylopectin, and with both amylopectin and glycogen higher substrate concentrations display an stabilizing effect upon the lyases.

The pH optimum is important if the lyase will be working together with other enzymes, for example, debranching enzymes.

Table 5 *The pH and temperature optima of the algal and fungal α-1,4-glucan lyases.*

Enzyme	pH optimum [a]	Optimal pH range	Temperature optimum
The algal lyases:			
GLq1-lyase	3.8	3.7-4.1	40 °C [b]; 45 °C [c]
GLs1-lyase	4.1	3.8-4.5	48 °C [b]; 50 °C [c]
The fungal lyases:			
MV-lyase	6.4	5.9-7.6	43 °C [b], 48 °C [c]
MC-lyase	6.5	5.5-7.5	37 °C [b]; 40 °C [c]

[a] Measured at their optimal temperature for 60 min. [b] Amylopectin and [c] glycogen are used as substrate, respectively.

3.4 Effect of alcohols on the stability and activity of the lyase (GLq1-lyase)

Several alcohols were found to increase 1,5AnFru yield and this could be due to their stabilization effect to the lyase or to that the catalysis of the lyase is more efficient in a slightly hydrophobic microenvironment. The optimal concentration of ethanol for 7 days' reaction was 5% (v/v) which increased the 1,5AnFru yield by 12%. For 10 days' reaction the optimal concentration was 3%, which increased 1,5AnFru yield by 16%, when tested at the following concentrations: 1, 3, 5, 7, 9, 11, 13, and 15%. 1-Propanol, 2-propanol and 1-butanol displayed similar effect as ethanol. When tested at an alcohol concentration of 1, 5, and 10%, the optimal concentration of 1-propanol was found to be 5%. At this optimal concentration, the yield of 1,5AnFru increased by 34% after 6 days' reaction (Table 6).

Table 6 *The effect of 1-propanol on the stability and activity of α-1,4-glucan lyase (GLq1-lyase) in terms of 1,5-anhydro-D-fructose production (μmol).*

1-Propanol concentration (%)	0	1	5	10
Reaction time				
1 day	84	80	115	107
3 day	261	280	367	307
6 day	451	530	605	456
10 day	689	803	853	583

Table 7 *α-1,4-Glucan lyase limitation of different types of α-glucans*[a]

Starch type	Limitation degree (%)		n	Standard deviation
	Average	Maximum		
Potato amylopectin	57.5	58.3	5	0.5
Oyster glycogen	55.9	57.0	5	0.6
Rabbit liver glycogen	56.5	58.0	4	1.2
Potato starch	48.2,	53.8	4	3.4
Corn starch	50.8	55.5	4	3.0
Dextrin 10	69.0	71.5	5	2.9

[a] All the α-glucans used were from Sigma; dextrin 10 was from Fluka and it was enzymatically hydrolyzed maize starch and had an average chain length of 10.32. Each of glucans (20 mg) was dissolved in 25 mM acetate buffer (pH 3.8) and incubated with GLq1-lyase in a final volume of 4 ml at 35 °C for 24 h with shaking. Samples were taken at 1, 2, 3, 4, 5, 6, 23, and 24 h for analysis.

3.5 α-1,4-Glucan lyase limitation

As known from β-amylase, the action of the lyase on amylopectin or on glycogen leads to the formation of a limit dextrin due to the high specificity toward α-1,4-glcuosidic bonds. Table 7 shows limitation degrees of the red algal GLq1-lyase on several types of α-glucans.

3.6 The effect of debranching enzymes on the yield of 1,5-Anhydro-D-fructose

As α-1,4-glucan lyase is highly specific toward α-1,4-glucosidic linkages and amylopectin and glycogen are glucose polymers linked by α-1,4- and α-1,6-linkages, a complete degradation of such substrates therefore needs the cooperation of the lyase and a debranching enzyme, such as pullulanase, which is more specific for short branched chains and isoamylase, which has a preference for longer branched glucans. As seen from Table 8, the inclusion of pullulanase in the reaction mixture increased the 1,5AnFru yield by about 18-27%, depending on whether soluble starch or amylopectin was used as substrate. Similar effect was observed when isoamylase is used (data now shown). As expected, the by-product Glc increased by up to 16 times in the presence of pullulanase (Boehringer Mannheim)(Table 8) in the case of amylopectin. These two debranching enzymes have been used industrially to increase maltose yield when β-amylase is used for maltose production.

Table 8 *The effect of pullulanase on the yields of 1,5AnFru and Glc*[a].

Substrate	Lyase	Pullulanase	AF Yield (%)	Glc Yield (%)
Soluble Starch	+	-	55	0.86
	+	+	73	3.88
Amylopectin	+	-	53	0.23
	+	+	80	3.68

+, -, indicates whether the related enzyme was included in the reaction. [a] The reaction mixture contained 2% (w/v) potato amylopectin (Sigma A-8515) or 2% soluble starch (Merck), GLq1-lyase and 0.36 units of pullulanase (Boehringer Mannheim) in a final volume of 0.3 ml. The reaction was carried out at 30 °C for 24 h. At the end of the reaction, samples were taken for analysis of 1,5AnFru and Glc.

4 PREPARATION OF 1,5-ANHYDRO-D-FRUCTOSE

4.1 Historical review

Chemical synthesis of 1,5AnFru was first reported by Prof. Lichtenthaler and coworkers at Darmstadt, Germany in 1980[7]. In 1988 a Japanese patent filed by Nippon Kayaku Co., Ltd, disclosed an alternative chemical synthesis procedure[8]. Enzymatic preparation of 1,5AnFru was reported by Prof. Akanuma and coworkers in 1986 by oxidation of 1,5-anhydro-D-glucitol (1,5AnGlc-ol) using an extract containing glucose 2-oxidase[9]. Multiple steps are involved in the chemical synthesis, whereas for the enzymatic method, even though it is quantitative conversion when pure enzyme is used, 1,5AnGlc-ol is expensive. In 1988 Prof. Baute and coworkers at Bordeaux, France, prepared 1,5AnFru from starch by using an extract from morels (*Morchella vulgaris and M. costata*) containing α-1,4-glcuan lyase[10]. For the first time a yield of 400-500 mg material was obtained but the purity was only 90% as non-purified enzyme preparation was used, which obviously contained, among others, Glc releasing enzymes. Due to the limited availability, little research was carried out on the chemistry and application of 1,5AnFru. Gram to kilogram scale of 1,5AnFru with a purity of 99% was first achieved in our laboratory in 1993 using the algal GLq1-lyase heterologously expressed in the fungus *A. niger* which has no detectable endogenous α-1,4-glucan lyase, [1-6] and this made it possible to explore the chemistry and application of 1,5AnFru.[11-13]

4.2 Preparation of 1,5AnFru using immobilized algal α-1,4-glucan lyase

Immobilization of the red algal GLq1-lyase was achieved by using succinimide-activated Sepharose (Affigel 15 gel, Bio-Rad Laboratories) and glutardiadehyde-activated silica affinity adsorbent (Boehringer Mannheim). The recovery of lyase activity after immobilization on Affigel 15 gel varied between 40 to 50%. The immobilized lyase showed stability even at pH5.5, which is coincident with the pH optimum of pullulanase. The operational stability of a column packed with the immobilized lyase was at least 16 days when operated at 24 °C at pH 5.5.

With the lyase immobilized on glutardiadehyde-activated silica, the recovery was 80-100%. However, the silica-immobilized enzyme was unstable when the column was operated at pH 3.8 or pH 5.5. It is possible that some lyase was adsorbed on the surface of the silica gel beads and was slowly released from the silica gel after each washing of the column; It may therefore be the adsorbed lyase that contributed to the high recovery rate and the decrease in the column operational stability; it may also be due to the unstability of silica gel.

4.3 Purification of 1,5AnFru

When maltose, malto-saccharides and linear α-1,4-glucans are used as substrates, the products are 1,5AnFru and a small amount of glucose (Table 8), as maltose or the maltose unit on the reducing end of a glucan gives rise to equal mole of 1,5AnFru and Glc. The separation of 1,5AnFru from Glc is achieved by chromatography using a calcium carbohydrate column with water as the eluent. Glc can also be removed by ion exchange after having been treated with glucose oxidase and catalase. Prof. Baute *et al.* used bakers' yeast to remove Glc in preparing their 400-500 mg 1,5AnFru. Yeast, however, releases both proteinaceous and non-proteinaceous substances during the fermentation, even though 1,5AnFru is not fermentable by bakers' yeast, it is not toxic to the yeast even at 0.5M of 1,5AnFru (our observation).

When branched substrate is used as substrate the product is 1,5AnFru (in over 50% yield, Table 7), and a limit dextrin. The separation of those two products can be achieved by ultrafiltration, for example, using Amicon YM10 membranes (cut-off 10,000). Alternatively, the un-reacted limit dextrin can be precipitated with ethanol. The removal of ethanol can be achieved by rotary evaporation.

References

1. S. Yu, M. Pedersén and L. Kenne, *Biochim. Biophys. Acta*, 1993, **1156**, 313.
2. S. Yu, T. Ahmad, M. Pedersén and L. Kenne, *Biochim. Biophys. Acta*, 1995, **1244**, 1.
3. S. Yu, T.M.I.E. Christensen, K.M. Kragh, K. Bojsen and J. Marcussen, *Biochim. Biophys. Acta,* 1997, **1339**, 311.
4. S. Yu, C.E. Olsen and J. Marcussen, *Carbohydr. Res.*, 1998, **305**, 73.
5. K. Bojsen, S. Yu, K.M, Kragh and J. Marcussen, *Biochim. Biophys. Acta*, 1999, **1430**, 396.
6. K. Bojsen, S. Yu and J. Marcussen, *Plant Mol. Biol.*, 1999 (In press).
7. F.W. Lichtenthaler, E.S.H. El Ashry and V.H. Göckel, *Tetrahedron Lett.*, 1980, **21**, 1429.
8. Nippon Kayaku Co. Ltd., Production process of 1,5-anhydro-D-fructose and its hydrate. 1988, Unexamined Japanese patent publication No. 63-72696.
9. T. Nakamura, A. Naito, Y. Takahashi and H. Akanuma, *J. Biochem.*, 1986, **99**, 607.
10. M-A. Baute, R. Baute and G. Deffieux, *Phytochemistry*, 1988, **27**, 3401.
11. S.M. Andersen, I. Lundt, J. Marcussen, I. Søtofte and S. Yu, *J. Carbohydr. Chem.* 1998, **17**, 1027
12. S. Yu, K. Bojsen, K. M. Kragh, M. Bojko, J. Nielsen, J. Marcussen, and T. M. I. E. Christensen, PCT Int. Appl. WO 9510616; *Chem. Abstr.*, 1995, **123**, 226024f.
13. S. Yu, K. Bojsen, and J. Marcussen, PCT Int. Appl. WO 9612026; *Chem. Abstr.*, 1996, **125**, 56375t.

Engineering Carbohydrate Modifying Enzymes

BIOENGINEERING OF AMYLASE AND XYLOSE ISOMERASE THERMOZYMES

J.G. Zeikus[1,2], A. Savchenko[2], D. Sriprapundh[2], and Claire Vieille[2]

[1]MBI International, Lansing, MI 48910 USA
[2]Department of Biochemistry, Michigan State University, East Lansing, MI 48824 USA

1 ABSTRACT

We have applied genetic engineering techniques to make improved starch processing enzymes from thermophilic and hyperthermophilic microbes. The goals of our research include identifying the molecular determinants for high stability and activity of amylases and xylose-glucose isomerase (XI). An α-amylase gene was cloned from *Pyrococcus furiosus* that was modified by site-directed mutagenesis to be three-fold more active and thirteen-fold more stable at 95°C than the commercial enzyme from *Bacillus licheniformis*. Amylopullulanase genes (*apu*) were cloned and their products were characterized from *Thermoanaerobacter ethanolicus* and *P. furiosus*. Nested deletions were used to identify distinct thermal stability and thermal activity regions in the *T. ethanolicus* enzyme. The temperature optimum of *T. ethanolicus* APU was shifted from 85 to 65°C without decreasing enzyme thermal stability. XI-encoding genes from *Thermoanaerobacterium thermosulfurigenes* (TTXI) and *Thermotoga neapolitana* (TNXI) were cloned, expressed, and characterized. Mutations were constructed that doubled the specific activity for glucose conversion to fructose by TTXI at 55°C and by TNXI at 97°C. Insertion of a proline a in TTXI surface loop region enhanced the enzyme thermal stability.

2 INTRODUCTION

The commercial processing of starch into glucose (i.e. fermentation feedstock) and high fructose corn syrup (i.e. corn sweetener) produces over $7 billion/year in sugar-based products and requires $200 million/year in thermostable processing enzymes. Starch is liquefied into maltodextrin syrups at 100°C with α-amylase, and it is then converted into dextrose by glucoamylase. Glucose is further isomerized into a 42% fructose syrup (Figure 1). To improve starch bioprocessing, we initiated genetic engineering studies of thermozymes derived from thermophilic and hyperthermophilic microbes [1-11]. The goals of our research include identifying the molecular determinants for high stability and activity of ∀-amylase and xylose-glucose isomerase (XI). Carbohydrate-processing thermozymes are more stable and active than mesozymes and offer higher sugar syrup concentrations and lower processing costs than conventional industrial starch processing enzymes [12-14].

Our approach is four fold: (1) to screen for different carbohydrate-modifying enzymes in thermophiles and hyperthermophiles; (2) to clone and over-express thermogenes in mesophilic hosts; (3) to perform site-directed or random mutagenesis to enhance thermozyme activity or to alter stability; and (4) to readily purify the thermozymes by a single chemical heat treatment step. The purpose of this report is to describe our recent research accomplishments.

3 AMYLASES

Industry is interested in a more stable, calcium-independent α-amylase with a lower pH activity range. We found this target enzyme in *P. furiosus*. Figure 2 compares the biochemical properties of *P. furiosus* and *B. licheniformis* (Bli) α-amylases. The *P. furiosus* (Pfu) enzyme was twice as active and 13 times more stable at 98°C than the commercial enzyme. Pfu and Bli α-amylase sequences are aligned in Figure 3. Notably, the thermozyme contains five cysteines and lacks one of the residues responsible for calcium binding that was found in Bli α-amylase. A structural model was developed of Pfu α-amylase domains A and B. In this model, Cys_{152}, Cys_{153}, and Cys_{165} are located on the surface of the molecule. Pfu α-amylase mutants were generated in which Cys residues were individually replaced by serine or alanine. Mutation Cys165Ser strongly decreased Pfu α-amylase's thermoactivity and thermostability (Figure 4, Figure 5). The other mutations had little effect on the enzyme's activity and stability. [Savchenko et al, manuscript in preparation]. We are in the process of determining the role of Cys165 in activity and stability.

When we screened for either α-amylase or pullulanase activity in thermophiles, we discovered a new enzyme amylopullulanase [2,6,15-17]. Amylopullulanase (Apu) has a single catalytic site responsible for the cleavage of both α-1-4 and α-1-6 bonds in starch. Recently, we characterized an extracellular amylopullulanase from *P. furiosus*. The general biochemical properties of this thermozyme are shown in figure 6. As expected for an enzyme from a hyperthermophile, its optimum temperature was 105°C. This Apu serves as a model to demonstrate thermozymes' extreme chemical stability. The enzyme remains 100% active in 0.5 M guanidine HCl. Such denaturant concentration would unfold most mesozymes but it stimulates Pfu Apu's activity.

Amylopullulanases are monomeric enzymes. We identified separate thermostability and thermoactivity domains in the *T. ethanolicus* Apu by constructing nested deletions in its gene. Figure 7 illustrates the effect of N- and C-terminus deletions on *T. ethanolicus* Apu thermal stability and activity. Notably, the Apu N324 mutant had a 20°C lower temperature activity optimum and was 30% more stable than the enzyme encoded by the entire gene.

4 XYLOSE-GLUCOSE ISOMERASES

Industry utilizes XIs for converting glucose syrups into a 42% fructose – 58% glucose mixture. A chromatographic step is used to further concentrate the syrup to 52% fructose – 48% glucose. This concentration step could be eliminated if XIs were active at 90°C because the glucose/fructose chemical equilibrium favors higher fructose concentrations at high temperatures. Figure 8 compares the thermal parameters of the industrial glucose isomerase from *B. licheniformis* (BLXI) with the enzymes from *T. thermosulfurigenes* (TTXI) and *T. neapolitana* (TNXI). BLXI has a high temperature optimum for an enzyme found in a mesophilic (and thermoduric) microbe. However, its thermal stability and thermal activity optimum are lower than those of TTXI or TNXI. TNXI is 50% unfolded at 120°C and it denatures before it unfolds, whereas BLXI is optimally active when it is 50% unfolded.

Thermozymes have evolved multiple mechanisms to achieve stability and activity at high temperatures. Hyperthermophilic enzymes often contain less Asn and Gln residues than their mesophilic counterparts: Asn and Gln are among the residues most sensitive to covalent destruction—they deamidate—at 100°C. Figure 9 shows type II XI Asn plus Gln content in relation to the source organism's optimal growth temperature and to the enzyme's optimal activity temperature.

We collaborated with Professor David Blow's laboratory (Imperial College, London, England) to determine TTXI and TNXI structures. Conventional criteria failed to explain why TNXI is significantly more stable than TTXI [Gallay et al, manuscript in preparation]. TNXI contains a few more prolines than TTXI. Two of them, Pro58 and Pro62, are located in a large loop, in which 14 hydrophilic residues surround Phe59. Crystallographic studies indicate that this loop (and Phe59, in particular) is involved in intersubunit interaction: it participates in the architecture of the neighboring subunit's active site. The roles of Pro58 and Pro62 in TNXI stabilization were tested by site-directed mutagenesis (Figure 10) [Sriprapundh et al, manuscript in preparation]. In TTXI, mutation Gln58Pro increased the enzyme's half-life at 85°C by approximately 50%. The reciprocal mutation Pro58Gln in TNXI decreased the enzyme's half-life at 95°C by approximately 35%. The enzyme's optimum temperature for activity also decreased from 95°C to 88°C. Mutations Ala62Pro (in TTXI) and Pro62Ala (in TNXI) were both destabilizing.

We previously engineered TTXI into a "true" glucose isomerase by site-directed mutagenesis (Figure 11). Figure 12 compares the specific activities of TTXI, TNXI, and our "best" site-directed mutants designed for increased activity on glucose. Mutation W138F/V185T more than doubled TTXI activity on glucose at 55°C and significantly enhanced it at 75°C or 85°C. Mutation V185T slightly decreased TNXI activity on glucose at 55°C, but it nearly doubled it at 97°C. This mutation made TNXI more active than any reported Type II XI.

5 TRENDS IN UNDERSTANDING STABILITY AND ACTIVITY OF STARCH PROCESSING THERMOZYMES

Thermozymes offer enzymes with higher stability and activity for starch processing. No single mechanism exists for higher stability of thermozymes. Stabilization mechanisms identified so far include: (a) less Asn and Gln residues (i.e., XI's or ∀-amylase); and (b) stabilization of intersubunit interactions by an additional proline in a loop (i.e., TNXI). As seen for other enzymes, a single point mutation can dramatically alter stability and activity (i.e., Cys165Ser in Pfu α-amylase). The molecular determinants for thermal stability and activity can be distinct for certain enzymes (i.e., *T. ethanolicus* Apu).

Thermozymes are now attracting a lot of attention due to their high stability and activity above 100°C. In the future, they may serve as a molecular scaffold to design enzymes that have high physiochemical stability, and are active on selected substrates and at specific temperature ranges.

Figure 1 *Commercial processing of starch to mono- and oligo-saccharides*

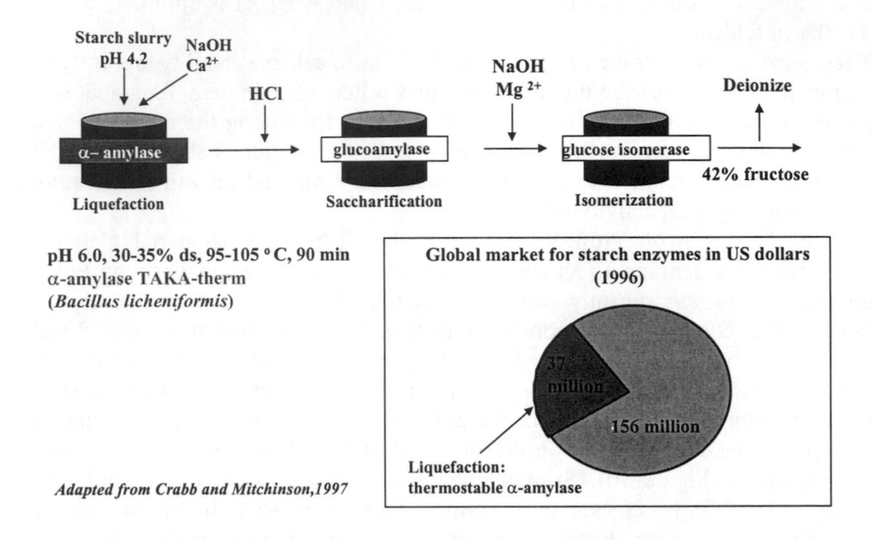

Figure 2 *Biochemical properties of P. furiosus and B. licheniformis α-amylases*

Enzyme	*B. licheniformis*	*P. furiosus*
Molecular weight	55,200	50,000
Specific activity (U/mg)	2,000 (at 90°C)	3,900 (at 98°C)
Optimal pH	7.0-8.0	5.5-6.0
Optimal temperature	90°C	100°C
Ca^{2+} requirement	Yes	No
Half life at 98°C	<1 h	13 h (no Ca^{2+})
End products	G1-G6	G2-G7

P. furiosus and *B. licheniformis* α- amylase sequences are 35.7% identical

Figure 3 *Biochemical properties of* *P. furiosus* *and* *B. licheniformis* *∀-amylases*

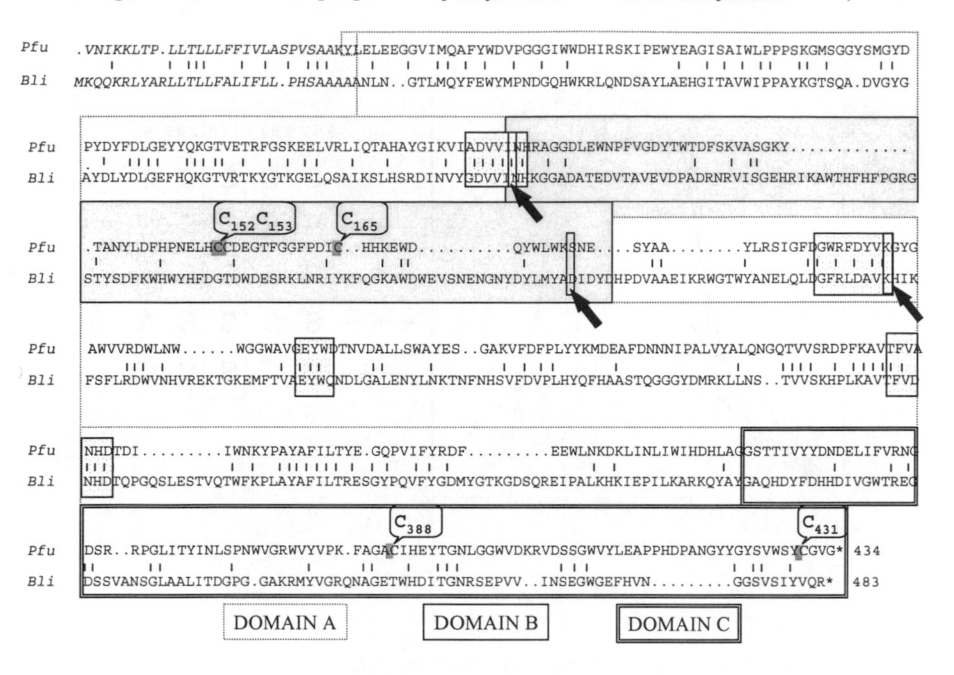

Figure 4 *Stability of* *Pfu* *∀-amylase wild-type and Cys mutants at 120°C*

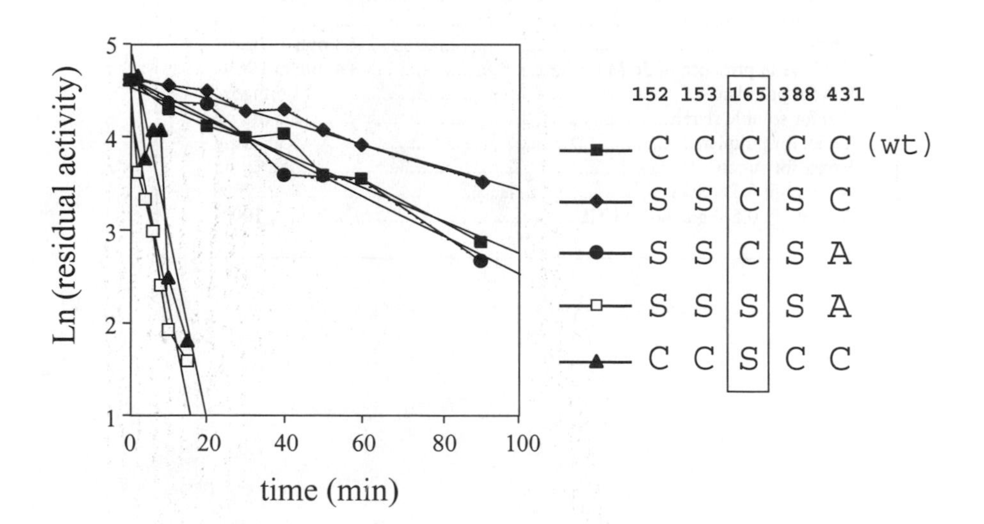

Figure 5 *Relative activity vs. temperature of Pfu ∀-amylase wild-type and Cys mutants*

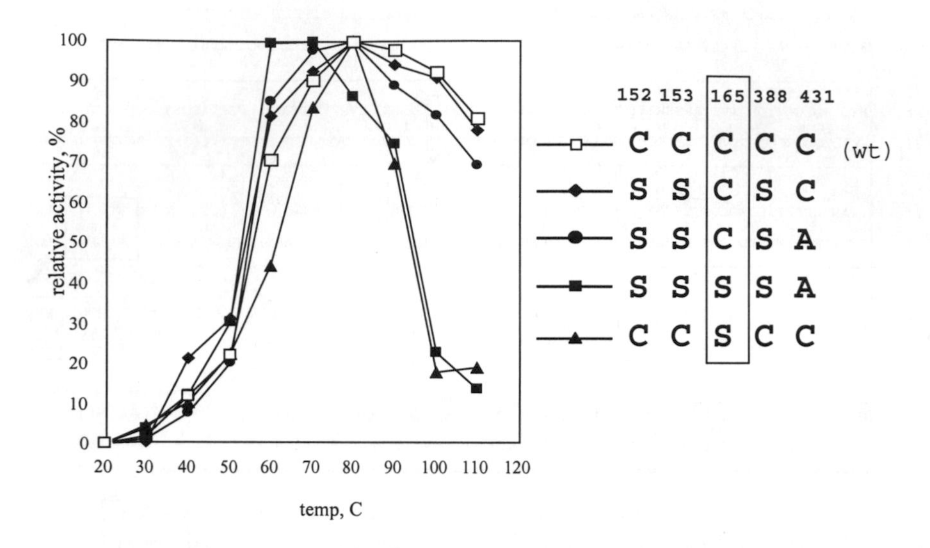

Figure 6 *General properties of P. furiosus amylopullulanase*

Optimum temperature for activity	105°C
pH range	> 80% activity at pH 4.5-7.0
Optimum pH for activity	pH 6.0
Half-life	7.5 min at 106°C
Half-life in presence of 5mM Ca^{++}	64 min at 106°C
Km for pullulan	1.6 mg/ml
Km for soluble starch	2.5 mg/ml
Vmax for pullulan	133 w/mg
Vmax for starch	88 U/mg
Activity in 0.5% SDS	75%
Activity in 0.5M guanidine HCL	100%

Figure 7 *Effect of N- and C-terminal deletions on T. ethanolicus Apu thermostability and activity*

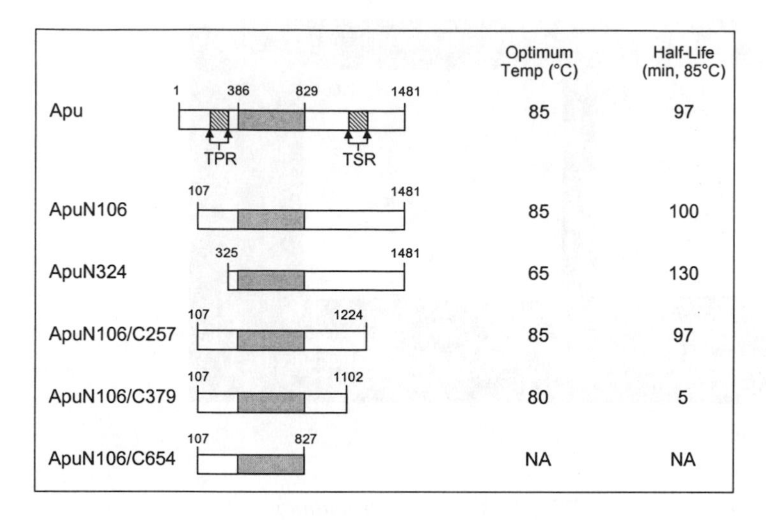

Figure 8 *Comparison of BLXI, TTXI, and TNXI thermal parameters*

Microorganism (optimal growth temp)	**T_{opt} (°C)**	**$t_{½}$ (min at 85°C)**	**T_m (°C)**	**E of inactivation (kCal/mol)**
B. licheniformis (37°C)	70	<1.0	74	260
T. thermosulfurigenes (60°C)	85	45	95	120
T. neapolitana (80°C)	95	330	120	11-29

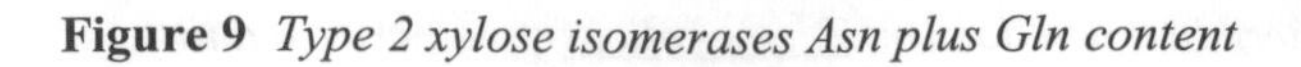

Figure 9 *Type 2 xylose isomerases Asn plus Gln content*

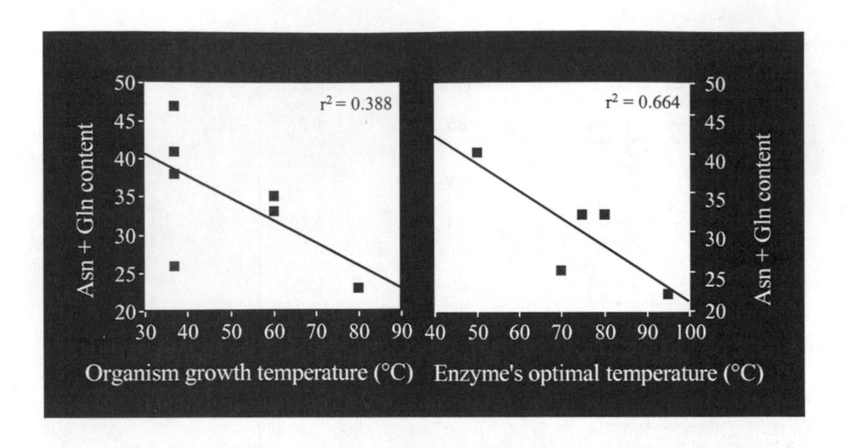

Figure 10 *Thermal parameters of TTXI, TNXI, and their mutants*

Enzyme	Optimum temperature (°C)	Half-life at 85°C (min)
TTXI	85	69
TTI Gln58Pro	85	99
TTXI Ala62Pro	74	3
TTXI Phe275Gly	78	5
		Half-life at 95°C (min)
TNXI	95	69
TNXI Pro58Gln	88	49
TNXI Pro62Ala	88	67
TNXI Gly275Phe	88	36

Figure 11 *Substrate preference change from xylose (Xyl) to glucose (Glc) associated with point mutations in the TTXI substrate-binding site*

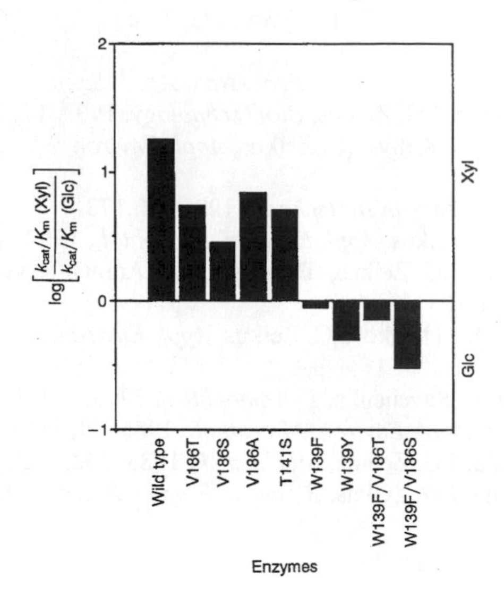

Figure 12 *Comparison of the specific activities of TTXI, TNXI, and their mutants on glucose*

	Specific Activity at 55°C (U/mg)	Specific Activity at 75°C (U/mg)	Specific Activity at 85°C (U/mg)	Specific Activity at 97°C (U/mg)
TTXI	3.82	10.65	13.57	N/A
V138F/V195T in TTXI	8.11	13.79	18.79	N/A
TNXI	3.75	6.46	17.48	26.02
V185T in TNXI	3.23	9.54	22.55	45.38

6 REFERENCES

1. G.-J. Shen, B.C. Saha, L. Bhatnager, Y.-E. Lee, J.G. Zeikus, *Biochem J.*, 1998, **254**, 835.
2. B.C. Saha, J.G. Zeikus, *Biotech. and Bioeng.*, 1989, **34**, 299.
3. C. Lee, L. Bhatnagar, B.C. Saha, Y.-E. Lee, M. Takagi, T. Imanaka, M. Bagdasarian, J.G. Zeikus, *Appl. Environ. Microbiol.*, 1990, **56**, 2638.
4. C. Lee, M. Bagdasarian, M. Meng, J.G. Zeikus, *J. Bio. Chem.*, 1990, **265**, 19082.

5. M. Meng, C. Lee, M. Bagdasarian, J.G. Zeikus, *Proc. Natl. Acad. Sci. USA*, 1991, **88**, 4015.
6. S.P. Mathupala, S.E. Lowe, S.M. Podkovyrov, J.G. Zeikus, *J. Bio. Chem.*, 1993, **268**, 16332.
7. M. Meng, M. Bagdasarian, J.G. Zeikus, *Proc. Natl. Acad. Sci. USA*, 1993, **90**, 8459.
8. M. Meng, M. Bagdasarian, J.G. Zeikus, *Bio/Technology*, 1993, **11**, 1157.
9. C. Vieille, J.M. Hess, R.M. Kelly, J.G. Zeikus, *Appl. Environ. Microbiol.*, 1995, **61**, 1967.
10. C. Vieille, J.G. Zeikus, *Trends in Biotechnol.*, 1996, **14**, 173.
11. G. Dong, C. Vieille, J.G. Zeikus, *Appl. Environ. Microbiol.*, 1997, **63**, 3577.
12. C. Vieille, D.S. Burdette, J.G. Zeikus, 'Biotechnology Annual Review', Elseivier Science, 1996, Vol. 2, p. 1.
13. G. Dong, C. Vieille, A. Savchenko, J.G. Zeikus, *Appl. Environ. Microbiol.*, 1997, **63**, 3569.
14. J.G. Zeikus, C. Vieille, A. Savchenko, *Extremophiles*, 1998, **2**, 179.
15. H.H. Hyun, J.G. Zeikus, *Appl. Environ. Microbiol.*, 1985, **49**, 1168.
16. B.C. Saha, S. Mathupala, J.G. Zeikus, *Biochem. J.*, 1988, **252**, 343.
17. S. Mathupala, B.C. Saha, J.G. Zeikus, *Biochem. Biophy. Research Comm.*, 1990, **166**, 126.

ENGINEERING OF CYCLODEXTRIN GLYCOSYLTRANSFERASE

L. Dijkhuizen, B.A. van der Veen, J. Uitdehaag*, B.W. Dijkstra*

Microbiology and *Biophysical Chemistry
Groningen Biomolecular Sciences and Biotechnology Institute (GBB)
University of Groningen, The Netherlands

1 INTRODUCTION

Many bacteria and fungi excrete enzymes that degrade starch to facilitate the uptake of carbohydrates into the cell. A number of bacteria belonging to the genera *Bacillus*, *Thermoanaerobacter*, *Thermoanaerobacterium*, *Clostridium*, *Micrococcus*, *Klebsiella*, are able to grow on starch using the extracellular enzyme cyclodextrin glycosyltransferase (CGTase; EC 2.4.1.19) for the initial attack on this polymeric substrate [1]. CGTases catalyse the formation of cyclodextrins from starch and related $\alpha(1\rightarrow4)$ linked glucose polymers via an intramolecular transglycosylation reaction. CGTase enzymes are functionally related to α-amylases, which hydrolyse starch into linear products. In contrast with α-amylases, CGTase enzymes preferentially add the "non-reducing end" glycosidic C4 hydroxyl group across the scissile $\alpha(1\rightarrow4)$ glycosidic bond, resulting in a glycosidic exchange and the formation of cyclodextrins. CGTase enzymes and α-amylases share about 30% amino acid sequence identity [2]. Bacteria employing a CGTase for starch degradation are able to subsequently metabolize the cyclodextrins produced as carbon- and energy sources for growth (Figure 1). This appears to involve a cell-associated cyclomaltodextrinase (CDase; EC 3.2.1.54), yielding glucose, maltose and maltotriose from cyclodextrins [1]. The further glucose metabolism proceeds intracellularly via the glycolytic pathway.

Cyclodextrins are cyclic glucose oligomers linked via α(1,4) glycosidic bonds (Figure 2). They possess a hydrophobic internal cavity formed by the glucopyranose CH groups and a hydrophilic exterior surface formed by primary and secondary hydroxyl groups [4]. Cyclodextrins are able to form inclusion complexes with many small hydrophobic molecules in aqueous solutions, resulting in changes in physical properties, e.g. increased solubility and stability, and decreased chemical reactivity and volatility (Figure 2) [5]. Cyclodextrins find an increasing use in industrial and research applications [6]. All CGTase enzymes studied thus far from different bacterial sources form a mixture of cyclodextrins consisting of 6, 7 or 8 glucose residues, α-, β- and γ-cyclodextrin, respectively. The relative proportion of the α-, β- or γ-cyclodextrins produced varies with the bacterial source of the enzyme [7].

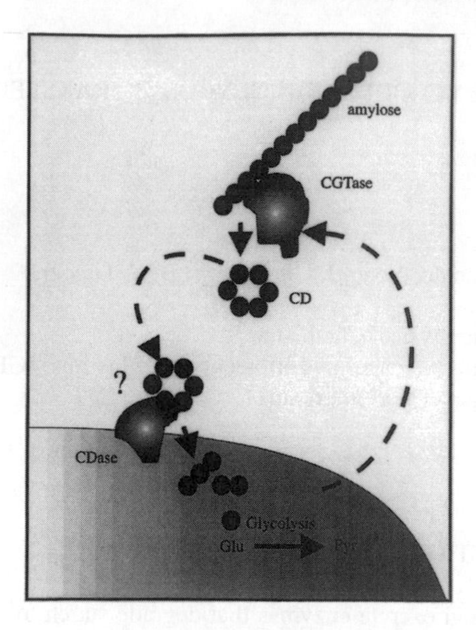

Figure 1. Bacterial growth on starch (amylose) involving an extracellular CGTase (yielding cyclodextrins, CD), a cell-associated CDase (yielding glucose, maltose and maltotriose), and intracellular glucose metabolism via the glycolytic pathway, yielding pyruvate (Pyr) [3].

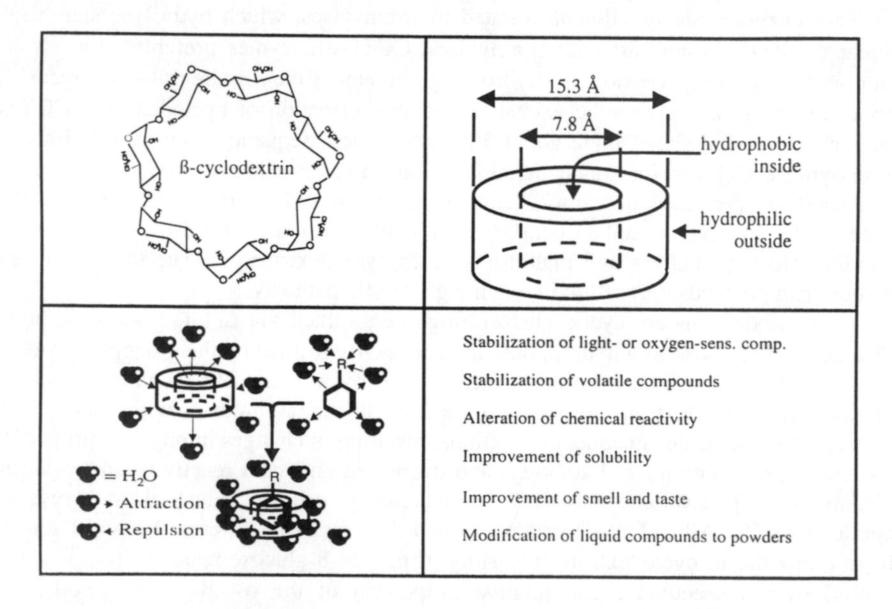

Figure 2. Structure and properties of β-cyclodextrin [3].

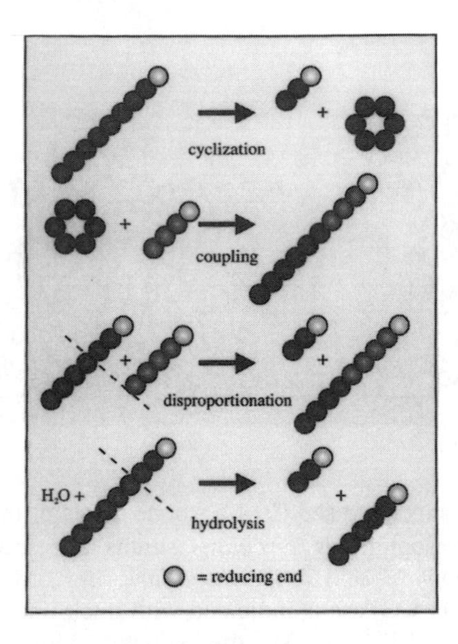

Figure 3. Reactions catalyzed by CGTase enzymes [3].

CGTase enzymes not only produce cyclodextrins from starch but also catalyze coupling, disproportionation and hydrolysis reactions (Figure 3) [8]. All CGTase enzymes studied are also rather sensitive to product inhibition by cyclodextrins. The industrial cyclodextrin production process, involving CGTase, thus could be improved considerably by the construction of mutant CGTase enzymes with improved product specificity and decreased product inhibition [9].

2. CGTase Three-Dimensional Structures

A number of CGTase proteins have been crystallized successfully, allowing determination of their three-dimensional structures. High concentrations of maltose are required for crystallization of the *B. circulans* strain 251 CGTase [10]. Three maltose binding sites were observed at the protein surface (Figure 4) [11], serving to identify amino acid residues strongly interacting with the starch substrate [12]. The maltose-dependence of CGTase crystallization is based on the proximity of two of the maltose binding sites to intermolecular crystal contacts [11]. A comparison of the crystal structures of the CGTases from the mesophiles *B. circulans* strain 251 [11] and strain 8 [13] (both at 2.0 Å resolution), and the thermophiles *B. stearothermophilus* [14] (at 2.5 Å resolution) and *T. thermosulfurigenes* EM1 [15] (at 2.3 Å resolution) shows that the Overall fold of these enzymes is highly similar (Figure 4) [16]; the hydrophobic core is well conserved and the two calcium binding sites are virtually identical. Significant

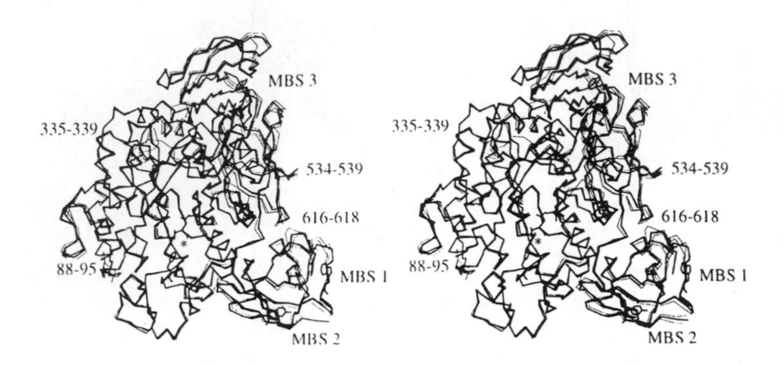

Figure 4. Stereo view of the superimposed Cα-backbone traces of the crystal structures of the CGTases from the mesophiles *B. circulans* strains 251 and strain 8, and the thermophiles *B. stearothermophilus* and *T. thermosulfurigenes* EM1. The backbone of the *T. thermosulfurigenes* EM1 CGTase is indicated with thick lines. Indicated are also the maltose binding sites (MBS1 to 3), the variable loop regions (using *T. thermosulfurigenes* EM1 CGTase numbering), and the active site (*) [15, 16].

differences between CGTases from thermophilic and mesophilic sources occur only at the surfaces of these proteins. Thermostable enzymes contain less flexible loops, have additional salt bridges, hydrophobic interactions and hydrogen bonds [15]. Especially loops 88-95, 335-339 and 534-539 (Figure 4) contribute with novel hydrogen bonds and apolar contacts to the stabilization of the *T. thermosulfurigenes* EM1 CGTase. The structural data donot provide direct insights in factors determining the different cyclodextrin product specificity of these CGTase enzymes.

All CGTases studied catalyze the four reactions presented in Figure 3. Incubation of CGTase with starch may not only yield cyclodextrins but also linear products. This is caused by the hydrolytic activity of the enzyme, followed by disproportionation- and coupling activity. The CGTase from *T. thermosulfurigenes* EM1 possesses a relatively high hydrolytic activity that results in production of cyclodextrins plus substantial amounts of glucose, maltose and maltotriose. When incubated with this CGTase, 35% of the starch is converted into cyclodextrins and 11% is converted into linear sugars [17]. The CGTase from *B. circulans* strain 251, however, possesses a relatively low hydrolytic activity, and formation of linear products is therefore observed hardly at all [18]. Comparison of the crystal structures of both CGTases revealed differences in the amino acid composition and in the conformation of the 88-95 loop (*T. thermosulfurigenes* EM1 CGTase numbering). In the CGTase from *B. circulans* strain 251 Tyr89 interacts with subsite -3 and -4 of a bound oligosaccharide substrate by a stacking interaction (see below). In the *T. thermosulfurigenes* EM1 CGTase there is no residue homologous to Tyr89 and this interaction therefore is no longer possible. Bound oligosaccharides thus lack an important stabilizing interaction, reducing cyclization activity of the CGTase from *T. thermosulfurigenes* EM1 by 50% compared to the cyclization activity of the CGTase from *B. circulans* strain 251. A decrease in cyclization activity implied an increase in hydrolytic activity, as was indeed observed in the experimental studies [15]. Apparently, some catalytic functionality of CGTase has been sacrificed for the sake of improved structural stability by modifying loop regions near the active site.

3. Catalytic Mechanism of Glycosylases

Three different catalytic mechanisms for hydrolysis by glycosylases have been proposed: the inverting mechanism, the retaining mechanism proceeding through an oxo-carbonium ion intermediate, and the retaining mechanism involving a covalent intermediate. In all glycosylases one to three invariant carboxylic acids are found in the active site cleft. For the CGTase of *T. thermosulfurigenes* EM1 these residues are Glu258, Asp230 and Asp329 (numbering without signal peptide).

In the proposed mechanism for inverting enzymes the glycosidic oxygen is initially protonated by a general acid catalyst, which is followed by a nucleophilic attack of a water molecule on the C1 atom of the sugar at subsite -1, activated by a carboxylate base, leading to inversion of the anomeric conformation. This mechanism is known as a single-displacement mechanism [19]; bond breaking and bond making both proceed in a single concerted step. The reaction rate depends on the concentrations of both nucleophile and substrate, kinetically known as a second order type of a substitution (S_N2).

The retaining reactions proceed via a double-displacement mechanism; the first step involves a similar protonation of the glycosidic oxygen by a general acid as in the inverting mechanism, creating an oxo-carbonium transition state [20] that subsequently collapses into an intermediate [21]. This intermediate is subsequently attacked by a water nucleophile in the second step, assisted by the base form of the acid catalyst. Each step inverts the configuration of the anomeric carbon. The two displacement steps therefore result in an overall retention of the configuration. For retaining enzymes the intermediate could either be an oxo-carbonium ion which is electrostatically stabilized by a carboxylate, or involves formation of a covalent bond, in which one of the catalytic aspartates (in some cases a glutamate) is presumed to act as a nucleophile. Kinetically the covalent-bond mechanism involves a second order substitution (S_N2) mechanism, while the conversion via a electrostatically stabilized oxo-carbonium ion is thought to proceed through a first order mechanism (S_N1).

Transglycosylases (e.g. CGTase) employ a similar reaction mechanism as described for retaining hydrolases. The second step of the reaction does not involve a water nucleophile, but the non-reducing end of another oligosaccharide is the nucleophile that attacks the intermediate, probably assisted by the base form of the acid catalyst.

The nature of the intermediate formed by retaining enzymes in these reactions initially was disputed. It is now generally accepted, however, that the reaction proceeds via a covalent intermediate. Clear evidence for a covalent glycosyl-enzyme intermediate in family 13 has been obtained from rapid trapping studies with natural substrates. Low-temperature ^{13}C NMR experiments have provided evidence for the formation of a β-carboxylacetal ester covalent adduct between maltotetraose and porcine pancreas α-amylase [22]. Conclusive evidence subsequently came from experiments involving trapping of a covalent intermediate with 4-deoxymaltotriosyl α-fluoride as a substrate in the inactive Glu257Gln mutant of *B. circulans* 251 CGTase (Glu 257 is the general acid catalyst in this enzyme) [23]. Two glycosylated peptides were identified in a proteolytic digest of the inhibited enzyme by means of mass spectrometry. Edman sequencing of these labeled peptides provided evidence for a covalent intermediate in CGTase.

It appears from the double displacement mechanism that for optimal catalysis of CGTase the nucleophile Asp230 must be unprotonated, whereas the general acid Glu258 must be protonated in the first step, but unprotonated in the second step of the reaction. A third catalytic residue, Asp329, has been found to be hydrogen bonded to Glu258 in the unliganded CGTase, thereby elevating the pKa of Glu258 and assuring its protonation. After substrate binding, this hydrogen bond is lost, making deprotonation of Glu258 possible. It has been suggested that this Glu258-Asp329 interaction is responsible for the broad pH optimum (pH 4-7) exhibited by CGTases [24]. However, mutations in *T. thermosulfurigenes* EM1 CGTase near the proton donor Glu258 suggest that the pH optimum curve of CGTase may be determined only by the protonation state of residue Glu258 [17]. Both the high and low slopes of the pH optimum curve could be manipulated by site-directed mutations close to Glu258. Changes in the environment of Glu258 before and after substrate binding can account for its broad pH optimum.

4. Cyclodextrin Product Specificity of CGTase Enzymes

It has been suggested that also the size of the aromatic amino acid (Phe or Tyr), present in a dominant position in the center of the active site cleft of CGTases (Tyr195 in CGTase of *B. circulans* 251), influences the preferred cyclodextrin size. Sin et al. [25] proposed a mechanism in which the starch chain folds around this residue. Substitution of this central amino acid Tyr188 by Trp in the *B. ohbensis* CGTase doubled the production of ε-cyclodextrin [26]. However, several other Tyr188 mutations in the same protein, as well as the substitution of Tyr195 of the *B. circulans* 251 CGTase by other amino acids [18], the mutation F191Y at the similar position in the CGTase of *B. stearothermophilus* NO2 [27], and the fact that the γ-CGTase of *B. subtilis* strain 313 contains leucine at this position [28], do not support this proposed mechanism.

New insights in factors determining CGTase cyclodextrin product specificity came from crystal soaking experiments which revealed the structure of a CGTase mutant from *B. circulans* 251 complexed with a maltononaose inhibitor (Figure 5) [16, 29]. This resulted in identification of 9 sugar binding sites (+2 to -7) in the active site cleft which make contact with the amylose chain. This amylose chain was not folded around the central amino acid, but was pointing away from this residue. This conformation may resemble a specific intermediate in cyclization for the formation of β-cyclodextrin, since β-cyclodextrin is the main product formed by the CGTase from *B.*

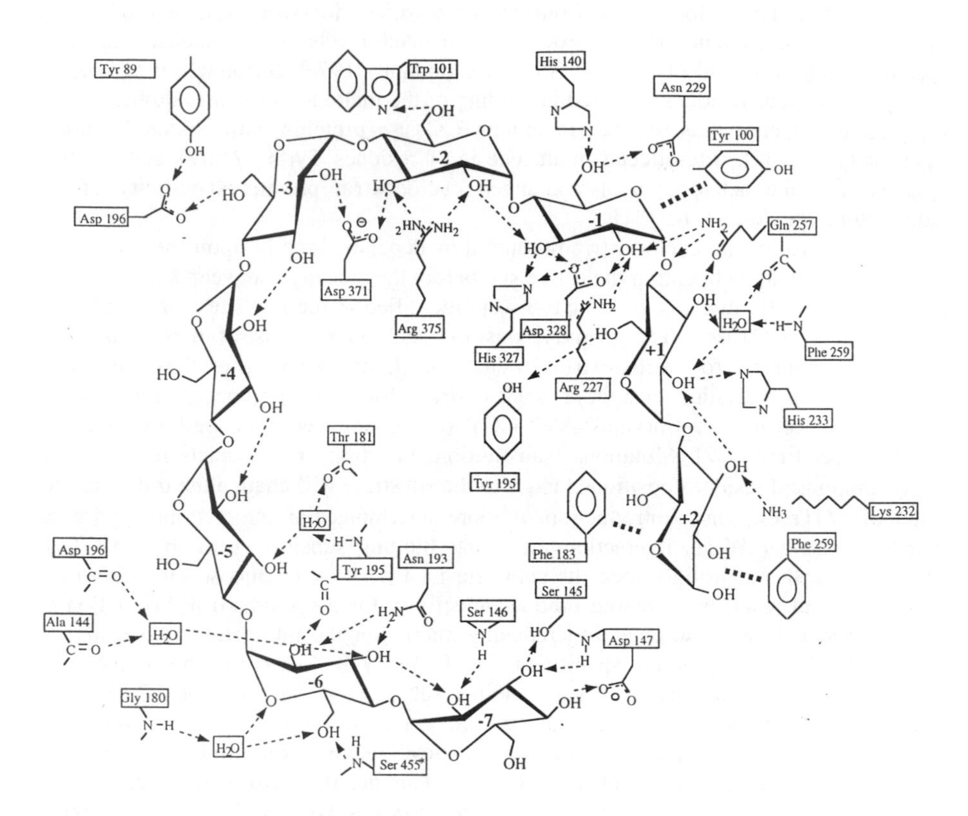

Figure 5. Schematic representation of the hydrogen bonds between the *B. circulans* 251 CGTase and a maltononaose inhibitor bound at the active site [16, 30].

circulans 251. A similar crystal soaking experiment with the α-CGTase from *T. thermosulfurigenes* EM1 resulted in the structure of a complex with a maltohexaose inhibitor, bound in a different conformation [17]. The maltohexaose conformation was more bent compared to the maltononaose conformation which was more straight. This suggests that indeed α-and β-CGTases bind their linear substrates differently. Also the structure of the *B. circulans* CGTase double mutant Y89D/S146P with a maltohexaose bound in the active site cleft revealed the existence of a specific intermediate in cyclization for the formation of α-cyclodextrin. This double mutant produced significantly more α-cyclodextrin compared to the wild-type enzyme. The cyclodextrin product ratio changed from 14:66:20 (α:β:γ) for the wild-type enzyme to 30:51:19 for the mutant enzyme. Mutations D371R and D197H of *T. thermosulfurigenes* EM1, aiming to hinder and to stabilize, respectively, the maltohexaose conformation, resulted in clear changes in cyclodextrin product ratios [17]. Mutant D371R produced enhanced levels of β- and γ-cyclodextrins: from 25:58:14 (α:β:γ) for wild type to 6:68:26 for mutant D371R. Mutant D197H produced enhanced levels of α-cyclodextrin: from 28:58:14 (α:β:γ) for wild type to 35:49:16 for mutant D197H. Introduction of specific changes in various residues involved in binding of the maltononaose or maltohexaose in the active site cleft thus resulted in mutant CGTase proteins with changed product specificity [17, 30]. Site-directed mutagenesis of residues Tyr89, Asn94 and Tyr101, located near the active site, also changed cyclodextrin product specificity of the alkalophilic *Bacillus* sp. I-5 [31].

The size of the cyclodextrin product thus depends directly upon the number of glucose units that are bound in the active site before the cleavage between subsite -1 and 1 takes place [29]. Subsites -6, -7 and -8 were identified as the key sites for the product specificity of CGTases. The identity, position and interactions of the amino acid residues responsible for cyclodextrin product specificity of CGTases thus have been identified. This detailed insight provides straighforward approaches for rational construction of mutant (thermostable) CGTase enzymes with desired cyclodextrin product specificity [32]. Mutations (substitution, insertion and/or deletion) in one or more amino acid residues positioned close to the substrate will change the α:β:(product ratio of CGTases. Thus, introduction of more intermolecular interactions (hydrogen bonding, van der Waals interactions) at sugar binding subsites -1 to -6 in CGTase (Figure 5) will increasingly lock the substrate in a position 6 glucose units from the active site and result in increased product selectivity for α-cyclodextrin. Formation of larger cyclodextrins is minimized by reducing intermolecular interactions at subsites -7 to -8. To increase product specificity for β- or γ-cyclodextrin, the number of intermolecular interactions at sugar binding subsites –6 and –7, or –7 and –8, respectively, is increased to lock the substrate in a position 7 or 8 glucose units, respectively, from the active site. Formation of α-cyclodextrin is minimized by reducing intermolecular interactions at subsites –1 to –6. This detailed information has allowed rational construction of thermostable CGTase enzymes with improved cyclodextrin product specificity.

5. REFERENCES

1. H. Bender, *Appl. Microbiol. Biotechnol.*, 1993, **39**, 714.
2. L. Dijkhuizen, D. Penninga, H.J. Rozeboom, B. Strokopytov, B.W. Dijkstra, 'Perspectives on Protein Engineering & Complementary Technologies' (M.J. Geisow, R. Epton, eds), Mayflower Worldwide Ltd, Birmingham, 1995, pp. 96.
3. L. Dijkhuizen, D. Penninga, H.J. Rozeboom, B. Strokopytov, B.W. Dijkstra, 'Proc. Carbohydrate Bioengineering Meeting Elsinore' (S.B. Petersen, B. Svensson, S. Pedersen, eds.), 1995, Elsevier, Amsterdam, pp. 165.
4. D. French, *Adv. Carbohydrate Chem.*, 1957, **12**, 12189.
5. W. Saenger, *Angew. Chem., Int. Ed. Engl.*, 1980, **19,** 344.
6. G. Wenz, *Angew. Chem., Int. Ed. Engl.*, 1994, **33,** 803.
7. G. Schmid, Tibtech, 1989, **7,** 244.
8. A. Nakamura, K. Haga, K. Yamane, *Biochemistry*, 1993, **32**, 6624.
9. S. Pedersen, L. Dijkhuizen, B.W. Dijkstra, B.F. Jensen, S.T. Jørgensen, *Chemtech*, 1995 (December), 19.

10. C.L. Lawson, J. Bergsma, P. Bruinenberg, G.E. de Vries, L. Dijkhuizen, B.W. Dijkstra, *J. Mol. Biol.*, 1990, **214,** 807.
11. C.L. Lawson, R. van Montfort, B. Strokopytov, H.J. Rozeboom, K.H. Kalk, G.E. de Vries, D. Penninga, L. Dijkhuizen, B.W. Dijkstra, *J. Mol. Biol.*, 1994, **236,** 590.
12. D. Penninga, B.A. van der Veen, R.M.A. Knegtel, S.A.F.T. van Hijum, H.J. Rozeboom, K.H. Kalk, B.W. Dijkstra, L. Dijkhuizen, *J. Biol. Chem.* 1996, **271,** 32777.
13. C. Klein, G.E. Schulz, *J. Mol. Biol.*, 1991, **217,** 737.
14. M. Kubota, Y. Matsuura, S. Sakai, Y. Katsube, *Denpun Kagaku*, 1991, **38,** 141.
15. R.M.A. Knegtel, R.D. Wind, H.J. Rozeboom, K.H. Kalk, R.M. Buitelaar, L. Dijkhuizen, B.W. Dijkstra, *J. Mol. Biol.* 1996, **256**, 611.
16. R.D. Wind, D. Penninga, J.C.M. Uitdehaag, R.M. Buitelaar, B.W. Dijkstra, L. Dijkhuizen, *FEMS Microbiol. Rev.*, 1999, in press.
17. R.D. Wind, R.D., J. Uitdehaag, R.M. Buitelaar, B.W. Dijkstra, L. Dijkhuizen, *J. Biol. Chem.*, 1998, **273**, 5771.
18. D. Penninga, B. Strokopytov, H.J. Rozeboom, C.L. Lawson, B.W. Dijkstra, J. Bergsma, L. Dijkhuizen, *Biochemistry*, 1995, **34**, 3368.
19. D.E.J. Koshland, *Biol. Rev.*, 1953, **28**, 416.
20. Y. Tanaka, T. Wen, J. S. Blanchard, E. J. Hehre, J. Biol. Chem., 1994, **269**, 32306.
21. J.D. MacCarter, S. G. Withers, *J. Biol. Chem.*, 1996, **271**, 6889.
22. B.Y. Tao, P.J. Reilly, J.F. Robyt, *Biochim. Biophys. Acta* 1989, **995**, 214.
23. R. Mosi, S. He, J. Uitdehaag, B.W. Dijkstra, S.G. Withers, *Biochemistry*, 1997, **36**, 9927.
24. B. Strokopytov, D. Penninga, H.J. Rozeboom, K.H. Kalk, L. Dijkhuizen, B.W. Dijkstra, *Biochemistry*, 1995, **34**, 2234.
25. K.A. Sin, A. Nakamura, H. Masaki, T. Uozumi, *Biosci. Biotechnol. Biochem.*, 1993, **57,** 346.
26. K.A. Sin, A. Nakamura, H. Masaki, Y. Matsuura, T. Uozumi, *J. Biotechnol.*, 1994, **32,** 283.
27. S. Fujiwara, H. Kakihara, K. Sakaguchi, T. Imanaka, *J. Bacteriol.*, 1992b, **174**, 7478.
28. K. Horikoshi, 'Proc. 4th Int. Symp. Cyclodextrins' (O. Huber, J. Szejtli, eds.), Dordrecht, Kluwer Academic Publishers, 1988, pp. 7.
29. B. Strokopytov, R.M.A. Knegtel, D. Penninga, H.J. Rozeboom, K.H. Kalk, L. Dijkhuizen, B.W. Dijkstra, Biochemistry, 1996, **35**, 4241.
30. D. Penninga, 'Protein engineering of cyclodextrin glycosyltransferase from *Bacillus circulans* strain 251', PhD thesis, Ponsen & Looijen BV, Wageningen, The Netherlands. http://www.biol.rug.nl/micfys/micfys.html/Ph.D.thesis, 1996.
31. Y.H. Kim, K.H. Bae, T.J. Kim, K.H. Park, H.S. Lee, S.M. Byun, *Biochem. Mol. Biol. Int.*, 1997, **41**, 227.
32. L. Dijkhuizen, L., B.W. Dijkstra, C. Andersen, C. Von der Osten, Cyclomaltodextrin glucanotransferase variants, International patent application WO 96/33267, 1996.

THE ENGINEERING OF SPECIFICITY AND STABILITY IN SELECTED STARCH DEGRADING ENZYMES

B. Svensson[1], K. S. Bak-Jensen[1], H. Mori[1], J. Sauer[1], M. T. Jensen[1], B. Kramhøft[1], T. E. Gottschalk[1], T. Christensen[2], B. W. Sigurskjold[2], N. Aghajari[3], R. Haser[3], N. Payre[4], S. Cottaz[4] and H. Driguez[4]

[1]Department of Chemistry, Carlsberg Laboratory, DK-2500 Copenhagen, [2]Department of Biochemistry, August Krogh Institute, University of Copenhagen, DK-2500 Copenhagen, [3]Institut de Biologie et Chimie de Protéines, C. N. R. S., F-69367 Lyon, [4]Centre de Recherche sur les Macromolécules Végétales, C. N. R. S., F-38401 Grenoble

1 INTRODUCTION

In recent years a growing number of three-dimensional structures and amino acid sequences of starch degrading and other enzymes active in the metabolism of polysaccharides have become available. Despite similarities in reaction mechanisms, the diversity in protein folds is impressive (1-4). Investigations in structure/function relationships using protein chemical and genetic engineering techniques have increased the knowledge on the mechanisms of action, enabling generalisations on numerous polysaccharide metabolising enzymes based on thorough insight established for others. Certain common concepts, however, remain poorly understood, such as i) the correlation between substrate cleavage pattern and fine architecture of the substrate binding site consisting of an array of consecutive subsites, ii) static and time-resolved description at atomic resolution of events in glycosidic bond cleavage and synthesis, iii) the interplay between binding and catalytic domains, and iv) the function of other non-catalytic domains. Moreover, the industrial application of polysaccharide degrading enzymes may depend on engineering of stability and function to be compatible with process conditions and on rational alteration of product and substrate specificities. The whole field progresses rapidly and provides many successful examples of basic and applied research going hand-in-hand.

2 RESULTS

This lecture will focus on advances in structure/function relationship investigations in i) glucoamylase from *Aspergillus niger*, an exoacting inverting enzyme of glycoside hydrolase family 15 having a catalytic $(\alpha/\alpha)_6$ barrel domain (CD) connected *via* an *O*-glycosylated Ser/Thr-rich linker to a starch binding domain (SBD; refs. 5, 6) and in ii) barley α-amylase, an endoacting retaining enzyme of the widely occurring $(\beta/\alpha)_8$-barrel starch metabolising enzymes of glycoside hydrolase family 13 (4, 7, 8). Issues addressed include i) the impact of the glucoamylase polypeptide linker (9, 10) on domain-domain interaction, function, and stability, ii) comparison in barley α-amylase of distinct isozyme properties (11, 12) including effects of salt-bridge engineering on stability (13), and iii) rational engineering at subsites +1/+2 (adjacent to the catalytic site), –6 and +4 (at the ends of the binding cleft; ref. 14).

2.1 Domain-domain interactions in glucoamylase from *Aspergillus niger*

Scanning tunnelling microscopy has suggested that CD and SBD in *A. niger* glucoamylase G1 are far apart with the connecting highly *O*-glycosylated linker adopting an extended conformation (15). Recently, heterobidentate inhibitors were synthesised that contain the potent inhibitor acarbose (16) connected via ethyleneglycol spacers of varying length to the cyclic starch analogue β-cyclodextrin, that binds to SBD (17). It was demonstrated by dynamic light scattering and isothermal titration calorimetry (ITC) that the two domains each bind a target moiety of the heterobidentate inhibitors regardless of their length varying from having none to having spacers 14, 36, and 73 Å long in their extended conformations, respectively. This strongly suggested that CD and SBD are close in the solution structure of glucoamylase G1 (17, 18). The heterobidentate inhibitors were found by ITC to bind with much lower affinity but with essentially the same enthalpy contributions as acarbose and β-cyclodextrin (18). Thus a severe entropy penalty had been paid by binding the double-headed inhibitors to glucoamylase G1 to occupy two sites in the same molecule, one at CD and one at SBD, respectively (18).

2.1.1 Linker region homologues of glucoamylase. Rather than varying the length of the double-headed inhibitor, the *O*-glycosylated linker (aa 440-508) could be changed by protein engineering. Earlier work indicated that truncation of the full length glucoamylase G1 within the linker was possible genetically at residue 481 (19) or proteolytically at residue 471 (20) resulting in a functional CD containing different portions of linker region. Like glucoamylase G2 (aa 1-512) (9, 21, 22) these truncated forms lack capacity to adsorb onto and degrade starch granules. Based on deletion analysis the most highly *O*-glycosylated stretch of the linker (9) was later speculated to be associated with stability (aa 485-512) and secretion (aa 466-483), respectively (23).

In the present linker variants, Ser468-Ser508 of *A. niger* glucoamylase was replaced by equivalent stretches of *Rhizopus oryzae* and *Humicola grisea* glucoamylases, representing different sequences shorter by 3 and 21 residues, respectively. The variants were produced in *Pichia pastoris* as established for *A. niger* glucoamylase G1 (24). Using non-natural sequences it was found that ≥11 amino acid residues were needed to replace Ser468-Ser508 to get a functional enzyme (data not shown). Both *Rh. oryzae* and *H. grisea* linker variant enzymes maintained the activity of *A. niger* glucoamylase toward maltose, isomaltose, maltooligosaccharides, and soluble and granular starches.

Table 1 *Properties of linker region variants of glucoamylase*

Enzyme	T_{50}[a] ℃	$[GuHCl]_{50}$[b] M	T_{D1}[c] ℃	T_{D2}[d] ℃
Wildtype glucoamylase G1	72	5.9	62.6	60.2
Rh. oryzae linker glucoamylase G1	67	5.0	60.0	57.4
H. grisea linker glucoamylase G1	66	5.0	60.5	56.1

[a]Temperature of 50% retained activity after 5 min at pH 4.5. [b]Guanidinium chloride concentration for 50% unfolding. DSC transition temperature values for irreversible[c] and reversible[d] unfolding

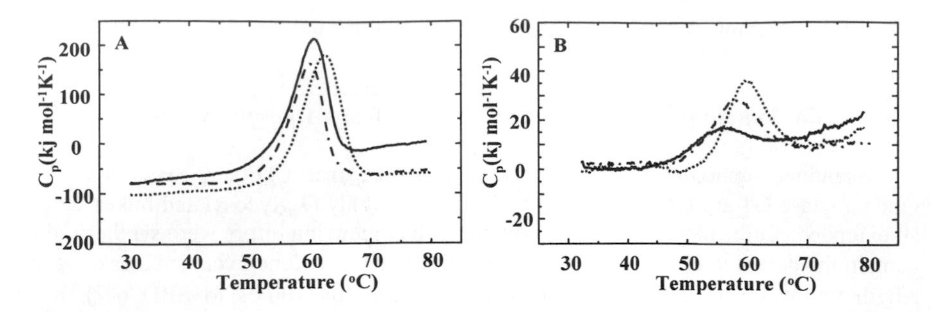

Figure 1 *A. DSC unfolding of wildtype glucoamylase G1 (•••), H. grisea (—), and Rh. oryzae linker variants (• - •). B. Reversible unfolding of SBD analysed after cooling*

However, differences from the *A. niger* enzyme appeared in binding of heterobidentate inhibitors and in stability at elevated temperature, in differential scanning calorimetry (DSC) (Figure 1) or in guanidinium chloride (Table 1; ref. 25).

2.1.2 Unfolding of wildtype glucoamylase by differential scanning calorimetry. Detailed investigations of wildtype glucoamylase G1 indicated that CD and SBD unfold separately (25, 26). While CD shows an apparent one step irreversible unfolding in DSC, the unfolding of SBD is reversible (25, 26). DSC was also performed on a truncated form containing CD (aa 1-471), the G2 (aa 1-512) and G1 forms, and two forms of SBD (prepared by T. P. Frandsen, Novo Nordisk) that contain residues 509-616 and 487-616. These studies indicated that G2, which has the full-length linker, was less stable than G1 and the truncated form. The C-terminal part of the linker region thus destabilises CD (26). SBD with no linker unfolds irreversibly, in contrast to SBD with part of the linker. The same C-terminal part of the linker apparently stabilised the SBD (25).

2.2 Stability and protein engineering of barley α-amylase isozymes

Until recently the insufficient amounts of barley α-amylase 1 and 2 (low and high pI isozymes, or AMY1 and AMY2) precluded detailed comparison of the isozymes with respect to structure/stability relationships. However, the established expression for AMY1 in *Pichia pastoris* (27) and the large-scale purification procedure from malt developed for the major isozyme AMY2 (14) made both isozymes accessible. It was reported previously that AMY2 has higher thermostability (28), and that EDTA most severely decreased the stability of AMY2, which has lower affinity for calcium ions than AMY1 (29). The three-dimensional structure is known of native AMY2 and of the acarbose complex (30, 31), and the 80 % sequence identity to AMY1 can guide protein engineering to alter the structure of AMY1 towards that of AMY2 (see 2.2.1 below). Recently, the crystal structure was solved of AMY1 (32), but the refinement is still in progress and the two crystal structures have not been used for interpretation of isozyme differences.

2.2.1 The stability of barley α-amylase isozymes 1 and 2 and salt-bridge engineering. It was recently shown that AMY2 has lower stability at high ionic strength than AMY1, and that AMY2 is the most stable isozyme in urea (13, 33). It was concluded therefore that electrostatic interactions stabilised AMY2. In contrast, screening of charges increased the stability of AMY1, which thus seemed to be destabilised by electrostatic interactions. Whereas addition of calcium chloride has a very small effect on the stability

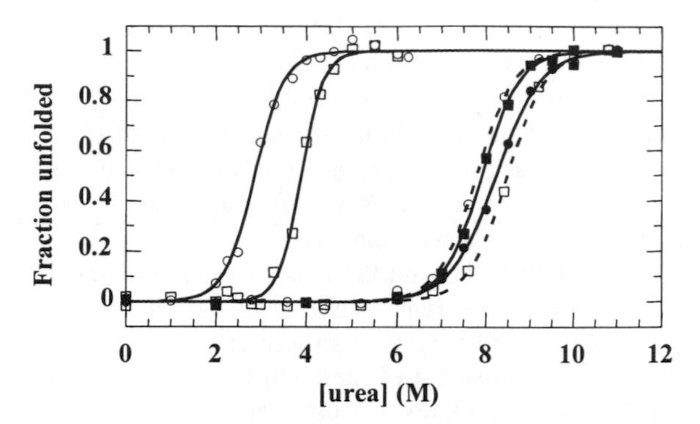

Figure 2 *Stability in 5 mM Ca^{2+} of AMY1 (—■—) and AMY2 (—●—); plus 1 M NaCl, AMY1 (- ■ -) and AMY2 (- ● -); or 10 mM EDTA, AMY1 (—□—) and AMY2 (—○—)*

of either isozyme, EDTA clearly destabilised AMY2 the most (Figure 2) in agreement with the tighter binding of calcium ions to AMY1 (29).

By 1D/3D structure comparison of AMY1 and AMY2 using the crystal structure of AMY2 it was evident that four AMY2 salt-bridges could not occur in AMY1, because one of the involved side chains was substituted (30). In the structure three of these four residues participate in electrostatic networks between structural elements (30). In order to assess the impact of the electrostatic interactions for the isozymes it was attempted to introduce the four AMY2 specific salt-bridges into AMY1 by site-directed mutagenesis (13). The AMY1 mutants Ala68→Lys, Asp97→Glu, Gln269→Asp, and Asn346→His, showed only minor variation in enzymic activity toward an oligosaccharide substrate and insoluble Blue Starch (data not shown) and small changes in stability. Moderate increase and decrease in thermostability were obtained for Asp97→Glu and Asn346→His AMY1, respectively (Table 2; ref. 14). The first connects the β→α loop 2 of the barrel domain to the protruding domain B, whereas the latter is between side chains in α-helix 8b of the barrel and the C-terminal domain (30). Of the four mutated residues the counterpart of $Asn346_{AMY1}$, i.e., $His344_{AMY2}$, has the smallest accessible surface area in the crystal structure (Table 2; ref. 13) in agreement with the lower stability of the mutant.

Table 2 *Effects of salt-bridge mutation on the stability of AMY1*

Enzyme	*T_{50}[a]* *°C*	*$[GuHCl]_{50}$[b]* *M*	*ASA[c] in AMY2* *(%)*
Wildtype AMY2	79.0	3.3	
Wildtype AMY1	75.5	3.6	
Ala68→Lys	75.5		52
Asp97→Glu	76.0		43
Gln269→Asp	n.d.		41
Asn346→His	74.5		26

[a]Temperature of 50% retained activity after 10 min at pH 6.7. [b]Guanidinium chloride concentration for 50% unfolding. [c]Available surface area

2.2.2 Engineering of barley α-amylase/subtilisin inhibitor sensitivity into an insensitive barley α-amylase isozyme hybrid. AMY2, the dominant isozyme in malt, is specifically inhibited by BASI the endogenous double-headed barley α-amylase/subtilisin inhibitor (34), that belongs to the soybean trypsin inhibitor family (35). Members of this β-trefoil fold protein family are involved in protein-protein interactions and their diverse specificities for cereal α-amylases, serine proteases of the subtilisin family, or receptors corresponding to the different growth factors in the family, are associated with different loops connecting β-strands in the supersecondary structure. The distinct inhibition and high affinity of BASI for AMY2 (34) and the lack of effect on AMY1 despite its 80% sequence identity with AMY2 are remarkable. This target enzyme selectivity was excellent motivation for mutational analysis of structural elements required in recognition of BASI. This further allows assessment of the contribution of individual side chains to the stability of the AMY2-BASI complex and may throw light on the structural basis of the isozyme specificity.

Isozyme hybrids of AMY1 and AMY2 were generated by *in vivo* homeologous recombination in *Saccharomyces cerevisiae*. The biochemical properties of a series of hybrids suggested that domain B which protrudes from the third β-strand of the $(\beta/\alpha)_8$ domain is involved in BASI sensitivity and several other isozyme specific properties (11, 36). Thus the isozyme origin of domain B was important for the hybrid properties. The AMY2-BASI complex was sensitive to charge screening (37) and tests on additional isozyme hybrids generated within the domain B (aa 89-152; AMY2 numbering; ref. 38) tentatively identified the stretch of aa 116-144 to carry a critical determinant for the complexation. This brought attention to sequence variations for AMY1 and AMY2 involving charged side chains in this short segment. Since $Arg128_{AMY2}$ was the only charged residue not conserved in AMY1, this residue and the succeeding *cis*-Pro129 from AMY2 (30), were introduced by site-directed mutagenesis into a BASI-insensitive AMY1/AMY2 hybrid, H161 (11), having the domain B of AMY1 origin. The resulting mutant, H161-$ThrLys130_{AMY1}$→ArgPro, was inhibited by BASI (Table 3). Interestingly, while the K_i = 18 nM of the H90-$Arg128_{AMY2}$→Gln mutant in a BASI sensitive hybrid H90 (K_i = 0.33 nM) confirmed a loss of affinity, the inhibition of this mutant was only very little sensitive to charge screening. In contrast, in the presence of salt K_i was 140 μM of H161-Thr129$Arg130_{AMY1}$→LysPro compared to K_i of 7 μM under optimal conditions for complex formation with BASI. This charge screening effect corresponds to loss of approximately 2 kcal × mol^{-1} in binding energy which is similar to the effect on BASI

Table 3 *Introduction of BASI–sensitivity in an inactive AMY1/AMY2 hybrid*

Enzyme	K_i^a *μM*	K_i^b *μM*	$\Delta\Delta G_i^c$ *kcal × mol⁻¹*	$\Delta\Delta G_i^d$ *kcal × mol⁻¹*
Wildtype AMY2	0.00022	0.0030		1.6
Wildtype AMY1	≥10³	n.d.		
Wildtype H161ᶜ	≥10³	n.d.		
H161-T129R/K130P	7	140	6.4	1.9
H90-R128Q	0.018	0.045	3.0	0.6

[a]In 40 mM Tris, pH 8.0, 5 mM $CaCl_2$. [b]In 40 mM Hepes, pH 6.8, 5 mM $CaCl_2$, 0.2 M NaCl. [c]$\Delta\Delta G_i = -RT\ln(K_{i,mutant}/K_{i,AMY2})$. [d]$\Delta\Delta G_i = -RT\ln(K_i^b/K_i^a)$. n.d. not determined

inhibition of AMY2 and the H90 parent. It seems therefore that a proper recognition site for BASI has been introduced by the mutation in the insensitive isozyme hybrid H161. In the AMY2-BASI crystal structure the proteins share a large interface of 2355 Å^2 (39) and among selected intermolecular contacts thus the charged hydrogen bond between side chains of Arg128$_{AMY2}$ and Ser77$_{BASI}$ was critical for complexation. Another important interaction identified by mutational analysis was between the conserved Asp142$_{AMY2}$ and Lys140$_{BASI}$ (data not shown). Apparently, few protein-protein contacts were crucial for formation and stability of AMY2-BASI, in spite of the large interface area. Recent studies on different systems reached the same conclusion that a small number of "hotspots" can determine vital properties of protein-protein complexes (40, 41).

2.2.3 Binding subsite mutagenesis to alter substrate specificity and action pattern of barley α-amylase. Amylolytic enzymes bind substrate glucosyl residues at an array of consecutive subsites that extends throughout the active site cleft. In typical endoacting enzymes such as α-amylases the substrate binding region comprises from five to eleven subsites. In barley α-amylase ten binding subsites were defined by subsite mapping. This analysis consists in measurement of the kinetics of the hydrolysis of a series of oligosaccharides and combination with the relative bond cleavage frequencies allows the calculation of binding affinities for individual subsites. AMY1 and AMY2 have six subsites accommodating glucosyl residues towards the non-reducing end from the bond to be cleaved and four towards the reducing end; the higher affinities are associated with subsites -6, -5, -2, and +1 (14, 42).

Difference Fourier analysis of the AMY2-acarbose complex revealed that three rings of the pseudotetrasaccharide inhibitor occupy subsites –1, +1, and +2 (31). The AMY2-acarbose structure was used to model complexes with longer substrates. Maltodecaose has in this way been proposed to interact with Tyr104 (AMY2 numbering) at subsite –6 and Tyr211 at subsite +4 (43) and the present mutations at these subsites add new information to data on previous binding site mutants (Figure 3). For these it was reported that substrate preferences could shift when residues were substituted at β→α binding loops 4 (Arg183GlyTyr in AMY1) and 7 (Phe286ValAsp in AMY1) of the $(\beta/\alpha)_8$ domain by various random mutagenesis approaches (7, 33). Thus while certain loop 4 mutants gained in activity towards short amylose of DP 17 and an oligosaccharide substrate

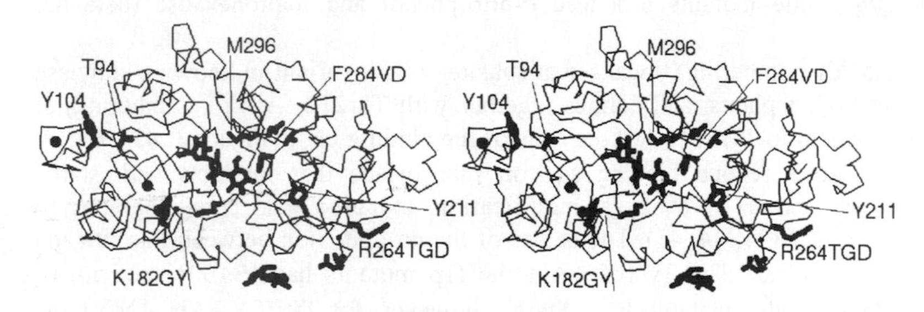

Figure 3 *Stereo view of substrate binding site mutations illustrated in the structure of AMY2-acarbose. Work on subsite –6 (Tyr104), -5 (Thr94), +1/+2 (Met296) and +4 (Tyr211) is reported here, while the Lys182GlyTyr, Phe294ValAsp and Arg264ThrGlyAsp mutations at equivalent sequences in AMY1 were described previously (7, 33). Three rings of acarbose are seen at the active site to fill subsites –1, +1, and +2 and a disaccharide unit occupies the surface binding site*

Table 4 *Enzymatic properties of AMY1 mutants at subsites +1/+2 and -5*

AMY1	*Blue Starch*	*Cl-PNPG$_7$*[b]		*Amylose DP 17*	
	U/mg	K_m	k_{cat}	K_m	k_{cat}
		mM	s^{-1}	*mg/mL*	s^{-1}
Wildtype[a]	2500	1.3	58	0.4	236
Cys95→Ala	4200	9.8	75	2.2	464
Met298→Ala	2800	6.2	43	0.5	208
Cys95→Ala/Met298→Ala	4900	n.d.		n.d.	

[a]From barley malt. [b]2-chloro-4-nitrophenyl β-D-maltoheptaoside

unable to span the entire binding site, respectively (7), loop 7 mutants had increased activity for the oligosaccharide, but decreased activity for amylose DP 17 (33).

Very recently, site-directed mutagenesis of Met298 at subsite +1/+2 in AMY1 was found to influence the substrate preference differently when combined with the mutation Cys95→Ala at or near subsite –5 (Table 4). Remarkably, with an oligosaccharide and amylose DP 17 substrate saturation of the AMY1 double mutants (only data for one double mutant, Cys95→Ala/Met298→Ala, are shown) were not approached at the highest substrate concentration compatible with the assays. In spite of this poor affinity for these shorter substrates, the double mutants have higher activity than wildtype AMY1 towards insoluble Blue Starch. The elevated activity was primarily a characteristic of Cys95→Ala AMY1. However, also for this single mutant, affinity decreased importantly for the soluble substrates, K_m being 5 – 7 times higher than for wildtype AMY1 (Table 4; ref. 7). The single Met298 mutants (only data for Met298→Ala are shown) have increased K_m for the oligosaccharide. The action pattern of these mutants towards oligosaccharides, as determined by the cleavage frequencies of individual substrate bonds, indicated that some of the double mutants preferably produced *p*-nitrophenyl maltoside and maltotetraose from *p*-nitrophenyl maltohexaoside, whereas Cys95→Ala AMY1 mainly released *p*-nitrophenyl glucoside and maltopentaose, and wildtype AMY1 and the Met298 single mutants produced *p*-nitrophenol and maltohexaose (data not shown).

Tyr105 (AMY1 numbering), localised at subsite –6 is invariant in known α-amylase sequences from higher plants, and defines together with Thr212 (AMY1 numbering) at subsite +4 (equivalent to Tyr211$_{AMY2}$) the ends of the binding cleft (Figure 3; 43). Barley α-amylase has also a second binding site on the surface that interacts with starch granules. This was shown by chemical modification, mutation, and crystallography to include Trp276Trp277 (31, 44, 45). The nature of the co-operation between this site and the active site is unclear. The Tyr105→Ala/Phe/Trp mutants had ≥ 100% activity of wildtype AMY1 towards insoluble Blue Starch, however, for Tyr105→Ala AMY1 the affinity for amylose DP 17 was reduced about 5-fold, while both activity and affinity were even more decreased for the oligosaccharide substrate. Tyr105→Trp/Phe AMY1 in contrast essentially retained or slightly increased activity for the soluble substrates (data not shown). Thr212→Pro/Tyr/Trp AMY1, at subsite +4, had ≤ 80% activity of AMY1 for insoluble Blue Starch, whereas k_{cat}/K_m for the two soluble substrates was 120 - 195 %. Compared to AMY1 the Thr212→Trp mutant resulted in improved substrate affinity for soluble substrates.

With respect to the action pattern the mutational changes at subsites -6 or +4 indeed affected the preferred productive binding modes of oligosaccharide substrates. Thus *p*-nitrophenyl maltohexaoside was cleaved to liberate mainly *p*-nitrophenyl maltoside, *p*-nitrophenyl glucoside, and *p*-nitrophenol and the corresponding non-labelled maltooligosaccharide products after substitution of Tyr105 with Ala, Phe, and Trp, whereas AMY1 produced essentially only *p*-nitrophenol and maltohexaose. The effect of mutation at subsite +4 was most pronounced on *p*-nitrophenyl maltopentaose hydrolysis that gave *p*-nitrophenyl maltotrioside and maltose as main products, while AMY1 liberated *p*-nitrophenyl maltoside and *p*-nitrophenyl glucoside in equal amounts together with maltotriose and maltotetraose.

The results demonstrate that rational engineering of action patterns is feasible and that it is possible to achieve predicted changes in preferred binding modes. Thus, when the strongly binding subsite –6 was changed, the maltohexaoside substrate was less frequently accommodated to bind at subsite –6, the substrate rather occupied several subsites on the leaving group side of the catalytic site in the productive complexes with these mutants. Similarly, when Thr212 at subsite +4 was mutated to the aromatic Tyr or Trp, productive binding of oligosaccharide substrates seemed favoured at the outer subsites +3 and +4. This supports a stacking interaction as modelled for maltodecaose to Tyr211 at subsite +4 in AMY2 (43). Subsite –6 might be envisaged to be involved in multiple attack on large soluble substrates due to its high affinity (10). However, preliminary analysis of Tyr105→Phe/Ala AMY1, showed no change from the degree of multiple attack of approximately 2 determined earlier for AMY1 (46).

3 CONCLUSION AND FUTURE DIRECTIONS

Recent protein engineering based investigations of structure/function relationships in amylolytic enzymes supported that the catalytic and starch binding domains of glucoamylase from *Aspergillus niger* influence each other both with respect to function, as demonstrated in the interaction with the heterobidentate inhibitors, and with respect to stability as shown by differential scanning calorimetric studies of intact wildtype G1 enzyme, truncated forms, and separate domains, as well as of linker homologue replacement G1 variants (17, 18, 25, 26).

Stability studies of barley α-amylase isozymes 1 and 2 suggested that the AMY2 was stabilised by electrostatic interactions, whereas AMY1 gained in relative stability in guanidinium chloride or at high concentrations of salt. However, attempts to introduce salt-bridges specific to AMY2 by site-directed mutagenesis of single residues in AMY1, resulted in only minor improvements of stability for some mutants, without approaching the superior stability of AMY2 at elevated temperature or in urea (13).

The specific activity of AMY1 was earlier differentially modulated by several mutations at β→α binding loops 4, 6, and 7 (7, 33). The present work demonstrated that the structure at specific subsites could be changed by site-directed mutagenesis to result in alteration of specificity and action pattern. Remarkably, certain mutants with low activity towards soluble shorter maltooligosaccharide substrates had higher activity than the wildtype enzyme towards insoluble starch.

In the future, three-dimensional structures would be strongly required in order to gain further insight in particular into the action of multidomain enzymes that like the fungal glucoamylases consist of separate linker-connected catalytic and binding domains. For amylolytic enzymes of glycoside hydrolase family 13, which represents more than 20 substrate and product specificities, it will be important to continue the engineering studies of the specificity by various methods applied to substrate binding loops and sequence motifs as guided by three-dimensional structures and multiple sequence alignment (4-8, 33). It is moreover important to identify functional roles of non-catalytic domains, present for example in the newly determined structures of the larger family 13 members with specificity for α-1,6 branch points and α-1,4 bonds adjacent to α-1,6 bonds (47, 48) as well as in the sucrose-utilising glucosyltransferases which belong to family 13 but have a circularly permuted catalytic domain (49). Especially in the case of cereal α-amylases, it will be important to elucidate the mechanism of recognition for the surface binding site (31, 44) and the possible interplay of this site and the active site. This may or may not be connected to events in the multiple attack mechanism which might be explored by a combination of protein engineering and substrate specificity analysis.

Acknowledgements
The present work was supported by the EU Biotechnology Programme (BIO4-CT98-0022), the Danish Research Councils' Committee on Biotechnology (grant no. 9502914), and a Novo Nordisk scholarship to KSBJ.

References

1. J. D. McCarter and S. G. Withers, *Curr. Opin. Struct. Biol.*, 1994, **4**, 885.
2. G. Davies and B. Henrissat, 1995, *Structure*, 1995, **3**, 853.
3. URL: http://afmb.cnrsmrs.fr/~pedro/CAZY/ghf-table.html
4. B. Svensson, *Plant Mol. Biol.*, 1994, **25**, 141.
5. T. P. Frandsen, H.-P. Fierobe, and B. Svensson, in *Protein Engineering in Industrial Biotechnology* (L. Alberghina, ed., Harwood Press) 1999, in press.
6. S. M. Southall, P. J. Simpson, H. J. Gilbert, G. Williamson, and M. P. Williamson, *FEBS Lett.*, 1999, **447**, 58.
7. I. Matsui and B. Svensson, *J. Biol. Chem.*, 1997, **272**, 22456.
8. H. Jespersen, E. A. MacGregor, M. R. Sierks and B. Svensson, *Biochem. J.*, 1991, **380**, 51.
9. B. Svensson, K. Larsen, I. Svendsen, and E. Boel, *Carlsberg Res. Commun.*, 1983, **48**, 554.
10. P. Coutinho and P. J. Reilly, *Protein Eng.*, 1994, **7**, 393.
11. K. W. Rodenburg, N. Juge, X.-J. Guo, M. Søgaard, J.-C. Chaix, and B. Svensson, *Eur. J. Biochem.*, 1994, **221**, 277.
12. R. L. Jones and J. V. Jacobsen, *Int. Rev. Cytol.*, 1991, **126**, 49.
13. M. T. Jensen, MSc Thesis, Roskilde University, 1998.
14. E. H. Ajandouz, J. Abe, B. Svensson, and G. Marchis-Mouren, *Biochim. Biophys. Acta*, 1992, **1159**, 193.
15. G. F. H. Kramer, A. P. Gunning, V. J. Morris, N. J. Belshaw, and G. Williamson, *Chem. Soc. Faraday Trans.*, 1993, **89**, 2595.
16. B. W. Sigurskjold, C. R. Berland, and B. Svensson, *Biochemistry*, 1994, **33**, 10191.
17. N. Payre, S. Cottaz, C. Boisset, R. Borsali, B. Svensson, B. Henrissat, and H. Driguez, *Angew. Chem. Int. Ed.*, 1999, **38**, 974.
18. B. W. Sigurskjold, T. Christensen, N. Payre, S. Cottaz, H. Driguez, and B. Svensson, *Biochemistry*, 1998, **37**, 10446.
19. R. Evans, C. Ford, M. R. Sierks, Z. Nikolov, and B. Svensson, *Gene*, 1990, **91**, 131.
20. B. Stoffer, T. P. Frandsen, P. K. Busk, P. Schneider, I. Svendsen, and B. Svensson, *Biochem. J.*, 1993, **292**,1 97.

21. B. Svensson, T. G. Pedersen, I. Svendsen, T. Sakai and M. Ottesen, *Carlsberg Res. Commun.*, 1982, **47**, 55.
22. B. Svensson, K. Larsen, and A. Gunnarsson, *Eur. J. Biochem*., 1996, **154**, 497.
23. C. B. Libby, C. A. G. Cornett, P. J. Reilly, and C. Ford, *Protein Eng*., 1994, **7**, 1109.
24. H.-P. Fierobe, E. Mirgorodskaya, T. P. Frandsen, P. Roepstorff, and B. Svensson, *Prot. Express. Purif.*, 1997, **9**, 159.
25. T. Christensen, PhD Thesis, University of Copenhagen, 1999.
26. T. Christensen, B. Svensson, and B. W. Sigurskjold, *Biochemistry*, 1999, **38**, 6300.
27. N. Juge, J. S. Andersen, D. Tull, P. Roepstorff, and B. Svensson, *Prot. Express. Purif.*, 1996, **8**, 804.
28. E. Bertoft, C. Andtfolk, and S.-E. Kulp, *J. Inst. Brew.*, 1984, **90**, 298.
29. D. S. Bush, L. Sticher, R. B. V. Huystee, D. Wagner, and R. L. Jones, *J. Biol. Chem.*, 1989, **264**, 19392.
30. A. Kadziola, J. Abe, B. Svensson, and R. Haser, *J. Mol. Biol.*, 1994, **239**, 104.
31. A. Kadziola, M. Søgaard, B. Svensson, and R. Haser, *J. Mol. Biol.*, 1998, **278**, 205.
32. N. Aghajari, X. Robert, B. Svensson, and R. Haser, 3rd Carbohydrate Bioengineering Meeting, Newcastle, April 1999, abstract 4.10.
33. B. Svensson, K. S. Bak-Jensen, M. T. Jensen, J. Sauer, T. E. Gottschalk, and K. W. Rodenburg, *J. Appl. Glycosci.*, 1999, **46**, 51.
34. J. Mundy, I. Svendsen, and J. Hejgaard, *Carlsberg Res. Commun.*, 1983, **48**, 81.
35. A. G. Murzin, A. M. Lesk, and C. Chothia, *J. Mol. Biol.*, 1992, **223**, 531.
36. N. Juge, M. Søgaard, J.-C. Chaix, M. F. Martin-Eauclaire, B. Svensson, G. Marchis-Mouren, and X.-J. Guo, *Gene*, 1993, **130**, 159.
37. J. Abe, U. Sidenius, and B. Svensson, *Biochem. J.*, 1993, **293**, 151.
38. N. Juge, K. W. Rodenburg, X.-J. Guo, J.-C. Chaix, and B. Svensson, *FEBS Lett.*, 1995, **363**, 399.
39. F. Vallée, A. Kadziola, Y. Bourne, M. Juy, K. W. Rodenburg, B. Svensson, and R. Haser, *Structure*, 1998, **6**, 649.
40. S. Atwell, M. Ultsch, A. M. De Vos, and J. A. Wells, *Science,* 1997, **278**, 1125.
41. A. W. Bogan and K. S. Thorn, *J. Mol. Biol.*, 1998, **280**, 1.
42. E. A. MacGregor, A. W. MacGregor, L. J. Mori, and J. E. Morgan, *Carbohydr. Res.*, 1994, **257**, 249.
43. G. André, PhD Thesis, University of Nantes, 1998.
44. R. M. Gibson and B. Svensson, *Carlsberg Res. Commun.*, 1987, **52**, 373.
45. M. Søgaard, A. Kadziola, R. Haser, and B. Svensson, *J. Biol. Chem.*, 1993, **268**, 22480.
46. B. Kramhøft and B. Svensson, in *Progress in Biotechnology* (A. Ballesteros, F. J. Plou, J. L. Iborra and P. J. Halling, Eds., Elsevier), 1998, **15**, 343.
47. Y. Katsuya, Y. Mezaki, M. Kubota, and Y. Matsuura, *J. Mol. Biol.*, 1998, **281**, 885.
48. S. Kamitori, S. Kondo, K. Okuyama, T. Yokota, Y. Shimura, T. Tonuzuka, and Y. Sakano, *J. Mol. Biol.,* 1999, **287**, 907.
49. E. A. MacGregor, H. M. Jespersen, and B. Svensson, *FEBS lett.*, 1996, **378**, 266.

MUTAGENESIS OF THE CATALYTIC CORE OF A STREPTOCOCCAL GLUCOSYLTRANSFERASE

R. R. B. Russell and V. Monchois

Department of Oral Biology
Dental School
University of Newcastle
Newcastle upon Tyne NE2 4BW

1 INTRODUCTION

In recent years, a very substantial amount of information has been gathered on the structure-function relationships of enzymes which degrade carbohydrate and there is now increasing interest in enzymes capable of synthesising polymeric carbohydrates, such as glycosyltransferases utilising phosphorylated precursors and glycosidases acting 'in reverse' as glycosynthases. However, the most substantial long-established industrial process in which a polysaccharide is produced is the synthesis of dextran from sucrose by the extracellular dextransucrase enzyme of the bacterium *Leuconostoc mesenteroides* B512F. Commercial dextran is an α1,6-linked homopolymer of glucose but it is known that different species of *L. mesenteroides* can produce glucans other than dextran which differ in their content of different types of branch. Since these glucans differ in their properties, there is interest in identifying the mechanisms by which the specificity of the glucansucrases is determined so that defined glucans can be produced. Furthermore, the `acceptor reaction' in which glucose is transferred not to a growing glucan chain but to an alternate acceptor such as maltose yields novel gluco-oligosaccharide products with commercial use.[1]

The other major group of organisms producing glucansucrases is the oral streptococci that colonise the tooth surface and lead to the accumulation of dental plaque and the process of dental caries. The importance of these enzymes in disease has stimulated much research into their properties and the search for means of inhibiting their activity either by chemical agents or via the immune response.[2]

Glucansucrases are classified as glucosyltransferases (EC. 2.4.1.5) and the streptococcal enzymes have generally been described in the literature as glucosyltransferases (GTF) or glucan synthases while the *Leuconostoc* enzyme is called dextransucrase (DSR), a term that is not appropriate when the product is not dextran. In this paper, we use the term glucansucrase to include all these enzymes. The amylosucrase produced by some species of *Neisseria* is another glucansucrase but since it does not share the structural features of *Leuconostoc* and streptococcal glucansucrases[3] it will not be discussed further here.

2 PROPERTIES OF GLUCANSUCRASES

Glucansucrases transfer glucose from sucrose to an acceptor molecule with the release of free fructose. The energy comes from scission of the high-energy bond in sucrose and it should be noted that no phosphorylated intermediates are involved. The principal acceptor is the growing glucan chain but glucose can also be transferred from the covalent glucosyl-enzyme intermediate to water (the net result being hydrolysis of sucrose) or to a small molecule acceptor. With sucrose as substrate, the acceptor present is fructose and so leucrose (5-O-α-D-glucopyranosyl-D-fructose) is formed. Maltose is also generally a good acceptor but different glucansucrases differ in their ability to utilise other acceptors. The glucans may contain α1-3, α1-6, α1-4 or α1-6-glycosyl linkages and different glucansucrases differ in the type of linkage, the length of chain and the degree of branching that they introduce. A single enzyme may form more than one type of linkage and the challenge is to identify those structural features that determine this specificity. In addition, the mechanisms of the transfer reaction is not yet understood and in particular there is the puzzle of how glucose is transferred to the reducing end of the glucan chain but to the non-reducing end of alternative acceptors. Robyt has proposed a 'two-site insertion' mechanism[4] but so far mutagenesis experiments have failed to find support for this model.

2.1 Structure of glucansucrases

Glucansucrases have posed considerable problems of biochemical purification, as frequently more than one enzyme is produced by an individual species and post-translational modification or degradation yields multiple forms of each. However, considerable advances have been made with the development for procedures for cloning and expressing individual genes in heterologous hosts. The nucleotide sequences of 14 streptococcal *gtf* genes are now known, along with 3 *dsr* genes from *Leuconostoc*.

All glucansucrases possess a common pattern of organisation; they are of high molecular weight, around 160 kDa, and have a signal peptide followed by a variable domain, a highly conserved core region of about 900 amino acids including the catalytic domain and a C-terminal glucan-binding domain covering about 400 amino acids.[2]

2.2 The variable domain

The variable domains of glucansucrases vary from 117 - 232 amino acids and multiple alignment of sequences has revealed that there is no conservation of identical or functionally-homologue residues. Furthermore, structural prediction with the PREDATOR programme[5] shows no common pattern of structural organisation. Even within those species that have multiple glucansucrases, each enzyme has a distinctive variable domain. The possibility that this hypervariability was a mechanism of immune evasion had been proposed for the oral streptococci but the fact that a similar pattern is seen in *Leuconostoc* makes this interpretation unlikely. The variable domain is not essential for glucansucrase function as functional truncated genes can be constructed which still express active enzyme so its purpose remains unknown. It is possible that the variable domain is a preprotein that is in some way involved in secretion, folding or localisation of the mature enzyme.

2.3 The catalytic domain

The central domain is c. 900 amino acids in length and strongly conserved, with c. 50% of amino acids being identical in all glucansucrases sequenced. Its catalytic function was first defined by amino- and carboxy-terminal deletions but has subsequently been shown to contain a number of residues that contribute to the enzyme mechanism, including the catalytic triad of carboxylic acids involved in the active site.[6,7,8] These residues (D415, D418 and E454 in GTF-I, Fig. 1) correspond to catalytic nucleophiles in other glycoside hydrolases of the amylase superfamily and indeed nearly all the motifs characteristic of amylases can be identified in glucansucrases[9] and for this reason they have been placed together in Family 70 on the CAZY database. The motifs, however, do not occur in the same order as in amylases but are cyclically permuted so that the components corresponding to the $(\beta/\alpha)_8$ barrel structure of amylases are predicted to occur in an order beginning with helix-3 and ending with β-sheet 3 (Fig. 1). It seems probable then, that glucansucrases contain a $(\beta/\alpha)_8$ barrel but multiple alignment by the ClustalW programme combined with secondary structure-prediction using the PREDATOR program shows that the region upstream of the proposed barrel is also highly structured, with alternating α-helix and β-sheets.

2.4 The glucan binding domain

The last one-third of the glucansucrases consists of a series of tandem repeats[10]. These differ in number and arrangement between different glucansucrases but all are characterised by conserved aromatic amino acids and glycine.[11] The domain is responsible for binding to dextran but until its full range of binding activity is defined it may be more appropriate to consider it as a carbohydrate recognition module since it can be expressed as an independent peptide and a closely-related glucan binding protein (GBP) is found in *S. mutans*.[12] In addition, similar repeat units are found in proteins of several other bacteria that bind to carbohydrate ligands other than glucans.[13]

3 GLUCOSYLTRANSFERASE I OF *STREPTOCOCCUS DOWNEI*

Streptococcus downei is one of the mutans group of oral streptococci. It expresses 4 GTF, including GTF-I which produces a water-insoluble glucan containing α1-3-glucosyl linkages.[10,14] We have recently described the construction of a series of truncated derivatives of GTF-I in which the signal peptide and variable domain are replaced by a $(His)_6$ tag fused to the start of the catalytic domain.[15] Expression of these constructs in *E. coli* followed by affinity chromatography on Nickel-Nitriloacetic acid agarose resin provides a rapid and efficient method for producing pure enzyme and we have demonstrated that the core domain of GTF-I alone (GTF-Ic) is fully capable of synthesising α 1-3 linked glucan. Sequential truncations of the glucan-binding C-terminal domain resulted in a slight loss of activity (Fig. 1) but this effect is

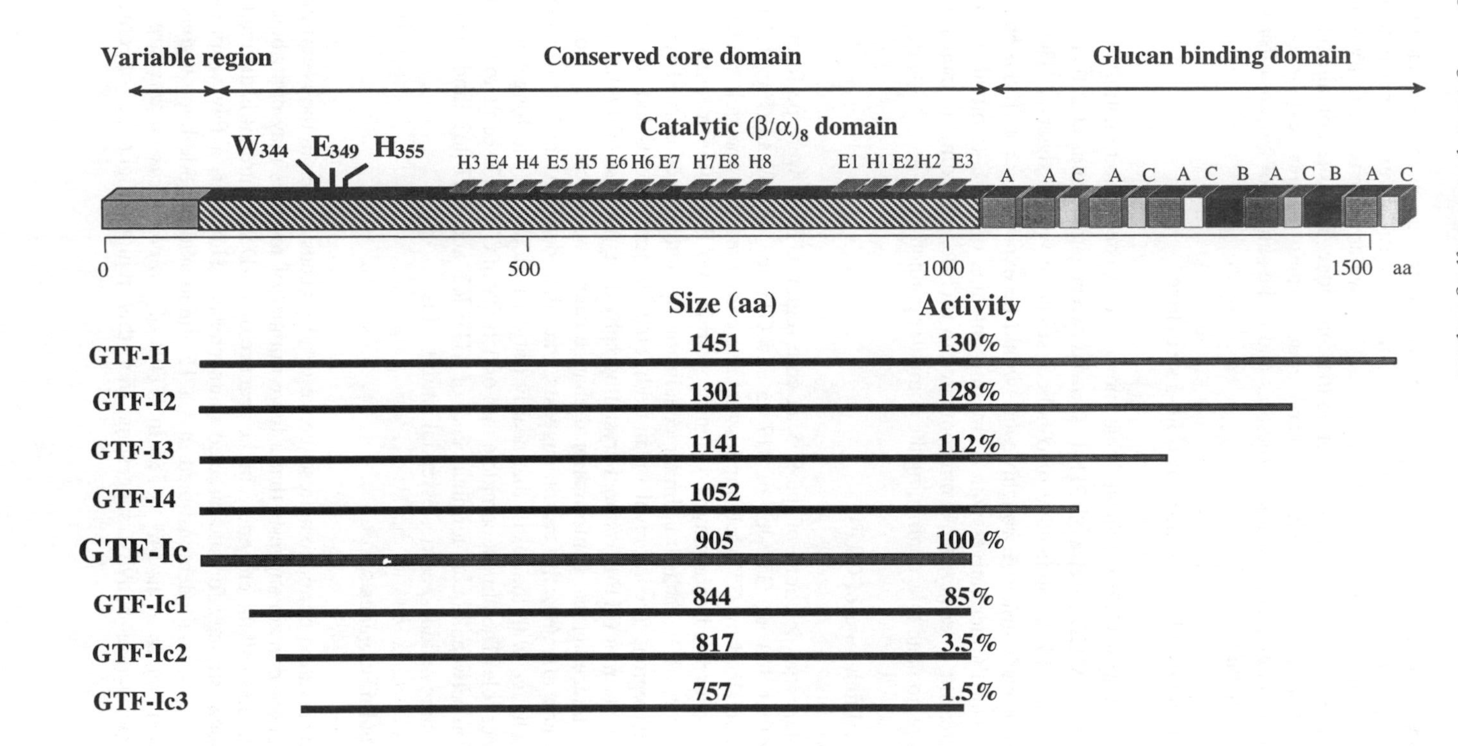

Figure 1. *Schematic representation of the structure of GTF-I and truncated derivatives.*

considerably less marked than has been reported for other glucansucrases that make soluble dextran products. The glucan binding domain does, however, appear to be required for the stimulatory effect of added Dextran T10, which increases activity of the full-length enzyme 1.5 fold but has no effect on the core alone. The core domain GTF-Ic thus represents a suitable model to carry out further studies exploring the structure/function relationships of glucansucrases and is the subject of current efforts to obtain a crystal structure.

3.1 Circular dichroism investigation of GTF-I structure

Computer prediction of GTF-I indicated that the core consisted of alternating α-helix and β-sheets. Far-UV CD spectra of GTF-Ic showed two negative bands at 210 and 222 nm with a peak at 215 nm indicative of α-helical structure with β sheets. The data indicated 30% of each structure, roughly what would be expected for a $(\beta/\alpha)_8$ barrel. The CD spectra of the derivatives with increasing amounts of the C-terminal domain showed an increasing intensity at 215 nm compared to GTF-Ic, indicating an increase in β-sheet in this region and thus confirming the computer prediction.

3.2 N-Terminal deletions of GTF-I

Successive N-terminal deletions of the 905 aa core region GTF-Ic were performed to obtain GTF-Ic1 of 844 aa, GTF-Ic2 of 817 aa and GTF-Ic3 of 757 aa (Fig. 1). The strong activity retained by GTF-Ic1 implies that the conserved tyrosine-rich stretch at the start of the conserved core plays no significant role into the action mechanism of GTF-I. Tsumori et al.[16] reached a similar conclusion by mutating conserved tyrosine residues in this region. The removal of an additional 27 amino acids led to almost complete inactivation of GTF-I. Without crystallography data, the possibility that such a small deletion induces global conformational change cannot be rejected but, as overall stability was not affected, the result suggests that this deletion leads to a local conformation change with the result that nearby critical residues are no longer able to play their proper role. The almost complete loss of activity in GTF-Ic3 seems to indicate that the region between the N-terminal end of GTF-Ic2 and GTF-Ic3, and rich in conserved aromatic residues, carries essential amino acids.

3.3 N-Terminal mutagenesis

The strong conservation of amino acid sequence in all glucansucrases in the N-terminal region of the core domain suggests that a large number of residues may contribute to functions common to all the enzymes. In the absence of a 3-D structure, it is difficult to define a rational strategy for site-directed mutagenesis. However, a few years ago, Dertzbaugh and Macrina[17] demonstrated that a 15-amino-acid peptide homologous to the region extending from aa 343 to 356 in GTFB of *Streptococcus mutans* strongly inhibited glucansucrase activity suggesting that this region might be of crucial importance. Of the 15-mer only 3 residues are conserved in all glucansucrases so, in order to investigate the role of this region, we carried out site directed mutagenesis of the homologous conserved residues in GTF-I, Trp-344, Glu-349 and His-355. These lie 100 amino acids upstream of the beginning of the putative $(\beta/\alpha)_8$ barrel domain containing the catalytic site. The levels of expression level of GTF-Ic and the three mutants were similar in *E. coli* and after purification with the N-terminal $(His)_6$ tag, no difference in stability was observed.

(...variable) SNIPSDLAKMSNVKQV**DGKYYY**YDQD**G**NVKKNFAVSVGEKIY**YF**DET**G**
AYKDTSKVEADKSGSDISKEETTFAAN**N**RAYSTSAENFEA**ID**NYL**TA**DS**WYRP**KS**IL**KD
GKT**W**TE**S**SK**D**DF**RPLL**MA**WWP**DTET**K**RN**Y**V**NY**MNKVVGIDKT**Y**TAETSQAD**L**TA**A**AELV
QAR**IE**Q**KI**TTEQN**T**K**WLR**EAISA**FV**KTQPQ**W**NGES**E**KPYDD**H**L**Q**N**GA**LK**F**D**N**QSDL**T**PD
TQ**S**N**YR**L**L**N**RTP**TNQTGSLDSRFTYNANDPLG**GYELLLANDVDNS**NP(...helix3)

Figure 2. *Sequence of the N-terminal part of the core domain of GTF-I. Amino acids conserved in all glucansucrases are shown in bold and the peptide identified by Dertzbaugh and Macrina*[17] *underlined.*

3.4 Effect of mutations on activity

Activity assays carried out with GTF-Ic and the variants showed that, in the presence of sucrose, all three mutations greatly affected the activity of the enzyme. With a single change to leucine at the Trp-344 position, only 3% of the initial activity was retained. Mutants carrying E349L and H355V changes retained 30 and 15% of the initial activity respectively. These mutations affected both synthesis of glucan and synthesis of oligosaccharide with maltose as acceptor to a similar extent, showing that these residues are important for both reactions.

3.5 Effect of mutations on mutan synthesis reaction

In order to determine whether the reduction of activity was due to changes in sucrose affinity for sucrose or in the turnover rate of the reaction, kinetic parameters were determined for wild type and mutant GTF-I. In the mutant with W344L substitution, the K_M value did not significantly change but the *kcat* value decreased nearly 25-fold. The strong effect of the W344L mutation on the *kcat* value suggests that Trp-344 may be involved into the turnover of the two reactions by stabilising the binding between sugar units and GTF-I.

The E349L mutation resulted in a change of the same magnitude in the two kinetic parameters (a 2.5-fold decrease for *kcat* and a 2-fold increase for K_M). Finally, the H355V mutant exhibited a slightly increased K_M value but decrease of 6.5-fold for the *kcat*. These results show that the reduction in activity is not primarily a consequence of reduced binding of the substrate sucrose.

None of the mutations had any influence on the nature of the glucan produced, which in all cases was over 95% α 1-3-linked but it was possible that the reduction in yield was due to an alteration in the destination of the glucose resulting from the splitting of sucrose. Therefore, to investigate how the mutations influenced the transfer ability of glucosyl moieties, yields of mutan production, hydrolysis and leucrose were assayed (Fig. 3). Replacement of the invariant residues Trp-344, Glu-349 and His-355 resulted in no change of the rate of hydrolysis, which stayed around 10%. The two mutants W344L and E349L exhibited similar yields of mutan and leucrose synthesis as the wild-type enzyme. In contrast, a significant change concerning these yields was observed with the H355V mutation. The yield of mutan increased from 62% to 75%, whereas the yield of leucrose decreased to 18% (Fig 3). His-355 of GTF-I thus appears in some way to be involved in determining the destination of the glucosyl residues coming from sucrose breakdown.

3.6 Effect of mutations on acceptor reaction in the presence of maltose

Maltose has been identified as a good acceptor for *Leuconostoc* glucansucrases and is the most commonly used acceptor in studies focused on oligosaccharide synthesis reaction. For GTF-Ic, maltose also appeared to be a good acceptor and in the presence of equimolar amounts of sucrose and maltose, the yield of mutan synthesis was reduced to about 10% and a number of oligosaccharides synthesised (Table 1). These can tentatively be identified as being mainly short linear gluco-oligosaccharides (GOS) ranging from DP3 (panose) to DP5 with minor components which may be non-linear glucooligosaccharides (nl-GOS) most probably containing α(1-3) glycosidic bonds, since these are the predominant type of linkage made by GTF-I.

The yield of oligosaccharides produced by the mutated variants was similar to that obtained with wild-type GTF-Ic enzyme and the two mutations W344L and E349L had no effect on the distribution of the oligosaccharides produced. In contrast, the H355V mutation influenced the distribution of the oligosaccharides and especially the ratio between nl-GOS and GOS: the yield of nl-GOS increased 2-fold, and the distribution between the three species of nl-GOS was also altered from the distribution obtained with other mutation or the wild type enzyme.

3.7 Effect of mutation on acceptor reaction in the presence of various types of acceptor

The effect of the mutations on the reaction with a range of different acceptor molecules was explored. In the presence of maltose, nigerose or isomaltose, the yield appeared not to be affected by the mutations and the variants exhibited a similar yield to that obtained with GTF-Ic. Reactions carried out using maltotriose and maltotetraose as acceptor resulted in similar yields for GTF-Ic and when mutations affected Trp-344 or Glu-349 but with the H355L mutation, the yield was significantly reduced (Table 1). This suggests that His-355 may play a role in a binding subsite necessary for oligosaccharide and glucan elongation.

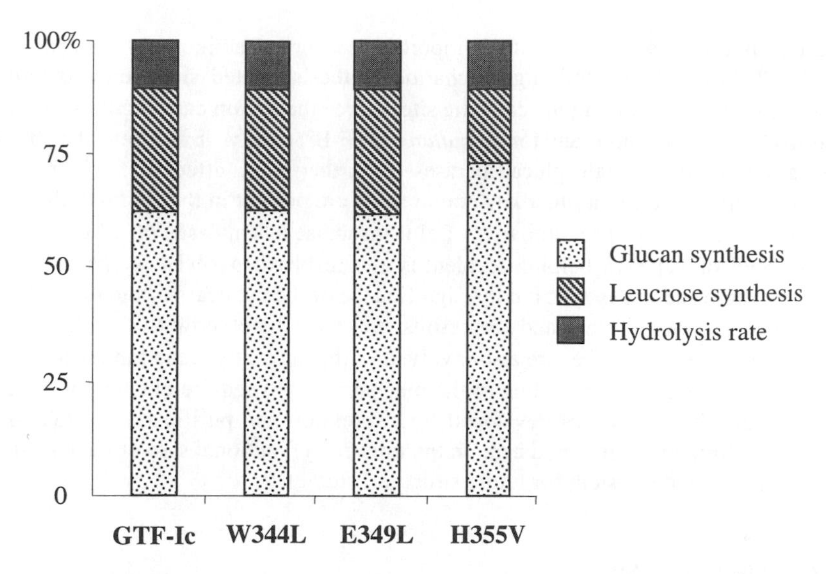

Figure 3. *Destination of the glucose moiety of sucrose from reaction of sucrose with GTFIc and mutated derivatives.*

	maltose	maltotriose	maltotetraose
GTFIc	92	69	34
W344L	93	66	33
E349L	93	68	33
H355V	96	60	20

Table 1. *Percentage of glucose derived from sucrose incorporated into oligosaccharides in the presence of different acceptors.*

4 CONCLUSION

The present investigation has identified important conserved amino acids in the core region of GTF-I for polymer and oligosaccharide synthesis located outside the putative $(\beta/\alpha)_8$ barrel domain containing the catalytic site. Since the region carrying these amino acids seems also to be important for *S. mutans* GTF-B activity, it is likely that these residues are required for all glucansucrases. Furthermore, effects of N-terminal deletions strongly suggest that other key amino acids are present in the first one-third of the core region of GTF-I. The similarity of glucansucrase to amylases in $(\beta/\alpha)_8$ barrel region has been of value in helping to identify the residues involved in the catalytic reaction site where sucrose is split but the significance of the cyclical permutation of the barrel elements is not yet clear and the results presented here show that parts of the enzyme outside the $(\beta/\alpha)_8$ barrel are also involved in the distinctive reaction mechanism of glucansucrases. Full understanding of the mechanism will require elucidation of the 3D-structure and the procedures developed for expression and purification of GTF-Ic, together with information presented here on the minimum functional size of the enzyme, provide a suitable model system for future structural studies.

5 ACKNOWLEDGEMENTS

This work was supported by Wellcome Trust grant 049554 and the European project BIOTECH CT98-0022 Alpha Glucan Active Designer Enzymes (AGADE).

6 REFERENCES

1. Z. Djouzi, Z., C. Andrieux, V. Pelenc, S. Somarriba, F. Popot, F. Paul, P. Monsan, and O. Szylit, 1995, *J. Appl. Bacteriol.*, 1995, **79**, 117.
2. S.M. Colby SM and R.R.B. Russell, *J. Appl. Microbiol. Symp.* 1997, **83,** 80S.
3. G.P. Potocki de Montalk, M. Remaud Simeon, R.M. Willemot, V. Planchot and P. Monsan, *J. Bacteriol.*, 1999, **181**, 375.
4. J. Robyt, 'Carbohydrate Bioengineering', Elsevier, Amsterdam, 1995, p.295.
5. D. Frishman and P. Argos, *Proteins* 1997, **27**, 239.
6. G. Mooser, 'The Enzymes', Academic Press, 1992, p. 187.
7. V.M. Monchois, M. Remaud Simeon, R. R. B. Russell, P. Monsan, and R. M. Willemot *Appl. Microbiol. Biotechnol.*, 1997, **48**, 465.
8. K.S. Devulapalle, S.D. Goodman, Q. Gao, A. Hemsley and G. Mooser, *Protein Sci.*, 1997, **6**, 2489.
9. E.A. MacGregor, H.M. Jespersen and B. Svensson B. *FEBS Lett.*, 1996, **378**, 263.
10. J.J. Ferretti, M.L. Gilpin and R.R.B. Russell, *J Bacteriol.*, 1987, **169**, 4271.
11. B.W. Wren, *Mol. Microbiol.*, 1991,5, 797.

12. J.A. Banas, R. R. B. Russell and J.J. Ferretti, *Infect. Immun.* 1990, **58**, 667.
13. P.M. Giffard and N.A. Jacques NA. *J. Dent. Res.* 1994, 73, 1133.
14. S.M. Colby, R. E. McLaughlin, J. J. Ferretti, and R. R. B. Russell. *Oral Microbiol. Immunol.*, 1999, 14, 27.
15. V. Monchois, M.A. Arguello-Morales and R.R.B. Russell, *J. Bacteriol.,* 1999, **181**, 2290.
16. H. Tsumori, T. Minami, and H. K. Kuramitsu, *J. Bacteriol.*, 1997, **179**, 3391
17. M.T. Dertzbaugh and F. L. Macrina, *Infect. Immun.*, 1990, **58**, 1509.

CONSEQUENCES OF REMOVING THE CATALYTIC NUCLEOPHILE IN *BACILLUS* 1,3-1,4-β-GLUCANASES: SYNTHASE ACTIVITY AND NOVEL PROPERTIES BY CHEMICAL RESCUE

Antoni Planas*, Josep-Lluís Viladot, Magda Faijes

Laboratory of Biochemistry, Institut Químic de Sarrià, Universitat Ramon Llull
Via Augusta, 390, 08017 Barcelona, Spain. E-mail: aplan@iqs.url.es

1 INTRODUCTION

The enzymatic depolymerization of 1,3-1,4-β-glucan –the major polysaccharide of endosperm cell wall of cereals– is an early event in the germination process, catalysed by endogenous glycosidases with three different specificities: 1,4-β-D-glucan 4-glucanohydrolase (EC 3.2.1.4), 1,3-β-D-glucan 3-glucanohydrolase (EC 3.2.1.39), and 1,3-1,4-β-D-glucan 4-glucanohydrolase (EC 3.2.1.73). The latter is the most efficient and has a strict specificity for hydrolysis of β-1,4 glycosidic bonds in 3-*O*-substituted glucopyranose units[1,2].

1,3-1,4-β-D-glucan 4-glucanohydrolases (1,3-1,4-β-glucanases) are also found in bacteria. Whereas the plant enzymes belong to family 17 of glycosyl hydrolases[3,4], the bacterial enzymes are members of family 16. A number of genes from different *Bacillus* species, *Fibrobacter succinogenes*, *Ruminoccocus flavofaciens*, *Clostridium thermocellum*, *Streptococcus bovis*, and *Orpinomyces*[5-7] have been cloned and the enzymes characterised. They are monodomain proteins with molecular masses of 25-30 kDa, active in a wide pH range, basic pI (8-9), and quite thermostable compared to the plant isozymes.

Important progress on the structure-function relationships in *Bacillus* 1,3-1,4-β-glucanase has been recently gained through mutational, crystallographic and kinetic studies. The development of series of tailor-made oligosaccharides of general structure [βGlc*p*(1→4]$_n$βGlc*p*(1→3)βGlc*p*–X, X= aryl, –[(1→4)βGlc*p*]$_m$–OMe as specific substrates[8-10] has allowed detailed kinetic characterisation of the *Bacillus licheniformis* enzyme[10-13]. It is a retaining *endo* glycosidase, with a binding site cleft composed of 6 subsites (four on the non-reducing end from the scissile glycosidic bond, and two on the reducing end). The crystal structures of the wild-type *B.licheniformis*[14] and *B.macerans*[15] enzymes, as well as several hydrids and circularly permuted proteins[16] characterised the jellyroll β-sandwich fold of family 16 glycosyl hydrolases, and guided the mutational analysis of substrate specificity and stability[17-19].

The mechanism of a retaining glycosidase involves a double-displacement reaction assisted by general acid-base catalysis[20,21] (Figure 1). In the first step (*glycosylation*), the amino acid residue acting as a general acid protonates the glycosidic oxygen with concomitant C-O breaking of the scissile glycosidic bond, while the deprotonated carboxylate functioning as a nucleophile stabilises the oxocarbenium intermediate through electrostatic interaction or by forming a covalent glycosyl-enzyme intermediate. The covalent nature of the intermediate may be general for most of the retaining glycosidases, except for lysozyme and glycosidases acting by substrate-assisted catalysis such as chiti-

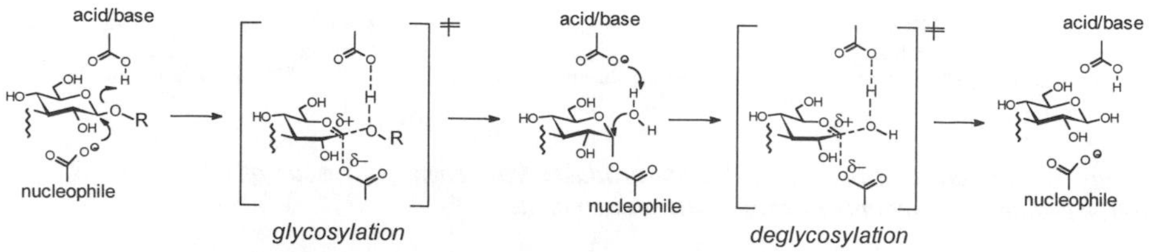

Figure 1 *Retaining glycosidase mechanism, involving a double-displacement reaction.*

nases. It is supported by a substantial body of evidence from kinetic isotope effects[21] and trapping experiments using substrate analogues, mainly 2-deoxy-2-fluoro glycosides[22] for which the covalent intermediate is detected by mass spectrometry and x-ray crystallography[23]. The second step of the mechanism (*deglycosylation*) involves the attach of a water molecule assisted by the conjugate base of the general acid to render the free sugar with overall retention of the anomeric configuration. For the same residue to act as a general acid in the first step and general base in the second, a significant shift in the pK_a of the carboxyl side chain must occur during the enzyme cycle, as demonstrated by NMR titration (*i.e. B.circulans* xylanase[24]).

For the 1,3-1-4-β-glucanase from *B.licheniformis*, the first *glycosylation* step is rate-limiting in the hydrolysis of aryl β-glycoside substrates[10], even for the highly activated 2,4-dinitrophenyl trisaccharide **1** (Figure 2), and kinetic pK_a values are 5.5 for the nucleophile residue and 7.0 for the general acid-base in the free enzyme, the latter shifted up to 7.3 in the E·S complex with an aryl trisaccharide and to 8.5 with the natural barley β-glucan substrate[12].

We report here further mechanistic studies on the hydrolase activity of the enzyme through kinetic analysis of the chemical rescue of inactive mutants at essential catalytic residues, and on the synthase activity of the enzyme using a nucleophile-less mutant.

2 HYDROLASE ACTIVITY

2.1. Chemical Rescue of the General Acid-Base Mutant

The catalytic residues of *Bacillus* 1,3-1,4-β-glucanases have been identified by mutational analysis[25] and X-ray crystallography[15]. For the *Bacillus licheniformis* enzyme, Glu138 is the general acid-base residue, and Glu134 the catalytic nucleophile, their functional role being assessed by a chemical rescue methodology of inactive alanine mutants on each essential residue[26]. Azide as exogenous nucleophile reactivated the mutants in a concentration-dependent manner using an activated 2,4-dinitrophenyl glycoside substrate (**1**, Figure 2). The E138A mutant yields the β-glycosyl azide product (**2**) arising from nucleophilic attack of azide to the glycosyl-enzyme intermediate, thus proving that Glu138 is the general acid-base residue. By contrast, azide reactivates the E134A mutant through a single inverting displacement to give the α-glycosyl azide product (**3**), consistent with Glu134 being the catalytic nucleophile.

The kinetics of E138A reactivation by added sodium azide are atypical among retaining glycosidases. Enzymes for which Ala or Gly mutants at the general acid-base residue have been kinetically studied in the presence of exogenous azide (*i.e. C.fimi* exo-glucanase/xylanase[27], *A. faecalis* β-glucosidase[28], *E. coli* β-galactosidase[29]) have shown that the azide anion participates as a nucleophile in the *deglycosylation* step by attacking the glycosyl-enzyme intermediate. In these systems, the *deglycosylation* step is rate-

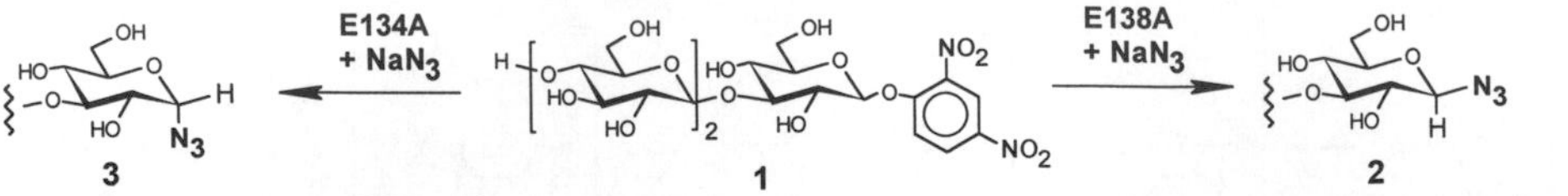

Figure 2 *Structure of the glycosyl azide products from chemical rescue of E138A and E134A 1,3-1,4-β-glucanase mutants by exogenous sodium azide.*

limiting for activated aryl glycosides. For excellent substrates such as 2,4-dinitrophenyl glycosides, k_{cat}/K_M values are not significantly slowed by the mutation, whereas for substrates requiring some general acid catalysis there is a significant rate reduction. When sodium azide was added as a nucleophile and 2,4-dinitrophenyl glycosides were used as substrates, both k_{cat} and K_M increased with azide concentration. In contrast, k_{cat}/K_M remained virtually constant. These results were interpreted as a firm evidence that the effect of azide was on the *deglycosylation* step: (a) glycosylation is not significantly compromised for the mutant, and (b) the parameter k_{cat}/K_M reflects the first irreversible step, formation of the glycosyl-enzyme intermediate, and this step is not affected by an exogenous nucleophile.

In our case for the *B. licheniformis* 1,3-1,4-β-glucanase, the results are quite different. The *glycosylation* step is rate-limiting with the 2,4-dinitrophenyl glycoside substrate, even for the E138A mutant, since a 2000-fold reduction in k_{cat}/K_M with the activated substrate is obtained relative to the wild-type enzyme. Moreover, K_M does not change significantly, whereas a large drop would be expected if *deglycosylation* becomes rate-limiting. Upon chemical rescue of the E138A mutant by azide, k_{cat} increases, but K_M experiences a large reduction to the μM range (Figure 3A), so the k_{cat}/K_M value largely increases with azide concentration approaching the wild-type value. It seems to indicate that azide not only has an effect on the *deglycosylation* step leading to the β-glycosyl azide product, but also has a major effect on a preceding step.

Activation of E138A by sodium azide is pH-dependent (Figure 3B). If azide behaves as a nucleophile in the second (*deglycosylation*) step, the lack of the general acid-base residue is expected to give a pH-independent k_{cat} value in the basic region. However, a kinetic pK_a of 7.7 is obtained, value very close to that determined for the wild- type enzyme with the same substrate.

Analysis according to the kinetic model describing the double-displacement reaction,

$$E + S \underset{}{\overset{K_1}{\rightleftharpoons}} ES \xrightarrow{k_2} EP \xrightarrow{k_3} E + P$$

where $k_{cat} = k_2{\cdot}k_3/(k_2+k_3)$, $K_M = k_3{\cdot}(k_2+k_{-1})/k_1{\cdot}(k_3+k_2)$, and $k_{cat}/K_M = k_1{\cdot}k_2/(k_2+k_{-1})$, allows to conclude that

a) the deglycosylation step is partially rate-limiting for the mutant even with an activated substrate,
b) azide participates in both steps: it activates *deglycosylation* by acting as a nucleophile that does not require general base assistance to lead to the β-glycosyl azide product, and it also has a large effect on a previous step, probably on *glycosylation* by increasing k_2,
c) the large decrease of K_M with increasing azide concentration results from a larger activation on k_2 than on k_3 by nucleophilic attack, and

The actual mechanism of azide reactivation besides its role as nucleophile is not yet clear and requires further analysis to establish the effect of the azide anion or its conjugated acid, hydrazoic acid, on the *glycosylation* step.

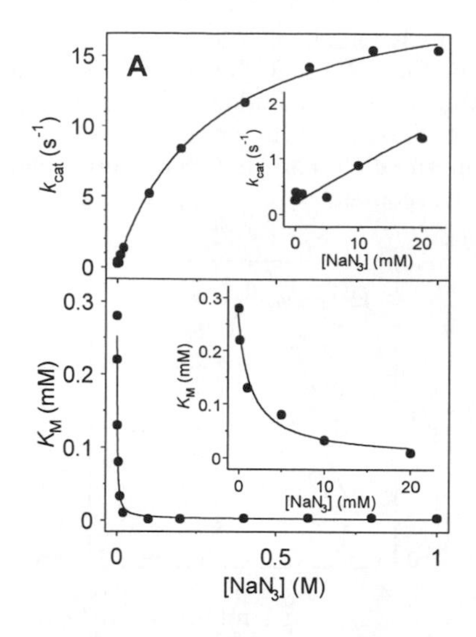

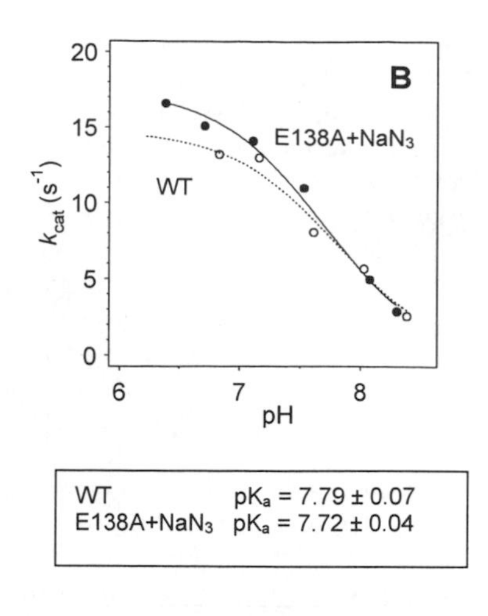

Figure 3 *Kinetics of the reactivation of E138A with substrate **1** in the presence of sodium azide. A) Dependence of the kinetic parameters on azide concentration at pH 7.2. B) pH dependence of k_{cat} at 1M azide. Conditions: citrate-phosphate buffer, 0.1 mM $CaCl_2$, I = 1.1 M, 30°C.*

Formate instead of azide is also able to reactivate some mutant glycosidases at the general acid-base residue with activated substrates for which *deglycosylation* is rate determining (*i.e. A. faecalis* β-glucosidase[28]). Formate probably acts as a general base in the second step leading to the hydrolysis product. However, formate has no activation effect on the 1,3-1,4-β-glucanase E138A mutant. Whether formate is not able to act as a general base or nucleophile in the *deglycosylation* step, or unable to have an effect on k_2 as azide does, remains to be evaluated.

The different behaviour of the E138A mutant 1,3-1,4-β-glucanase as compared to other retaining glycosidases for which *deglycosylation* in rate determining for activated aryl glycosides, indicate that other factors operate on the *glycosylation* step leading to the glycosyl-enzyme intermediate, since this step is rate-limiting for the reaction with activated substrates not requiring general acid catalysis either on the wild type or on the mutant.

2.2 Reactivation of the Nucleophile Mutant by Formate: Detection of a Covalent Glycosyl-Enzyme Mimic

Removal of the catalytic nucleophile by mutation to alanine (E134A) produces an inactive glycosidases whose activity is rescued by azide to give an α-glycosyl azide product. The inverting reaction is the result of a single displacement by azide which is able to bind in the cavity created by removal of the glutamate side chain acting as the catalytic nucleophile in the wild-type reaction.

Formate resembles more closely the structure of the excised glutamate in the E134A mutant. When added as an exogenous nucleophile to the reaction with an activated aryl glycoside, it restores the hydrolytic enzyme activity in a concentration-dependent manner. The first hydrolysis product obtained is the β-saccharide, thus resulting in an overall retention reaction[26]. Whereas this situation is also observed in other glycosidases[28,30], suggesting that the formate anion may play the role of the removed nucleophile, no direct evidence supports the participation of formate as nucleophile.

By ^{1}H-NMR monitoring of the reactivation of the 1,3-1,4-β-glucanase E134A mutant with the 2,4-dinitrophenyl trisaccharide **1** in the presence of added sodium formate, a long-lived intermediate was detected, tentatively assigned to an α-glycosyl formate adduct[26]. We proof here the structure of this intermediate and its kinetic competence to undergo further hydrolysis and transglycosylation reactions.

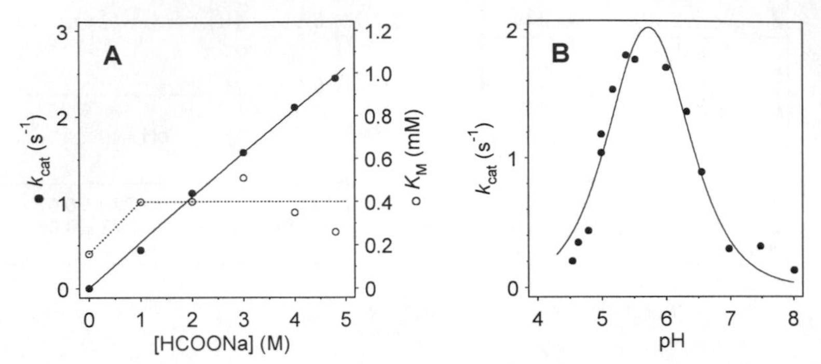

Figure 4 *Reactivation of E134A + substrate **1** by sodium formate. A) Dependence of the kinetic parameters with formate concentration at pH 7.2. B) pH dependence of k_{cat} in the presence of 2 M HCOONa. Conditions: citrate-phosphate buffer, 0.1 mM $CaCl_2$, I = 2.2 M, 30 ºC.*

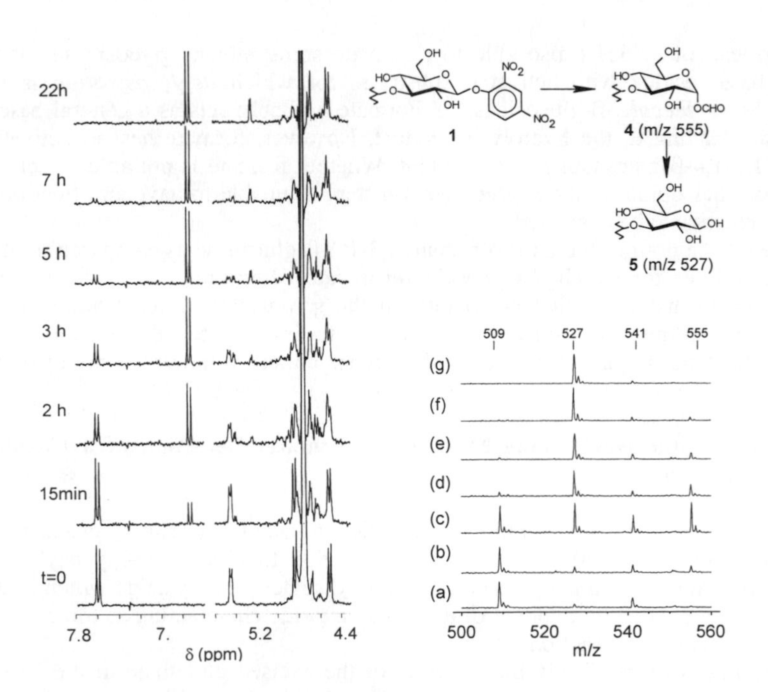

Figure 5 *Monitoring of the reaction of E134A with substrate **1** in the presence of 2 M sodium formate. A) ^{1}H-NMR in D_2O, pD 7.25, 25ºC. δ (ppm): 4.66 (H-1β, **5**), 5.23 (H-1α, **5**), 5.40 (H-1, **4**), 5.44 (H-1, **1**). B) MALDI MS. Reaction in H_2O, pH 7.2, 25ºC, a-g: 0, 1, 3, 9, 13, 24 and 48 h.*

a) UV monitoring.- Figure 4A summarises the kinetic parameters as a function of formate concentration when monitoring the release of the 2,4-dinitrophenol aglycon by UV spectrophotometry. At the highest formate concentration assayed, k_{cat} is increased 3000-fold relative to the formate-free reaction, and it is only 300-fold lower than the k_{cat} value for the wild-type enzyme with the same substrate. Since K_M does not change too much ($\leq$ 2-fold) the same large reactivation is seen on k_{cat}/K_M.
b) ^{1}H-NMR monitoring.- The transient intermediate accumulates in sufficient concentration to be detected by ^{1}H-NMR (Figure 5A). The new anomeric proton at δ 5.40 ppm and *J* 7.5 Hz is consistent with an α-glycosyl formate adduct. It is sufficiently stable to dissociate from the enzyme because it accumulates up to 2 mM concentration for a reaction with 10 mM initial substrate concentration and 11.4 μM E134A enzyme, with a half-life of 2-3 hours. The rate of formation and disappearance of this intermediate (to yield the β-saccharide) is enzyme-concentration dependent, proving that its hydrolysis is at least partially enzymatic.
c) MALDI-TOF MS monitoring.- To assess that the exogenous formate is incorporated in the intermediate, the reaction was monitored by MALDI mass spectrometry. Figure 5B shows the recorded spectra for the time course of the reaction of E134A (12 μM) with 10 mM substrate **1** in the presence of 2 M sodium formate in D_2O buffered at pD 7.25. At zero time the expected peak of the 2,4-dinitrophenyl trisaccharide substrate (**1**) at m/z 693 $[M+Na]^+$ is not observed, rather fragmentation peaks arising from photochemical reactions of the chromogenic aglycon due to the laser excitation appear at m/z 509 and 541. At increasing reaction time, a new peak at m/z 555 first increases and then decreases, whereas a peak at m/z 527, corresponding to the hydrolysis product, increases and is the only final reaction product. The 555 (m/z) peak is assigned to the glycosyl formate intermediate. When the reaction is carried out with ^{13}C-labeled formate, the transient peak appears at m/z 556, and when an equimolar mixture $H^{12}COONa + H^{13}COONa$ is used as exogenous nucleophile, two peaks of equal intensity at m/z 555 and 556 are observed.
d) Kinetic pK_a of enzyme-bound formate.- The pH dependence of k_{cat} for the reactivation reaction of E134A with the 2,4-dinitrophenyl trisaccharide **1** in the presence of sodium formate gives a bell-shaped curve with two pK_a's at 5.5 and 6.0 (Figure 4B). The kinetic pK_a in the acidic limb (5.5) is then assigned to the enzyme-bound formate in the E·S complex. This value is similar to that determined for the catalytic nucleophile in the wild-type reaction (5.5 in the free enzyme, < 5 in the E·S complex[12]), consistent with the function of the formate anion playing the role of the excised catalytic nucleophile.

In summary, the exogenous formate acts as a nucleophile in the enzyme-catalysed reaction. We here detect and identify for the first time an α-glycosyl formate adduct from an unmodified sugar that mimics the proposed glycosyl-enzyme intermediate in the reaction of retaining glycosidases, providing further evidence for the covalent nature of the intermediate on natural substrates, in addition to the trapping experiments with substrate analogues such as the 2-deoxy-2-fluoro glycosides where the 2-position is perturbed as compared to natural glycoside substrates.

3 SYNTHASE ACTIVITY

3.1 Wild-type enzyme: kinetically-controlled transglycosylation

The two main approaches to glycoside-catalysed synthesis of glycosidic linkages are direct reversal of hydrolysis (equilibrium-controlled synthesis), and trapping of the glycosyl-enzyme intermediate (kinetically-controlled synthesis). The latter strategy on retaining glycosidases has been widely exploited for the synthesis of target oligosaccharides, involving the use of aryl glycosides, glycosyl fluorides and oligosaccharides as glycosyl donors[31]. The approach depends on the more rapid trapping of an activated glycosyl-enzyme intermediate by the glycosyl acceptor than by water. Although under the right conditions glycoside formation is favoured kinetically, hydrolysis either of the intermediate or of the resulting product is favoured thermodynamically, and practical yields generally range from 20 to 40%, even with the help of organic cosolvents to reduce the extend of the hydrolytic reaction.

The wild-type 1,3-1,4-β-glucanase from *Bacillus licheniformis* has been fruitfully used for the synthesis of 1,3-1,4-β-D-glucooligosaccharides as novel substrates for kinetic and subsite mapping studies of bacterial and plant β-glucanases[32-34,13]. Yields were up to 40% with β-glycosyl fluoride donors when excess of acceptor and organic cosolvents were used, but no transglycosylation reactions were detected with aryl β-glycoside donors.

3.2 Nucleophile-less Mutant (E134A): Glucansynthase activity

In the kinetically-controlled approach, the actual glycosyl donor is the α-glycosyl-enzyme intermediate. Therefore, building up or providing a more efficient α-glycosyl donor may improve the transglycosylation reaction. In order for the enzyme to bind an activated α-donor, the catalytic nucleophile residue has to be removed. As shown in the formate reactivation, the nucleophile-less (E134A) mutant is able to bind an activated glycosyl derivative with α-configuration, thus mimicking the structure of the glycosyl-enzyme intermediate. Two different approaches towards the design of a more efficient glucansynthase are here preliminarily explored (Figure 6):

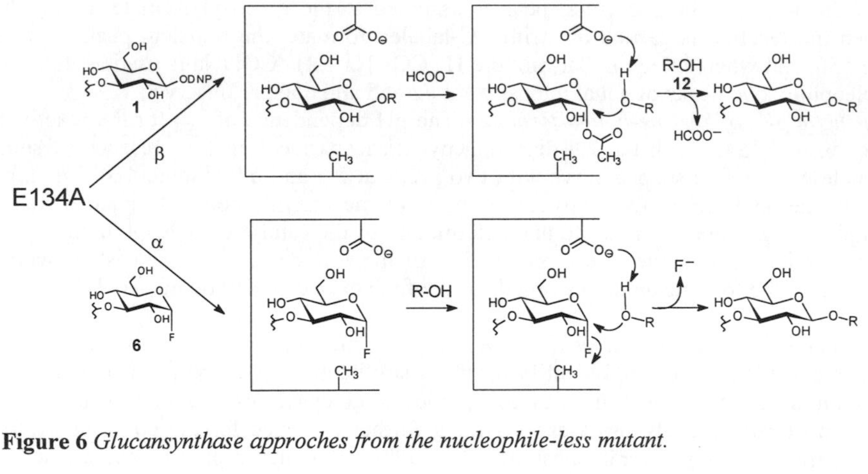

Figure 6 *Glucansynthase approches from the nucleophile-less mutant.*

3.2.1.In situ generation of the α-glycosyl donor. Incubation of the E134A mutant with the 2,4-dinitrophenyl β-glycoside **1** in the presence of sodium formate, followed by addition of an excess of acceptor (4-methylumbelliferyl β-cellobioside **12**, Figure 6, up), gives transglycosylation products of different degrees of polymerisation. The reaction progress was monitored by HPLC and MALDI-TOF spectrometry showing that a pentasaccharide was the major transglycosylation product, and higher oligosaccharides (hepta, nona, and undecasaccharides) were only detected as trace by-products (< 2%). However, because formate was used in large excess, the formed glycosides are subsequently hydrolysed by the formate-assisted hydrolase activity of E134A. The potential of this approach, *in situ* generation of an α-donor from an activated β-glycoside, requires further optimisation by modulating the amount of formate and the extend of the reaction to achieve high practical yields. Nevertheless, this is a promising approach that parallels the kinetically-controlled transglycosylation activity of the wild-type enzyme.

3.2.2. Direct binding of the α-glycosyl donor. Following the 'glycosynthase' approach first introduced by the Withers group on a β-glucosidase[35], the E134A 1,3-1,4-β-glucanase mutant showed that the methodology also applies to *endo* glycosidases[36]. An α-glycosyl fluoride mimics the structure of the glycosyl-enzyme intermediate upon binding to the nucleophile-less mutant, and is able to act as a glycosyl donor for transglycosylation reactions (Figure 6, down). The obvious advantage of this process with respect to the formate-assisted transglycosylation from a β-glycoside or the kinetic approach using the wild-type enzyme is that the condensation product cannot be hydrolysed by the mutant enzyme, and yields rise to 90%. Besides the practical interest of this methodology for enzyme-catalysed oligosaccharide synthesis, it provides a new system for the analysis of subsites selectivity and protein-carbohydrate interactions in *endo* enzymes, when combined with subsite mapping and inhibition studies on the hydrolytic activity of the wild-type and mutant enzymes.

Condensation between α-laminaribiosyl fluoride **6** and 4-methylumbelliferyl β-D-glucopyranoside **7** (Table 1) in a 1 to 5 acceptor excess gives a single trisaccharide product in 85% yield after preparative chromatographic purification. The synthase activity of E134A is regio and stereospecific for a newly-formed β-1,4 glycosidic bond, thus showing the same strict specificity as the hydrolase activity of the wild-type enzyme. Figure 7A plots the time course of the reaction at different enzyme concentrations, yields reaching 90% after 24 h with 4 μM E134A concentration at 35°C. The optimum pH under these conditions is 7-7.5 (Figure 7B), the pH profile showing a kinetic pK_a on the acidic limb (≈6.4). The residue Glu138, the general acid/base in the wild-type enzyme, may play the role of general base in the condensation reaction, selectively enhancing the nucleophilicity of the 4-OH group in the Glc*p* acceptor. A downward shift in its pK_a with respect to the wild-type is expected in the nucleophile-less mutant in agreement with the pK_a tuning of the general acid/base residue during the hydrolase catalytic cycle (see introduction). Therefore, the observed pH profile is consistent with a general base-catalysed condensation involving Glu138.

Kinetic parameters for the condensation of **6** with 4-methylumbelliferyl β-cellobioside (**12**) were determined from initial reaction rates at constant acceptor and varying donor concentrations (Figure 7C). No saturation is observed up to 5 mM, corresponding to a rather high K_M value, in agreement with the K_M of 17 mM for the wild type-catalysed hydrolysis of the homologous 4-methylumbelliferyl β-laminaribioside substrate[12]. The calculated k_{cat}/K_M value (donor) is 13.6 $M^{-1}s^{-1}$.

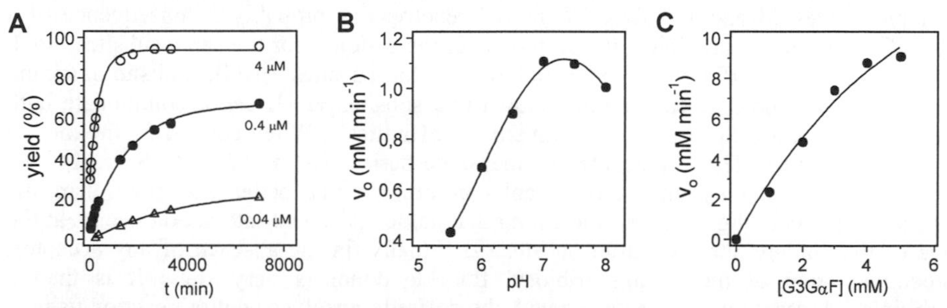

Figure 7 *Kinetics of transglycosylation catalysed by E134A 1,3-1,4-β-glucanase with the α-glycosyl fluoride donor* **6**. *50 mM maleate buffer, 0.1 mM* $CaCl_2$*, 35°C. A) Time course at different enzyme concentrations with acceptor* **7**, *pH 7. B) pH profile for the condensation* **6**+**7**. *C) Michaelis kinetics for the condensation* **6** + **12**, *[acceptor] = 7 mM, [Enz] = 3 μM, pH 7.0.*

Table 1 *Glucansynthase activity of the E134A 1,3-1,4-β-glucanase mutant. Relative initial rates of transglycosylation (R) with the α-laminaribiosyl fluoride donor (**6**) and different acceptors. 50 mM maleate buffer pH 7.0, 0.1 mM $CaCl_2$, [donor] = 3 mM, [acceptor] = 6 mM, [E134A] = 3 μM, 35°C. n.r.: no reaction.*

	Acceptor (monosaccharides)	R		Acceptor (disaccharides)	R
7	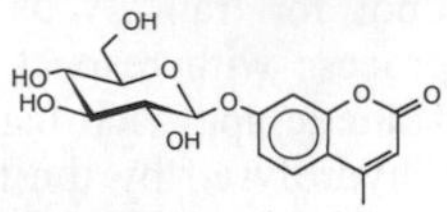	1.00	**12**	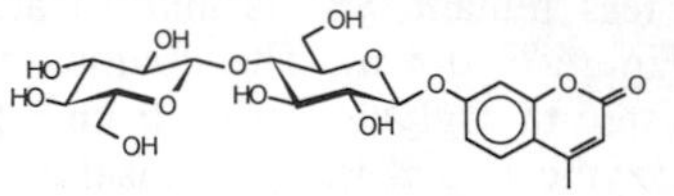	1.37
8	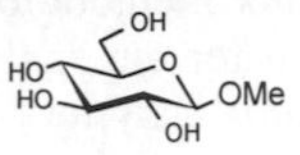	< 0.1	**13**	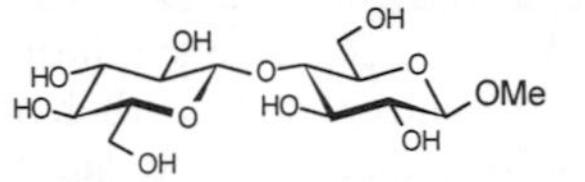	1.12
9	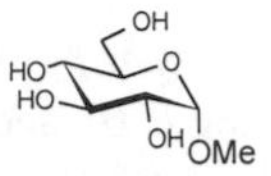	n.r.	**14**	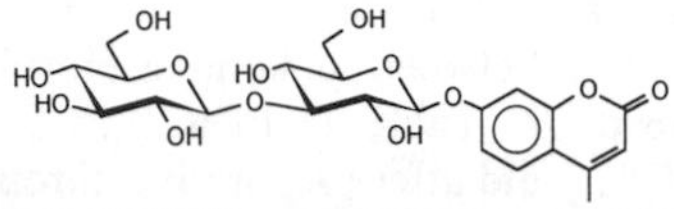	0.34
10	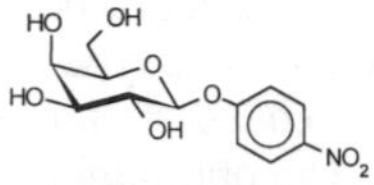	n.r.	**15**	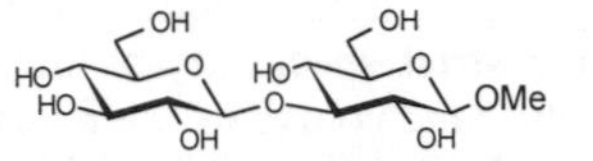	0.16
11	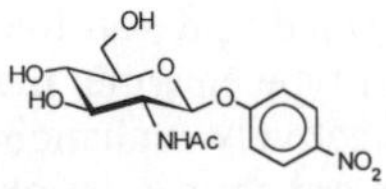	n.r.	**6**	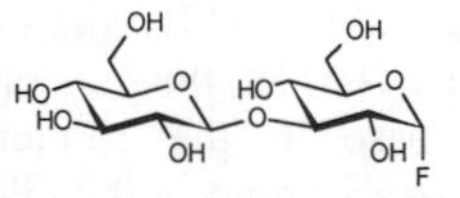	0.11

Table 1 summarises the selectivity for mono- and disaccharide acceptors using **6** as glycosyl donor. Initial rates of transglycosylation were determined by HPLC monitoring and results are expressed as relative initial rates to the 4-methylumbelliferyl β-D-glucopyranoside **7**. The methyl β-glucoside **8** is a poor acceptor whereas **7** reacts readily. Therefore subsite +I has low affinity and the acceptor must occupy subsite +II, as also seen from the high rates with disaccharides. The α-glucoside **9** does not react at all, neither sucrose nor maltose, indicating that an α-linkage is not accepted. As expected, the galactoside **10** and 2-acetamide-2-deoxy glucoside **11** are not glycosylated.

On disaccharides, the cellobiosides **12** and **13** are better acceptors than the laminaribiosides **14** and **15**. This differential reactivity is probably a consequence of a competition of the laminaribioside acceptor with the α-donor for the same subsites –II/-I, rather than higher affinity of the cellobiosides for subsites +I/+II. Subsite mapping analysis on the wild-type enzyme has shown that subsites +I/+II can accommodate both laminaribiosyl and cellobiosyl units with similar affinities[13]. When comparing the methyl and 4-methylumbelliferyl glycosides of the same series, **13** *vs.* **12** and **15** *vs.* **14**, the disaccharides containing an aromatic aglycon are slightly better acceptors than the corresponding methyl glycosides, indicating that some additional interaction occurs at the edge of the binding site cleft after subsite +II. Finally, in the absence of any acceptor, autocondensation of the α-laminaribiosyl fluoride donor is very slow; it is then a negligible side reaction not interfering with the desired transglycosylation reaction using a 2:1 excess of acceptor.

ACKNOWLEDGMENTS

This work was supported by Grant BIO97-0511-C02-02, CICYT, Madrid, Spain. J-LL. V. and M. F. acknowledge the pre-doctoral fellowships from Generalitat de Catalunya, and Instituto Danone, respectively.

REFERENCES

1. F.W. Parrish, A.S. Perlin and E.T. Reese, *Can. J. Chem.*, 1960, **38**, 2094
2. B.A. Stone and A.E. Clarke, 'Chemistry and Biology of (1→3)-β-glucans', La Trobe University Press, Australia, 1992
3. B. Henrissat and A. Bairoch, *Biochem. J.*, 1993, **293**, 781
4. G.J. Davis and H. Henrissat, *Structure*, 1995, **3**, 853
5. R. Borriss, *Current Topics in Mol. Genet.*, 1994, **2**, 163
6. M.S. Ekinci, S.I. McCrae, and H. J. Flint, *Appl. Environ. Microbiol.*, 1997, **63**, 3752
7. H. Chen, X.-L. Li, and L.G. Ljungdahl, *J. Bact.*, 1997, **179**, 6028
8. C. Malet, J.L. Viladot, A. Ochoa, B. Gállego, C. Brosa and A. Planas, *Carbohydr. Res.*, 1995, **274**, 285
9. C. Malet, J. Vallés, J. Bou and A. Planas, *J. Biotechnol.*, 1996, **48**, 209
10. A. Planas, O. Millet, J. Palasí, C. Pallarés, M. Abel, and J.Ll. Viladot, *Carbohr. Res.*, 1998, **310**, 53
11. C. Malet, J. Jiménez-Barbero, M. Bernabé, C. Brosa, and A. Planas, *Biochem. J.*, 1993, **296**, 753
12. C. Malet and A. Planas, *Biochemistry*, 1997, **36**, 13838
13. J.Ll. Viladot, H. Driguez, and A. Planas, in preparation
14. M. Hahn, J. Pons, A. Planas, E. Querol and U. Heinemann, *FEBS Letters*, 1995, **374**, 221
15. M. Hahn, O. Olsen, O. Politz, R. Borriss and U. Heinemann, *J. Biol. Chem.*, 1995, **7**, 3081
16. U. Heinemann, J. Aÿ, O. Gaiser, J.J. Müller and M.N. Ponnuswamy, *Biol. Chem.*, 1996, **377**, 447
17. O. Politz, O.Simon, O. Olsen, and R. Borriss, *Eur. J. Biochem.* 1993, **216**,829
18. J. Pons, E. Querol and A. Planas, *J. Biol. Chem.*, 1997, **272**, 13006
19. K. Piotukh, V. Serra, R. Borris, and A. Planas, *J. Biol. Chem.* submitted
20. D.E. Koshland, Jr., *Biol. Rev.* 1953, **28**, 416
21. M.L. Sinnott, *Chem. Rev.*, 1990, **90**, 1171
22. S.G. Withers, R.A.J. Warren, I.P. Street, K. Rupitz, J.B. >Kempton, and R. Aebersold, *J. Am. Chem. Soc.*, 1990, **112**, 5887
23. G.J. Davis, L. Mackenzie, A. Varrot, M. Dauter, A.M. Brzozowski, M. Schülein, and S.G. Withers, *Biochemistry*, 1998, **37**, 11707
24. L.P. McIntosh, G. Hand, P.E. Johnson, M.D. Joshi, M. Körner, L.A. Plesniak, L. Ziser, W.W. Wakarchuk, and S.G. Withers, *Biochemistry*, 1996, **35**, 9958
25. M. Juncosa, J. Pons, T. Dot, E. Querol and A. Planas, *J. Biol. Chem.*, 1994, **269**, 14530
26. J.Ll. Viladot, E. de Ramón, O. Durany, and A. Planas, *Biochemistry*, 1998, **32**, 11332
27. Q. Wang, D. Trimbur, R. Graham, R.A.J. Warren, and S.G. Withers, *Biochemistry*, 1995, **34**, 14554
28. A.M. MacLeod, D. Tull, K. Rupitz, A.R.J. Warren, and S.G. Withers, *Biochemistry*, 1996, **35**, 13165
39. J.P. Richards, R.E. Huber, C. Heo, T.L. Amyes, and S. Li, *Biochemistry*, 1996, **35**, 12387
30. M. Moracci, A. Trincone, G. Perugino, M. Ciaramella, and M. Rossi, *Biochemistry*, 1998, **37**, 17262
31. C.H. Wong, G.M. Whitesides, 'Enzymes in Synthetic Organic Chemistry', Tetrahedron Organic Chemistry Series, Pergamon, Oxford, 1994, Vol. 12, Chapter 5
32. J.-Ll. Viladot, V. Moreau, A. Planas, and H. Driguez, *J. Chem. Soc., Perkin Trans. I*, **1997**, 2383
33. J.L. Viladot, B. Stone, H.Driguez, and A. Planas, *Carbohydr. Res.*, 1988, **311**, 95
34. M. Hrmova, G.B. Fincher, J.L. Viladot, A. Planas and H. Driguez, *J. Chem. Soc., Perkin Trans. I*, **1998**, 3571
35. L.F. Mackenzie, Q. Wang, R.A.J. Warren, and S.G. Withers, *J. Am. Chem. Soc.*, 1998, **120**, 5583
36. C. Malet and A. Planas, *FEBS Lett.*, 1998, **440**, 208

ENGINEERING THE CATALYTIC AND BINDING PROPERTIES OF THE CELLOBIOHYDROLASES FROM *Trichoderma reesei*

T.T. Teeri[1], C. Divne[2], T. A. Jones[2], G. Kleywegt[2], A. Koivula, M. Linder, J. Ståhlberg[2], I. von Ossowski, G. Wohlfahrt and J.-Y. Zou[2]

VTT Biotechnology and Food Research, Box 1500, FIN-02044 VTT, Finland
[1]Current address: Royal Institute of Technology, Department of Biotechnology, SE-10044 Stockholm, Sweden, [2]Department of Molecular Biology, Uppsala University, Box 509, SE-75124 Uppsala, Sweden

1 INTRODUCTION

Filamentous fungi can degrade of crystalline cellulose very efficiently and cellobiohydrolases are the key components of their cellulolytic enzyme systems. The soft rot fungus *Trichoderma reesei* produces two cellobiohydrolases, Cel7A and Cel6A (formerly called CBHI and CBHII, respectively[1]), both composed of distinct catalytic and cellulose-binding domains. Their catalytic domains have very different overall structures but both have a long, tunnel shaped active site promoting so called processive mode of action. The two enzymes degrade cellulose crystals starting from the reducing ends (Cel7A) or from non-reducing ends (Cel6A) (reviewed in[2]). Similar pairs of cellobiohydrolases with opposite chain end specificities have been discovered in bacterial cellulolytic enzyme systems.[3,4] Endoglucanases homologous to Cel7A, Cel6A and other cellobiohydrolases have more open active sites allowing random cuts in the middle of the cellulose chains. Studies with the cellulose-binding domain (CBD) of Cel7A have revealed an exchange rate of binding on crystalline cellulase that allows for its proposed processive mode of action[5]. In contrast, the CBD of Cel6A, seems to bind to the substrate more tightly, and is not readily reversible[6]. The topology of the active sites and the two-domain structures of the two cellobiohydrolases are discussed below in view of their high activities against crystalline cellulose.

2. THE CATALYTIC DOMAINS

The catalytic domain structures of both *T. reesei* cellobiohydrolases have been determined in complex with a number of oligosaccharides and ligands[7-10]. The striking finding was that both enzymes have active sites buried in tunnels extending deep inside their catalytic domains. The roof of each active site tunnels is provided for by long surface loops, two in the case of Cel6A and four in the case of Cel7A. The tunnel of Cel6A is thus shorter, providing binding sites for 4 glucose units inside the tunnel while Cel7A can bind as many as 10 glucose units[7,9]. The glucan chains bind in these active sites through hydrogen-bonding networks and aromatic stacking interactions with tryptophan residues. The catalytic residues are placed to break the bond between the subsites -1/+1 in Cel7A and +1/-1 in Cel6A. In Cel7A, a structure in complex with an oligosaccharide filling the entire length of the tunnel reveals a massive twist in the chain between the subsites -4/-3[9]. In Cel6A a similar but less severe twist is observed between the subsites +1/+2.[7] It is possible, although still speculative that the twists help to push the chain in only one direction in the active site tunnels, thus promoting the processivity.

It seemed extraordinary that cellobiohydrolases, which are the enzymes most active on crystalline cellulose, had active site structures that seem to exclude easy access to the glucan chains of crystalline cellulose. This dilemma has since inspired several investigations addressing both the structure and the properties of the native substrate as well as the mechanisms by which the cellobiohydrolases access their crystalline substrate. Based on our structural and biochemical data, we have proposed that a single glucan chain end enters the tunnel from one end ("entrance"), threads through the entire tunnel for bond cleavage in the far end with the products leaving from the opposite, "exit" end of the tunnel.[7-11]. According to an alternative hypothesis, the initial contact occurs by sporadic opening of the active site loops[12,13]. This would allow the chain to be first bound along the length of the channel where after the loops would close, leading to the observed processive action. Structure determination of the cellobiohydrolase catalytic domains clearly shown that the substrate binds in the active sites of Cel7A and Cel6A in opposite orientations[7-10]. Therefore, Cel6A releases cellobiose from the non-reducing end of the glucan chain while the opposite holds for Cel7A. However, the structures alone cannot resolve how the substrate enters into the tunnel.

2.1. Endwise action of Cel6A

Several lines of evidence are consistent with our model of endwise action of Cel6A. Firstly, as compared with the endoglucanase Cel7B (formerly EGI), Cel6A is very slow to reduce the degree of polymerisation of bacterial microcrystalline cellulose and is thus unlikely to introduce significant numbers of internal cuts to the chains[14]. Secondly, our recent mutagenesis study shows that removal of a tryptophan residue (W272) at the very entrance of the active site tunnel reduces the activity of Cel6A selectively on insoluble substrates, but not on amorphous or soluble ones[11]. Inspection of the three dimensional structures shows that this residue is in a position where it is readily exposed to the solvent but yet close to the tunnel entrance. It could thus easily act as the first contact with a cellulose chain end helping to direct it further into the active site tunnel. Finally, an analysis of the soluble product profile reveals that, in addition to its main product, cellobiose, Cel6A produces low but clearly detectable amounts of cellotriose from crystalline cellulose[11]. Our interpretation is that this is a direct consequence of the both the structure of a cellulose chain, in which every glucose unit is rotated 180° relative to its neighbours and the tunnel shaped active site of cel6A, which limits the rotation of the cellulose chain once bound[15]. If Cel6A initiated crystalline cellulose degradation by an obligatory endolytic cleavage (illustrated in Figure 1C), the first catalytic event would create a single cut in the middle of a long cellulose chain sitting on the crystal surface. The products of this cleavage would thus be two polymers. The non-reducing end of one of these would be attached to the enzyme while the other would still be bound to the cellulose crystal by hydrogen bonds. In this case, no soluble products should be released. After this first cleavage, the enzyme would automatically be in register to produce its main product, cellobiose, and no cellotriose should be released in the beginning of the hydrolysis. If, on the other hand, endwise action was preferred by Cel6A, the chain could be introduced in the tunnel entrance in two alternative orientations. In the first orientation, the sugar orientation at the –1 site would be optimal for the production of cellobiose (Figure 1A). However, owing to the symmetry of the cellulose chain, it can equally well enter the active site when rotated 180° relative to the first orientation. In this case, production of cellobiose is not possible since the sugar at the –1 site has wrong orientation. (Figure 1B). Since glucose is not produced from the non-reducing end of a polymer by Cel6A, the first product of the alternative orientation would be cellotriose, which is clearly observed. The presence of cellotriose among the early products by Cel6A on crystalline cellulose thus implies that endwise action occurs but does not completely exclude the possibility of rare internal cutting due to loop opening. Our recent structural study revealed that the tip of one of the active site loops is able to undergo conformational changes that can open but also tighten the active site tunnel over the site of bond cleavage[10]. Whether these movements are substantial enough to open the active site and to allow initiation in the middle of a chain on some substrates remains to be seen upon further experiments.

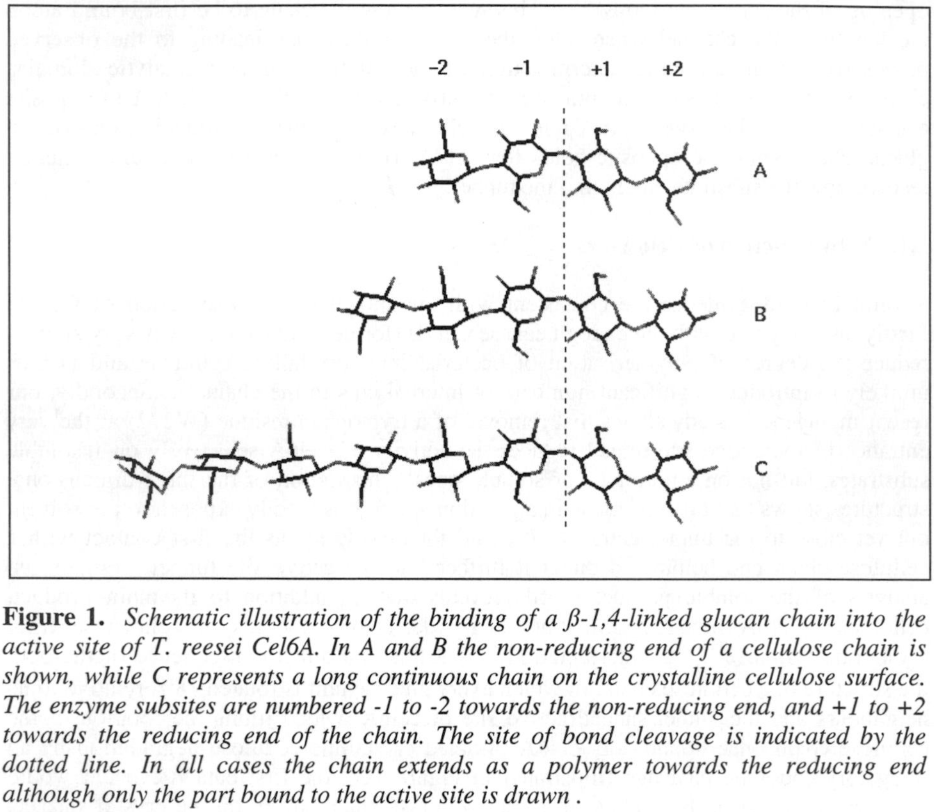

Figure 1. *Schematic illustration of the binding of a ß-1,4-linked glucan chain into the active site of T. reesei Cel6A. In A and B the non-reducing end of a cellulose chain is shown, while C represents a long continuous chain on the crystalline cellulose surface. The enzyme subsites are numbered -1 to -2 towards the non-reducing end, and +1 to +2 towards the reducing end of the chain. The site of bond cleavage is indicated by the dotted line. In all cases the chain extends as a polymer towards the reducing end although only the part bound to the active site is drawn .*

2.2 Engineering the "exo" –loop of Cel7A

The long active site tunnel of Cel7A is formed by four long loops, which serve to exclude the bound sugar chain almost completely from the solvent[8]. One of these, the so called 'exo' –loop covers the catalytic residues and should be flexible in order for the enzyme to be able to initiate cellulose degradation in the middle of the cellulose chains[9]. In our attempts to obtain direct experimental evidence for or against the endwise action of Cel7A, we searched for a site where a disulphide bridge could be introduced to prevent eventual change of conformation of this loop. Ideally the exo-loop should be disulphide bonded with another loop on the opposite wall of the tunnel. However, suitable positions for the introduction of free cysteins close enough in space to each other could not be identified in these two loops. Instead, replacement of two aspartic acid residues within the 'exo' -loop by cysteine residues led to the formation of a disulphide bridge freezing the structure of the loop (see Figure 2). X-ray structure determination of the mutant confirmed the formation of the disulphide bridge and showed no changes in the conformation of the loop relative to the wild type Cel7A (our unpublished data).

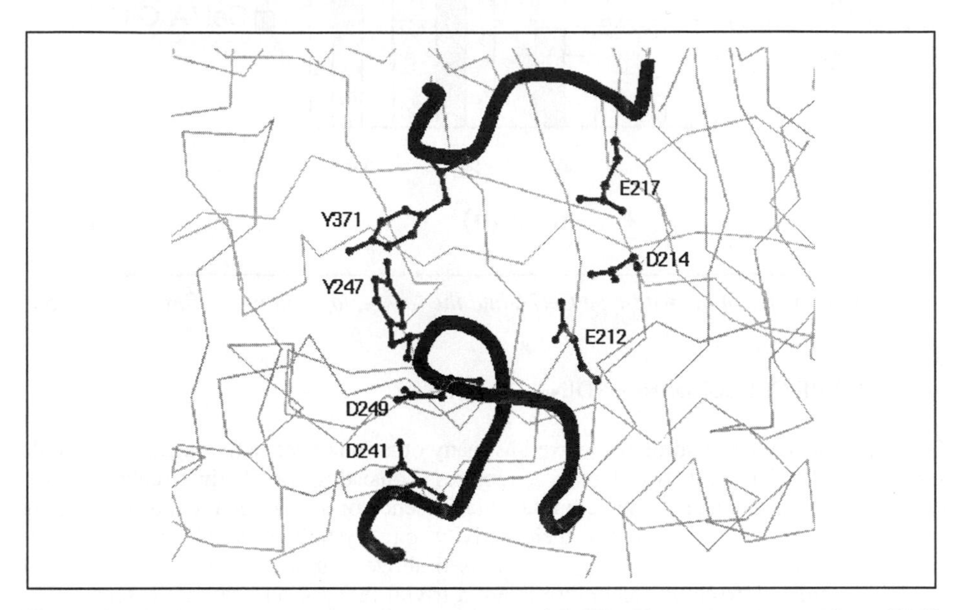

Figure 2. *The loop structure over the active site of Cel7A. The active site residues, E212, D214 and E217 are labelled. Two loops fold over the active sites and close the tunnel by an interaction of the tyrosines Y247 in the 'exo' –loop and Y371 in the loop on the opposite side of the tunnel wall. In order to freeze the structure of the exo –loop, the residues D241 and D249 were changed to cystein residues, which resulted in the formation of a disulphide bond.*

In spite of fixing the loop structure, no significant changes were observed in the activity of the loop mutant relative to the wild type Cel7A on the soluble para-nitrophenyl-lactoside (pNP-Lac) (Table 1) nor on bacterial microcrystalline cellulose (BMCC) (Figure 3). We therefore believe that this loop is not likely to undergo large conformational changes required for opening of the active site tunnel for initial endoactivity upon cellulose degradation.

Table 1. *Activities of the wild type and loop mutated Cel7A on para-nitrophenyl-Lactoside.*

Enzyme	K_{cat} (s^{-1})	K_M (mM)	K_{cat} /K_M (mM^{-1}*s^{-1})
Cel6A	0.093	0.41	0. 23
C-C	0.082	0.44	0.19

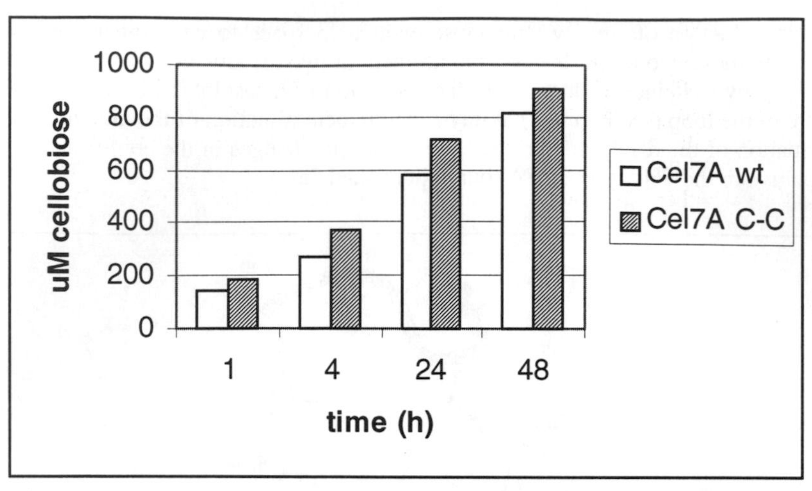

Figure 3. *Activities of the wild-type Cel7A and the 'exo'-loop mutant on BMCC at pH 5.*

3. THE CELLULOSE-BINDING DOMAINS

Both *T. reesei* cellobiohydrolases as well as many other enzymes active against insoluble substrates share typical two-domain structures consisting of distinct catalytic and substrate-binding domains[16]. In cellulases, the presence of a functional cellulose-binding domain (CBD) clearly promotes their activity on crystalline cellulose while it is unnecessary on soluble substrates. As discussed above, initiation of the degradation of the highly ordered crystalline cellulose is not a trivial task for an enzyme. It seems thus logical that a CBD is needed to bind the catalytic domain on the substrate surface for as long as it takes to achieve productive binding. However, it is not equally clear whether the CBD is needed thereafter in the process. In the processive cellobiohydrolases the CBD should have an exchange rate allowing the enzyme to slide along the cellulose surface during the progress of the hydrolysis. Our earlier data indicates that the Cel7A CBD indeed has an exchange rate compatible with its apparent catalytic rate, and should thus not limit the progression of the catalytic domain[5]. Moreover, a double CBD with two different fungal CBDs joined by a relatively long linker, showed high affinity due to synergistic binding of the two distinct CBDs[17]. In analogy, it seems that the tight binding observed with intact two-domain cellulases may result from synergistic yet dynamic binding interactions of the two functional domains. However, the question is not quite as simple as that since some other CBDs have been shown to exhibit binding behaviour which is hard to describe as anything but irreversible[6,18].

3.1 Engineering the elution of the Cel7A CBD

One of the motivations to study cellulose-binding domains is that they provide interesting 'tags' for the immobilisation of other biomolecules onto the cheap cellulose-based columns[19]. We and others have shown that the binding of such fusion proteins is governed by the binding properties of the CBD. The binding of the CBDs depends mainly on the ring-ring stacking interactions of three or more aromatic residues and the exposed glucose rings at the cellulose surface. The most common residues responsible for the binding interaction are tryptophans, followed by tyrosines and sometimes even phenylalanins. In principle the imidazole ring of histidine should also be able stack onto the glucose rings but in practice histidines are almost never seen to mediate protein-carbohydrate interactions. Histidine has a pK_a at the moderate pH of about 6.5, and changing the pH might influence its interactions with the sugar rings. In a recent study[20] we decided to explore this possibility by replacing some of the tyrosines in the Cel7A CBD by histidines and examine whether this would facilitate elution of the CBD by pH (Figure 4).

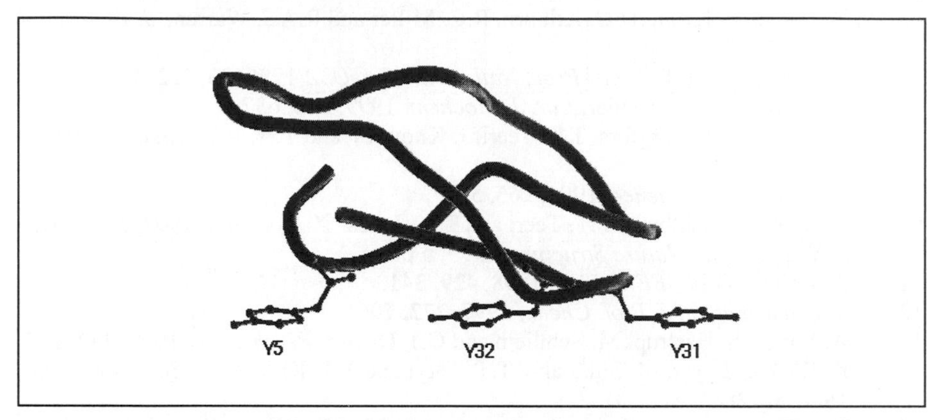

Figure 4. *Structure of the Cel7A CBD, showing the side chains of the three tyrosine residues involved in cellulose binding. Histidines were introduced to positions 5 and 31.*

Out of the two positions, which tolerate amino acid replacements, a histidine at position 5 practically inactivated the Cel7A CBD. On the other hand, a histidine replacing the tyrosine, Y31, led to a CBD which bound to the cellulose nearly as well as the wild type CBD in high pH, but much weaker at the pH below 6.5 (Figure 5). This shows that histidine is indeed capable of the required stacking interaction. Since the interaction is sensitive to pH, it fails to provide a general solution for binding required for the cellulolytic microbes living in the ever-changing natural environments, and perhaps selected against. It is, however, ideal for applied purposes providing a clearcut, stepwise elution of the CBD.

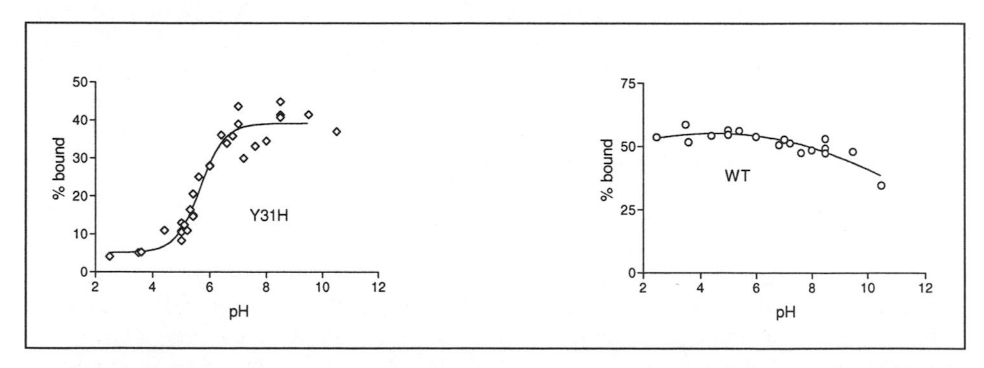

Figure 5. *Titration curves showing the binding of the wild type Cel7A CBD and the Y31H mutant as a function of pH.*

Acknowledgement
Dr. S. Raza is thanked for helping to produce Figures 1 and 2.

References

1. B. Henrissat, T.T. Teeri and R.A.J. Warren, *FEBS Lett.* 1998, **425**, 352.
2. T.T. Teeri, Trends Biotechnol. 1997, **15**, 160.
3. B. Barr, Y-L. Hsieh, B. Ganem and D.B. Wilson, *Biochemistry* 1996, **35**, 586.
4. N. Gilkes, E. Kwan, D.G. Kilburn, R.C. Miller and R.A.J. Warren, *J. Biotechnol.*, 1997, **57**, 83.
5. M. Linder and T.T. Teeri, *Proc. Natl. Acad. Sci. USA* 1996, **93**, 12251.
6. G. Garrard and M. Linder, *Eur. J. Biochem.* 1999, **262**, 637.
7. J. Rouvinen, T. Bergfors, T.T. Teeri, J. Knowles and T.A. Jones, *Science*, 1990, **249**, 380.
8. C. Divne *et al.*, *Science*, 1994, **265**, 524.
9. C. Divne, J. Ståhlberg, T.T. Teeri and T.A. Jones, *J. Mol. Biol,* 1997, **275**, 309.
10. J.-Y. Zou *et al.*, *Nature Structure*, 1999, n press.
11. A. Koivula *et al., FEBS Lett,.* 1998, **429,** 341.
12. S. Armand *et al., J. Biol. Chem.,* 1997, **272,** 706.
13. A. Varrot, S. Hastrup, M. Schülein and G.J. Davies, *Biochem. J.,* 1999, **337**, 297.
14. K. Kleman-Leyer, M. Siika-aho, T.T. Teeri and T.T. Kirk, *Appl. Env. Microbiol.*, 1996, **62**, 2883.
15. T.T. Teeri *et al., Biochem. Soc. Transact.*, 1998, **26**, 173.
16. M. Linder and T.T. Teeri, *J. Biotechnol.*, 1997, **57**, 15.
17. M. Linder, I. Salovuori, L. Ruohonen and T.T. Teeri, *J. Biol. Chem.*, 1996, **271**, 21268.
18. A.L. Creagh, E. Ong, E.J. Jervis, D.G. Kilburn and C.A. Haynes, *Proc. Natl. Acad. Sci. U.S.A.,* 1996, **93**, 12229.
19. M. Linder, T. Nevanen, K. Söderholm, O. Bengs and T.T. Teeri, *Biotechnol. Bioeng.,* 1998, **60**, 642.
20. M. Linder and T.T. Teeri, *FEBS Lett.*, 1999, **447**, 13.

Subject Index